Thin-Film Transistor Reliability

Authored by

Meng Zhang

College of Electronics and Information Engineering
Shenzhen University
Shenzhen, China

&

Mingxiang Wang

Department of Microelectronics
Soochow University
Suzhou 215006, China

Thin-Film Transistor Reliability

Authors: Meng Zhang & Mingxiang Wang

ISBN (Online): 978-981-5322-61-3

ISBN (Print): 978-981-5322-62-0

ISBN (Paperback): 978-981-5322-63-7

need for a court order if at any point you breach any terms of this License Agreement. In no event will any delay or failure by Bentham Science Publishers in enforcing your compliance with this License Agreement constitute a waiver of any of its rights.

3. You acknowledge that you have read this License Agreement, and agree to be bound by its terms and conditions. To the extent that any other terms and conditions presented on any website of Bentham Science Publishers conflict with, or are inconsistent with, the terms and conditions set out in this License Agreement, you acknowledge that the terms and conditions set out in this License Agreement shall prevail.

Bentham Science Publishers Pte. Ltd.
80 Robinson Road #02-00
Singapore 068898
Singapore
Email: subscriptions@benthamscience.net

BENTHAM SCIENCE

CONTENTS

FOREWORD

It is with great pleasure that I introduce "Thin-Film Transistor Reliability", a comprehensive exploration into the intricate world of thin-film transistor (TFT) reliability. This book, authored by Prof. Meng Zhang and Prof. Mingxiang Wang, delves deep into the essential aspects of TFT technology, shedding light on the critical factors that influence their reliability. The journey embarked upon in this book takes the reader through a meticulous study of TFTs, from their fundamental principles to the intricate details of reliability analysis methods and stress-induced degradation mechanisms.

The chapters within this book serve as a roadmap guiding readers through the evolution of TFT technology, the various applications in modern electronics, and the challenges posed by environmental factors and stress-induced degradation. By dissecting common defects, exploring reliability analysis techniques, and discussing strategies for improvement, this book equips readers with a comprehensive understanding of TFT reliability.

I commend Prof. Meng Zhang and Prof. Mingxiang Wang for their dedication and expertise in crafting this insightful masterpiece. May this book inspire further exploration and advancements in the realm of TFT reliability, shaping a more robust and reliable future for electronic devices.

Guangcai Yuan
Vice President of BOE Technology Group Co., Ltd.
Beijing
China

PREFACE

Welcome to the exciting world of "Thin-Film Transistor Reliability". This book delves into the intricate details and challenges associated with the reliability of thin-film transistors (TFTs), providing a comprehensive overview of their structure, fabrication processes, and applications in modern electronics. It explores the various degradation mechanisms that affect TFT performance and presents analysis methods to assess their reliability.

Chapter 1 serves as an introduction to TFTs, giving you a solid foundation by discussing their overview, development history, classification, and comparison. It also explores the various applications of TFTs in modern electronics, such as active-matrix displays, sensors, and other circuits. Additionally, this chapter sheds light on the reliability of TFT technology, highlighting the degradation processes that can occur in active-matrix displays, sensors, and other applications. Chapter 2 delves deeper into the reliability issues of TFTs, focusing on the common defects found in silicon-based TFTs and metal oxide TFTs. It explores the different types of defect states and typical degradation mechanisms that affect the performance of TFTs. In Chapter 3, various reliability analysis methods for TFTs are explored. From degradation analysis in different regions to CV curve analysis, low-frequency noise analysis, and thin-film quality analysis, this chapter provides a comprehensive overview of the techniques used to evaluate and analyze the reliability of TFTs. Simulation analysis techniques are also discussed, including TCAD simulation and thermal simulation. Chapter 4 examines the degradation induced by DC voltage stress in TFTs. It covers gate bias stress, hot-carrier effects, and self-heating effects, discussing their impact on both silicon-based TFTs and metal oxide TFTs. Chapter 5 shifts the focus to AC voltage stress-induced degradation in TFTs. It explores the degradation models and behavior of poly-silicon TFTs and metal oxide TFTs under AC stress, highlighting the dependence on waveform elements. A comparison between the two types of TFTs is provided as well. In Chapter 6, circuit-level stress-induced degradation in TFTs is discussed. This chapter covers AC degradation under DC bias, bipolar AC degradation, and ultra-fast AC degradation, shedding light on the impact of different stress conditions on the performance and reliability of TFTs. Chapter 7 explores the effects of environmental factors on TFT reliability. It discusses the influence of temperature, illumination, and moisture on the performance and degradation of TFTs. Finally, Chapter 8 presents strategies and methods for improving the reliability of TFTs. It discusses the implementation of special structures in TFTs and explores other improvement methods, providing valuable insights into enhancing the overall reliability of these electronic devices. In the concluding chapter, Chapter 9, the key findings and insights presented throughout the book are summarized, providing a comprehensive overview of TFT reliability. The future directions and potential areas of research in the field of TFT reliability are also discussed.

This book aims to provide researchers, engineers, and students with a comprehensive understanding of TFT reliability and the tools to assess and enhance it. It combines theoretical insights with practical knowledge, making it an invaluable resource for anyone involved in the field of electronics. I hope that readers will find this book informative, inspiring, and a catalyst for further advancements in the reliability of TFT.

Finally, we would like to extend our deepest appreciation to our dear students, Mr. Zhendong Jiang, Mr. Guanming Zhu, Mr. Yiming Song, Mr. Yunyang Wang, Mr. Xindi Xu, Ms. Yuwei Zhao, Mr. Pengfei Liu, Mr. Bin Wang, Mr. Qingcan Su, Mr. Zihan Wang, Mr. Feilian Chen, Mr. Mingjun Zhang, Mr. Ruipeng Shen, and Mr. Ming Guo for their invaluable assistance, from collecting research materials to assisting with illustrations and figures. We also would like to express our deep gratitude to the National Natural Science Foundation of China (Grant Number: 62274111) and Shenzhen Municipal Research Program (Grant Number: SGDX20211123145404006) for their generous support throughout the writing process.

Meng Zhang
College of Electronics and Information Engineering
Shenzhen University
Shenzhen, China

&

Mingxiang Wang
Department of Microelectronics
Soochow University
Suzhou 215006, China

CHAPTER 1

An Overview of Thin-Film Transistors

Abstract. This chapter introduces the fundamental concepts of thin-film transistors (TFTs), outlining their development, classification, and comparison. It delves into the various applications of TFTs in modern electronics, particularly highlighting their role in active-matrix displays, sensors, and other circuits. The chapter also addresses the reliability of TFT technology, focusing on the degradation processes that occur in different applications and the importance of understanding these for the advancement of electronic devices.

Keywords: Application, Development history, Reliability, Thin-film transistors (TFTs).

1.1. INTRODUCTION

Currently, active matrix (AM) display technology is the most mainstream display technology. Thin-film transistors (TFTs), as the core components of AM display technology, greatly influence every aspect of display performance. This chapter will begin with an overview of the development history, classification, and applications of TFTs. It will then provide a detailed discussion of their basic structure, including an introduction to several common types of TFTs, such as polycrystalline silicon (poly-Si) and metal-oxide (MO) TFTs. Following this, the chapter will explore the applications of TFTs in AM displays and extend the discussion to their use in sensors and other areas. The final section will underscore the critical importance of reliability in these various applications.

1.2. OVERVIEW OF THIN-FILM TRANSISTORS

With the rapid development of the Internet of Everything, artificial intelligence, and the 5G era, the electronic information industry has also stepped onto the fast track of soaring development alongside them. Among these advancements, the semiconductor industry's growth has attracted increasing attention. Today, the display FP industry has become an important strategic and foundational industry in the new generation of electronic information fields (Fig. **1**).

Currently, multiple technological pathways are advancing in tandem with innovative technologies springing up rapidly. The established active matrix liquid crystal display (AMLCD) [1, 2] takes the lead, while the high-quality active matrix organic light emitting diode (AMOLED) [3] serves as a premium technological successor. Additionally, micro light emitting diodes (Micro-LEDs) [4], laser projections, and electronic paper are expanding the display realm for specific applications. The domain of display technology has progressively moved beyond conventional settings like televisions, mobile devices, and surveillance systems to encompass areas such as smart homes, educational technology, community smart systems (Fig. **2**), autonomous vehicles, remote healthcare, virtual reality, and augmented reality.

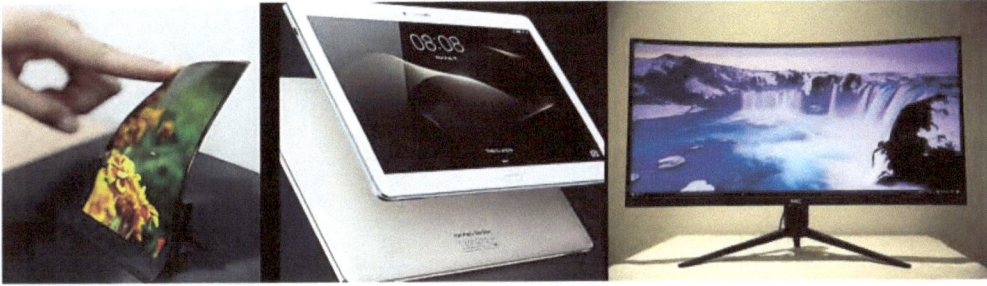

Fig. (1.1). Displays in mobile phones, tablets, and computers.

TFTs are electronic components extensively applied in the realm of display technology, commonly utilized as switches to regulate current flow. Serving as the cornerstone of AM driving systems, TFTs are essential in both pixel arrays [5] and driving circuits [6], exerting a profound influence on the quality and manufacturing efficiency of display devices. Their structure, manufacturing processes, characteristics, and reliability are critical to the performance of displays. Therefore, research into TFTs has become a focal point in the field of display technology.

1.2.1. The Development History of Thin-Film Transistors

The origins of TFT development can be traced back to 1925 when Julius Edgar Lilienfeld (Fig. **3**) innovatively proposed the concept of the effect transistor (FET) and applied for a patent for the FET invention five years later [7]. However, due to the limitations of the theories and conditions at the time, Lilienfeld did not recognize that the active layer in an FET must be made of semiconductor materials.

It wasn't until 1935 that Heil explicitly proposed the crucial theory that the active layer of an FET must be made of semiconductor materials [7]. However, the requirements for the electronic manufacturing process in solid-state devices greatly exceeded the capabilities of the production techniques available at that time, preventing the concept from advancing beyond the theoretical stage and resulting in no finished devices being produced. It was not until the 1950s that the production of FETs truly became a reality. The world's first junction, FET, was proposed by Shockley (Fig. **4**) in 1952 and successfully fabricated by Dacey and others in 1954 [8].

Fig. (1.2). Data visualization display in smart communities.

Fig. (1.3). Julius Edgar Lilienfeld (1882.4.18-1963.8.28).

Fig. (1.4). William Shockley (1910.2.13-1989.8.12).

The prototype of the TFT was proposed by P.K.Weimer in 1960, as shown in Fig. (**1.5**). He also successfully fabricated a top-gate top-contact structured TFT [9]. At that time, polycrystalline cadmium sulfide (CdS) was used as the active layer material, and the insulating layer was made of silicon dioxide (SiO_2). This marked the true beginning of TFT research.

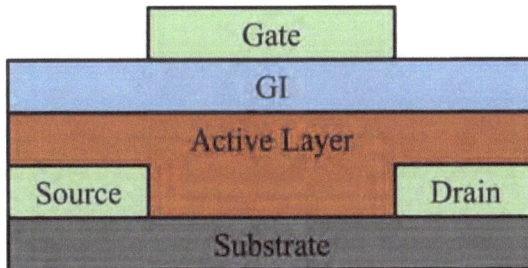

Fig. (1.5). Schematic of the basic structure of TFT proposed by P.K. Weimer et al. in 1960.

In 1964, Klasens and Koelmans successfully fabricated TFTs with an insulating layer of tin oxide (SnO_2) and an active layer of aluminum oxide (Al_2O_3), using aluminum (Al) as the electrode [10]. They utilized evaporation to create SnO_2 thin films, demonstrating the potential of this material as a transistor channel. This research marked the birth of tin oxide TFTs and laid the groundwork for subsequent studies and applications. By 1968, Boise and Jacobs had developed the first TFT with an active layer of zinc oxide (ZnO), inaugurating the development of MO TFTs [11].

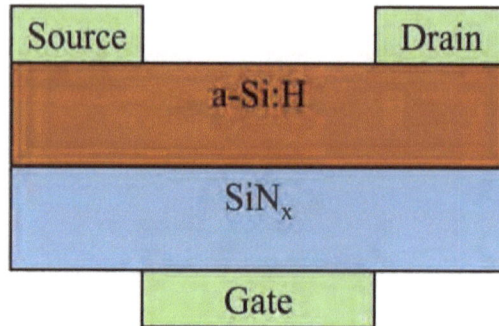

Fig. (1.6). Schematic of amorphous silicon TFT.

In 1979, P. LeComber and W. Spear from the University of Dundee, along with their colleagues, developed the amorphous silicon (a-Si) TFT, marking a breakthrough in TFT research. They created the first functional TFT made from hydrogenated amorphous silicon (a-Si:H) and a silicon nitride gate insulating layer (Fig. **1.6**). The a-Si TFT was soon recognized as more suitable for large-area AMLCD. This advancement spurred Japan's commercial development of AMLCD panels based on a-Si TFTs.

Although a-Si TFTs can generally meet the technical requirements of AMLCDs, their relatively low field-effect mobility [< 1 cm^2/(V·s)] prevents them from driving high-resolution LCDs. Moreover, emerging flat panel display (FPD) technologies such as AMOLED set higher standards for the electrical characteristics of TFTs, which a-Si TFTs cannot achieve. The poly-Si TFT technology, invented in 1980 by Deep and colleagues at IBM Corporation in the United States, effectively filled this technological gap. The mobility of low-temperature poly-Si TFTs obtained through excimer laser annealing (ELA) can even reach tens to hundreds of times that of a-Si.

However, the high production costs and poor uniformity in the fabrication of poly-Si have hindered its widespread application in the field of FPDs. In 1986, Tsumura and his team [12] pioneered the use of polythiophene as a semiconductor material for the fabrication of organic thin-film transistors (OTFTs), marking the inception of OTFT technology. During the 1990s, the employment of organic semiconductor materials as the active layer emerged as a novel research focus. Benefiting from advantages in manufacturing processes and cost, OTFTs are regarded as a prospective display technology with potential applications in the driving of LCD and OLED systems. However, due to poor reliability and low mobility, OTFTs remain mainly in the laboratory research stage.

Starting from the 1980s, extensive and in-depth research was conducted on TFTs [13], ushering in an era of rapid development for these technologies. In 1982, the Japanese company Sanyo successfully developed devices using a-Si TFTs to drive LCDs. Subsequently, poly-Si and microcrystalline silicon TFTs were also developed, establishing silicon-based TFTs as the core component in display technology and securing a dominant position in the display market [14-16].

Since the beginning of the 21st century, with the rapid advancement of technology and the increasing demands for display quality, the drawbacks of traditional silicon-based TFTs, such as poor light transmittance, instability under certain conditions, and poor uniformity in large-area applications [17], have become increasingly apparent. Consequently, more researchers are turning their attention to MO TFTs [18]. In 2003, Wager and Hoffmann, among others, published their latest findings on ZnO TFTs [19]. In 2004, Fortunato and colleagues also reported on TFT devices based on ZnO active layers. That same year, the research group led by Hosono at the Tokyo Institute of Technology first used amorphous MO semiconductor materials as the channel layer materials for TFTs, fabricating indium gallium zinc oxide (IGZO) TFTs [20, 21]. In 2010, Samsung Electronics developed 70-inch and ultra-high-resolution 3D large TVs driven by a-IGZO TFTs, signaling the gradual maturation of MO TFTs in display applications [22]. The evolution of TFTs is summarized in Fig. (**1.7**).

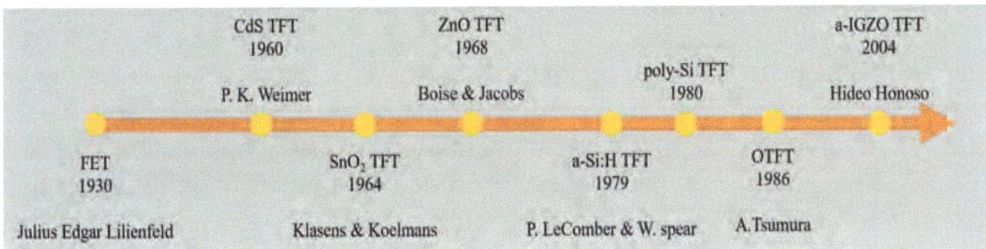

Fig. (1.7). Evolution of TFTs.

1.2.2. Classification and Comparison of Thin-Film Transistors

The performance of a TFT is critically influenced by its active layer, as different active layer materials can lead to significant variations in device performance and parameters. Consequently, the commercialized TFTs are typically categorized into three main types based on their active layers: a-Si TFTs, low-temperature poly-Si (LTPS) TFTs, and MO TFTs. As shown in the Table (**1**) below, the key performance comparisons of three typical TFTs are shown.

Table 1. Comparison of key performance parameters of typical TFTs.

AMOLED Panel	Poly-Si TFT	a-Si:H TFT	Oxide TFT
Active Channel	Poly-Si	Amorphous Si	Amorphous IGZO
TFT uniformity	Poor	Good	Good
Pixel circuit	Complex (5T +2C *etc.*)	Complex (4T + 2C, 5T + 2C)	Simple/Complex (2T + 1C/6T + 2C)
Channel mobility	~100 cm^2/Vs	< 1 cm^2/Vs	> 10 cm^2/Vs
TFT type	PMOS(COMS)	NMOS	NMOS
TFT mask steps	5 ~ 11	4 ~ 5	5 ~ 7
Cost/Yield	High/Medium	Low	Low/Medium

The primary characteristics of these TFT categories are as follows:

1.2.2.1. Amorphous Silicon Thin-Film Transistors

The a-Si films possess a high number of structural defects, such as dangling and broken bonds, which impede the diffusion pathways of internal charge carriers [23], resulting in low mobility and failing to meet industrial application standards. A-Si:H films, however, exhibit better semiconductor properties, with conductivity that can be enhanced by doping with either donors or acceptors. Due to the simplicity of the process, low-temperature processing, compatibility with large areas, uniformity, and low cost, a-Si:H TFT backplanes have been widely used in FPD manufacturing. Yet, as technological advancements demand higher picture quality from display devices, the shortcomings of a-Si:H TFTs become apparent. The primary reasons are the low hole mobility of a-Si:H and the need for improved stability under bias, which is not conducive to high-resolution displays. Additionally, light within the visible spectrum can easily affect electron transport, leading to rapid degradation of the device's electrical performance, significantly impacting the application of a-Si TFTs in large-area, low-cost, and flexible electronic devices [24].

1.2.2.2. Low-Temperature Polysilicon Thin-Film Transistor

Compared to a-Si TFTs, LTPS TFTs exhibit significantly higher mobility rates, ranging from tens to hundreds of $cm^2V^{-1}s^{-1}$. This satisfies the requirements for many high-resolution and high pixels per inch (PPI) display scenarios in AMLCDs or AMOLEDs, enhancing the application value of poly-Si over monocrystalline silicon [25, 26]. Furthermore, LTPS TFTs are also suitable for fabricating multiplexers and shift registers, offering good stability and high driving capabilities. Poly-Si films can be directly deposited in the polycrystalline phase or recrystallized from the amorphous Si phase through various annealing methods, such as solid phase crystallization (SPC) [27], metal-induced crystallization (MIC) [28], and ELA [29]. Among these, advanced laser annealing techniques can achieve high crystallinity. However, they are costly and result in large grain sizes. Therefore, materials with smaller grains are required to improve uniformity. Meanwhile, a common drawback of poly-Si materials is their large off-state current (I_{off}), which significantly increases power consumption.

1.2.2.3. Metal-Oxide Thin-Film Transistors

Recently, MO has emerged as another highly promising competitor, aiming to completely replace silicon in the backplanes of FPDs. MOs can be broadly categorized into two types [30]: one type consists of binary oxides formed by a single cation, such as ZnO [31] or SnO_2 [32], while the other type consists of oxides formed by multiple cations, such as indium zinc oxide (IZO) [33], zinc-tin oxide (ZTO) [34], indium-gallium oxide (IGO) [24], IGZO [34, 36], and indium-tin-zinc oxide (ITZO) [37-39]. Due to their high mobility, low-temperature processing, device compatibility, high transparency to visible light, and low-cost manufacturing, MO TFTs have attracted increasing attention from both academia and industry in recent years [39, 41]. They serve as ideal switching devices for large-area, high-frame-rate AMLCD/AMOLED technologies, flexible displays, and other applications.

1.2.3. Application of Thin-Film Transistors

Currently, the application of TFTs is mainly focused on FPDs. As one of the mainstays of the electronics industry, FPDs are widely used in areas such as mobile phones, tablets, computers, televisions, automobiles, and monitoring devices, greatly enriching people's lives and work. AMLCDs, as the mainstream technology for FPDs, still occupy nearly ninety percent of the market share. Meanwhile, AMOLED displays have developed rapidly in recent years, with their proportion in

FPDs continuously increasing. AMOLED, with its advantages of active emission, low power consumption, high image quality, wide viewing angles, and slim design, is particularly noteworthy for its flexible display capability, making it better suited for the future direction of FPD products.

In addition to their applications in FPDs, TFTs, with their transparent and flexible characteristics, can be embedded in packaging, mirrors, books, and devices to achieve low-cost, one-time data transmission functions [42]. They are commonly used in scenarios such as passive radio frequency identification (RFID) and near-field communication (NFC) tags. Furthermore, circuits [43], data storage [44], single-use sensors for the medical field [45], and sensor systems [46] based on TFTs have been successfully implemented by various research teams. Although these technologies have not yet been widely commercialized, with the increasing maturity of various technologies, cost reduction, and the rapid development of the Internet of Things, TFT applications in these fields are expected to shine brightly in the near future.

1.3. BASIC STRUCTURE AND FABRICATION PROCESSES OF THIN-FILM TRANSISTORS

In this section, the basic structure of TFTs is first introduced. Subsequently, from the perspectives of a-Si TFTs, poly-Si TFTs, and MO TFTs, the common structures of the devices are presented, along with some special structures appearing in current research.

1.3.1. Basic Structure of Thin-Film Transistors

A TFT mainly consists of four main components: the semiconductor active layer, gate dielectric layer, source-drain electrodes, and gate electrode. Generally, TFT structures are classified based on the relative positions of the active layer and three-terminal electrodes, resulting in top-gate top-contact, bottom-gate top-contact, top-gate bottom-contact, and bottom-gate bottom-contact structures, as shown in Fig. (**1.8**). Each structure has its own advantages and disadvantages, and the choice should be made based on practical considerations.

In industrial manufacturing, a-Si:H TFTs and MO TFTs commonly utilize bottom-gate structures [47, 48]. However, a-Si:H is susceptible to light, so opaque metal bottom-gate electrode materials are often used in device structure design to block the backlight layer of the display. The use of bottom-gate structures in MO TFTs facilitates the production and design of large-sized LCDs [49]. LTPS TFTs commonly employ top-gate structures to ensure uniform crystallization of the active layer at high temperatures, reducing damage to other interfaces in subsequent processes and thus improving device performance [50].

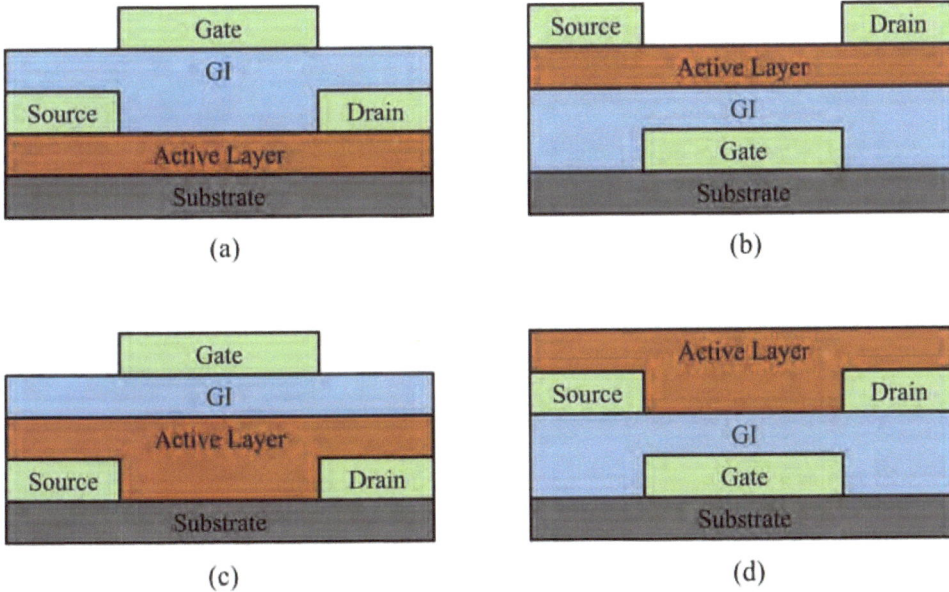

Fig. (1.8). Schematic diagrams of common TFT structures: (a) top gate top contact, (b) bottom gate top contact, (c) top gate bottom contact, and (d) bottom gate bottom contact.

Additionally, to address specific issues in engineering applications, structures such as vertical channels and dual-gate structures have been developed, as shown in Fig. (1.9). Vertical structure devices can achieve ultra-short channels compared to conventional structure devices [51]. Moreover, due to the vertical transmission of carriers between the source and drain electrodes, the device's ability to resist mechanical stress is greatly enhanced, ensuring maximum electrical performance, which is crucial for enhancing the electrical performance of devices in flexible devices [52]. Dual-gate structures, as the name suggests, involve the modulation of carrier concentration in the device channel by two gate electrodes, controlling the device's on and off states [53, 54]. Such a structure significantly improves the gate's control over the semiconductor device channel, resulting in clear advantages over conventional structure devices in parameters such as threshold voltage (V_{th}), subthreshold swing (SS), field-effect mobility (μ_{FE}), enhancing the device's static performance [55, 56].

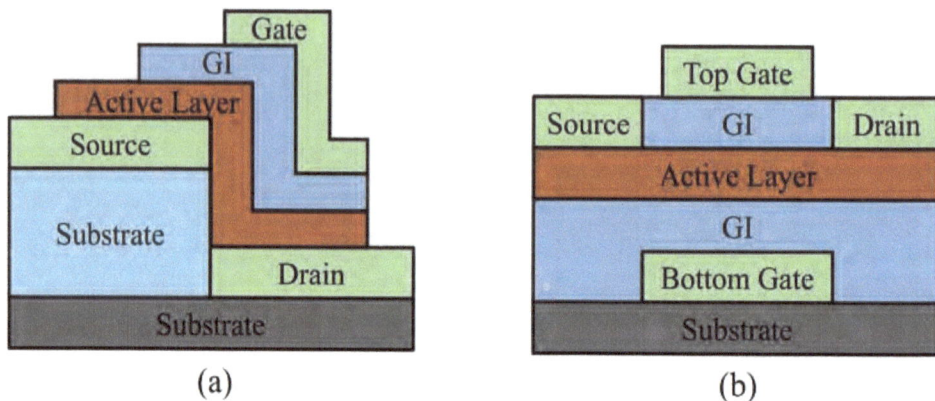

Fig. (1.9). Schematic diagrams of TFT with special structures: (a) vertical channel and (b) double gate structure.

1.3.2. Amorphous Silicon Thin-Film Transistors

The a-Si material is the earliest material used for the active layer of TFT devices, and the a-Si:H material is the most widely used. This is because the atomic arrangement of atoms in the structure of a-Si material lacks periodicity in space, resulting in a certain number of dangling and broken bonds. These dangling and broken bonds act as the defects that hinder the diffusion path of its charge carriers, leading to a low mobility of the a-Si material. The low mobility obviously limits the practical application of a-Si materials [57]. To address this limitation, researchers introduced a certain amount of hydrogen atoms during the formation of the a-Si film, saturating the dangling and broken bonds in the a-Si material, thereby improving its mobility [58]. The a-Si:H was discovered as a semiconductor material with application potential in the 1970s and quickly gained widespread application in TFTs due to its high current switching ratio and low off-state leakage current. However, with the advancement of technology and the increasing demand for display screen quality, the drawbacks of a-Si:H TFTs have gradually become apparent. The very low hole mobility of a-Si:H, poor stability under bias conditions, inability to achieve high-resolution display, and high sensitivity to light within the visible spectrum lead to severe degradation of the device's electrical performance, severely limiting the application of a-Si TFTs in large-area, low-cost, and flexible electronic device fields [59]. The schematic diagram of common a-Si:H TFT structures is shown in Fig. (**1.10**).

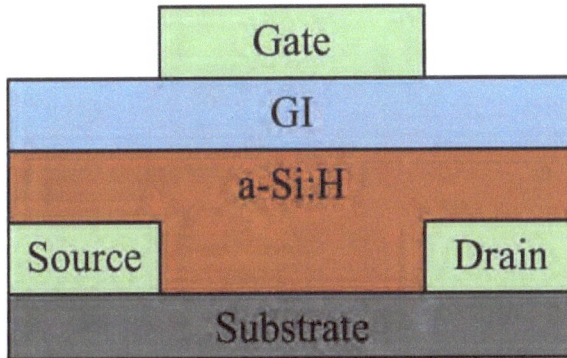

Fig. (1.10). Schematic diagram of common a-Si:H TFT structures.

1.3.3. Polycrystalline Silicon Thin-Film Transistors

Compared to a-Si, poly-Si exhibits a mobility that is several hundred times greater. This enhanced mobility endows LCDs fabricated from poly-Si with superior resolution and faster response times, as well as a higher aperture ratio, thereby increasing its practical value. Poly-Si can be categorized into high-temperature poly-Si (HTPS) [60] and LTPS [61] based on the fabrication process. The fabrication temperature for HTPS is relatively high, exceeding 600°C, which may lead to poor substrate stability during the annealing process, rendering it unsuitable for glass substrate and flexible substrates. Conversely, LTPS is produced at lower temperatures but entails higher manufacturing costs and complex processes, resulting in devices with inferior uniformity. This hinders its application in large-area displays.

Poly-Si TFTs typically adopt a top-gate, self-aligned process. The process starts by depositing an a-Si layer on a thermally oxidized silicon substrate. Many crystallization methods are then used to convert the a-Si into polycrystalline silicon (poly-Si), which is a main process in the fabrication of Poly-Si TFTs. After this, a gate oxide layer is usually prepared using low-pressure chemical vapor deposition (LPCVD), followed by gate material deposition and patterning. Source and drain regions are formed through self-aligned doping, and dopants are activated by high-temperature annealing. Finally, contact holes are opened, and metal electrodes are deposited to complete the TFT structure. Because the crystallization process of polycrystalline silicon directly affects the performance of the device, different crystallization processes are mainly introduced in the subsequent introduction.

1.3.3.1. Crystallization of Poly-Si Film

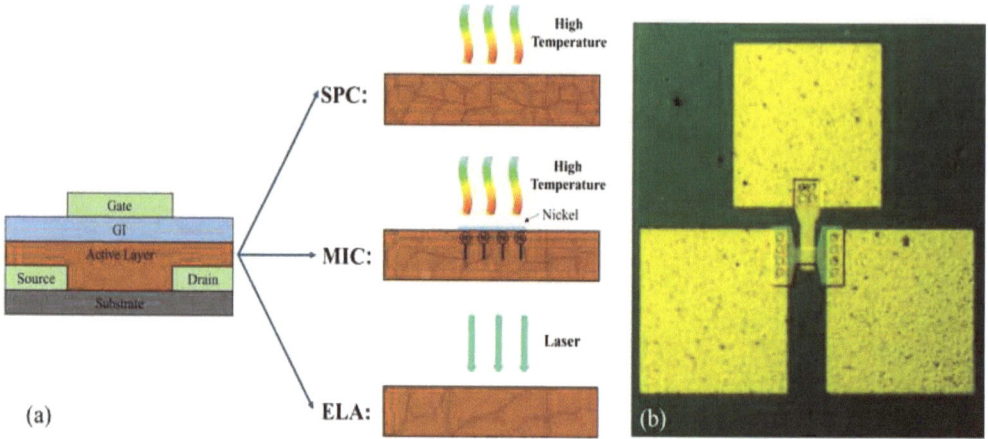

Fig. (1.11). (a) Schematic diagram of three crystallization processes in LTPS TFTs. (b) Microscopic image of a poly-Si TFT.

For LTPS TFTs, as depicted in Fig. (**1.11**), the common poly-Si thin-film deposition methods can be categorized into SPC [62], MIC [63], and ELA [64]. Below is a detailed introduction to each of these crystallization techniques:

Solid Phase Crystallization

This is the most straightforward of the three techniques for depositing poly-Si thin films. It is characterized by a simple production process, broad applicability, and rapid growth rate, making it widely used in semiconductor and photovoltaic industries. However, the primary drawback of this technique is the small grain size, high defect density, and inferior quality of the LTPS thin film, which in turn leads to high V_{th}, low μ_{FE}, and large SS in the resulting TFTs. The fabrication process mainly includes the following steps:

1) Amorphous Silicon Deposition: Techniques such as chemical vapor deposition (CVD) or physical vapor deposition (PVD) are commonly used to deposit a-Si film atoms onto the substrate. In CVD, silane gases (*e.g.*, trichlorosilane, trimethylchlorosilane, *etc.*) are used as silicon sources, which generate active silicon atoms through pyrolysis reactions at high temperatures and deposit them onto the seed to form a silicon layer. In PVD, silicon atoms are physically detached from a solid silicon source (*e.g.*, through thermal evaporation or sputtering) and deposited onto the seed.

2) Heat Treatment: After the deposition of a-Si, a heat treatment process is employed to crystallize the silicon atoms into poly-Si. The heat treatment conditions involve controlling parameters such as temperature and time, as well as adjusting the atmosphere.

3) Crystal Growth: During the heat treatment, the deposited silicon atoms begin to crystallize into poly-Si grains. The growth process involves diffusion, nucleation, and crystal growth of the grains, which are influenced by various factors, including temperature, atmosphere, and deposition rate.

4) Recrystallization: As the poly-Si grows, crystalline boundaries between the grains gradually form. To minimize the impact of these boundaries on the crystal properties, a recrystallization treatment is often performed, employing methods such as increasing the temperature or applying stress to realign the grains, thereby reducing the number and energy of the crystalline boundaries.

Metal-induced Crystallization

Metal-induced crystallization (MIC) is a widely utilized method for transforming a-Si into poly-Si. This process involves depositing a metal film on an a-Si thin film and utilizing the interaction between the metal and silicon under specific heat treatment conditions to facilitate the crystallization of a-Si. Below is a detailed description of the MIC process:

1) Substrate Preparation: Initially, a substrate, typically made of glass, quartz, or silicon, is prepared. An a-Si thin film is then deposited onto the substrate using techniques such as plasma-enhanced (PECVD) or LPCVD.

2) Metal Deposition: A metal film is deposited onto the surface of the a-Si thin film. Commonly used metals include nickel (Ni), Al, or cobalt (Co), with the thickness of the metal film ranging from a few to several tens of nanometers.

3) Heat Treatment: The sample is subjected to heat treatment in a high-temperature furnace. During this process, the metal reacts with the a-Si to form a metal silicide. The temperature for this reaction typically ranges from 450°C to 600°C, and the duration can vary from minutes to hours.

4) Metal-Silicide Reaction: The reaction between the metal and a-Si generally occurs in two stages. The first stage involves metal diffusion, where metal atoms begin to diffuse into the a-Si, forming an intermediate compound of metal silicide. This stage usually takes place at lower temperatures. The second stage is the crystallization phase, where, in the presence of metal silicide, the a-Si starts to crystallize towards the location of the metal silicide. The metal silicide provides seed nuclei for crystallization, guiding the growth of poly-Si crystals and resulting in poly-Si.

5) Crystallization Process: During the crystallization process, the a-Si thin film gradually transforms into poly-Si. The post-crystallization poly-Si thin film possesses higher crystal quality and crystallinity, making it suitable for the fabrication of various electronic devices.

6) Post-Treatment: After crystallization, further processing is often required, such as removing unreacted metal residues and optimizing the crystal structure of the poly-Si.

Excimer Laser Annealing

ELA is one of the most prevalent crystallization techniques for LTPS TFTs. During the ELA process, an excimer laser is used to irradiate an a-Si thin film, locally heating it to high temperatures, thereby inducing crystallization. By controlling the laser's energy and scanning speed, one can regulate the size and shape of the crystal grains, as well as the lattice structure of the thin film. Below are the primary steps involved in ELA:

1) Substrate and a-Si Thin Film Preparation: Initially, a substrate, commonly made of glass, quartz, or silicon, must be prepared. An a-Si thin film is then deposited onto the substrate, which can be achieved through methods such as PECVD or LPCVD.

2) Laser Irradiation: A laser beam is generated by a laser system and focused onto specific areas of the a-SI thin film's surface. The energy density of the laser beam is sufficiently high to produce localized heating on the a-Si surface.

3) Local Heating: Upon irradiation, the absorbed laser energy causes local heating of the a-Si, raising its temperature to approach or exceed the crystallization temperature.

4) Crystallization Process: Once the a-Si reaches a certain temperature threshold, its molecules begin to rearrange and form a poly-Si structure. Under the influence of laser irradiation, poly-Si grains gradually grow on the surface of the a-Si. Since the laser irradiation is localized, only the irradiated areas undergo crystallization, while the remaining areas retain their amorphous state.

5) Crystal Growth and Grain Size Control: During crystallization, the growth direction of the crystals is constrained by the laser beam, allowing control over the size and orientation of the poly-Si grains by adjusting parameters such as the scanning speed, power density, and irradiation duration of the laser beam. Proper selection of these parameters enables precise control over the grain size of the poly-Si thin film.

6) Post-Treatment: After the crystallization is complete, further processing may be necessary, such as removing any unreacted a-Si residues and optimizing the crystal quality of the poly-Si.

In summary, among the three silicon thin-film crystallization techniques, SPC is the simplest to operate but suffers from the highest density of grain boundaries in the poly-Si film, which significantly impacts the electrical performance of the devices. MIC requires stringent control over the temperature and duration of the MIC process. Excessive temperatures or prolonged times may lead to thermal damage of the a-Si or uneven grain growth, resulting in a slightly better grain boundary density compared to SPC. ELA benefits from its excellent directionality and the local nature of the laser beam, yielding the lowest grain boundary density in the poly-Si film among the three techniques, thereby offering the best electrical performance for the devices.

1.3.3.2. Special Structure of Poly-Si Thin-Film Transistors

Lightly Doped Drain

As a common form of degradation in TFTs, hot carrier (HC) degradation typically leads to a significant deterioration of the on-state current (I_{on}) in poly-Si TFTs, posing a substantial challenge to the reliability of poly-Si devices. To address this issue, a structure known as the lightly doped drain (LDD) has been proposed [65]. The LDD structure incorporates a lightly doped region between the drain and the active layer. The presence of this structure significantly reduces the high electric field emanating from the drain, thereby effectively suppressing HC degradation in poly-Si TFTs.

However, it is important to note that, in order to achieve the goal of suppressing the HC effect, the LDD structure is typically designed without vertical overlap with the gate electrode. Due to the lighter doping in this region, the number of carriers is significantly reduced, which in turn moderately inhibits the device's I_{on}. It can be said that the LDD structure is a trade-off between sacrificing some device performance for enhanced reliability [66].

Bridged-Grain Structure

A novel TFT structure, referred to as bridged-grain (BG), has been proposed to enhance the electrical characteristics of MIC and SPC TFTs, enabling their performance to rival that of ELA TFTs [67]. This structure is suitable for driving high-resolution displays and OLED displays. Known as BG-TFT technology, it offers a significantly lower production cost compared to ELA and superior stability to MO TFT, indicating a broad potential for application [68].

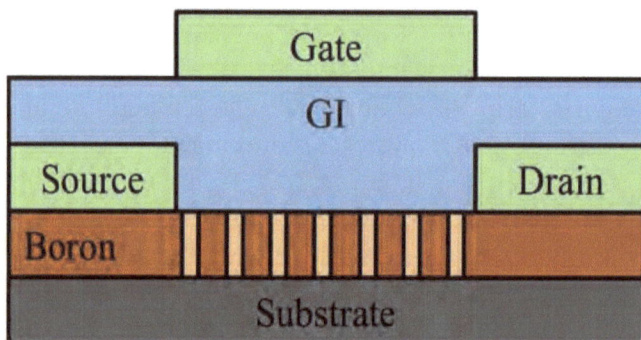

Fig. (1.12). Schematic diagram of BG poly-Si TFT structure.

The schematic of the BG-TFT structure is depicted in Fig. (**1.12**). Narrow regions of highly doped BG areas are uniformly distributed along the channel length. The doping type of the BG regions matches that of the source and drain regions, effectively dividing the active layer under the gate into a series of shorter channels. The distance between adjacent BG regions is intentionally designed to be short, allowing a single TFT with this channel length to exhibit a degree of short-channel effects (SCE). This innovative structure fully leverages the advantages of SCE and multi-junction effects, leading to substantial improvements in the TFT's V_{th}, SS, on-off current ratio, and μ_{FE}. Additionally, the device's HC reliability, self-heating (SH) reliability, and negative bias temperature instability (NBTI) are greatly enhanced.

Taking a p-type poly-Si TFT as an example, the channel is N-doped, while the source, drain, and BG regions are P-doped. With the incorporation of BG regions, the channel becomes P-N-P-···N-P···P-N-P, enabling the BG poly-Si TFT to achieve the following unique effects:

(1) In the off state, the BG-TFT exhibits a low leakage current, similar to a multi-gate TFT.

(2) In the on state, the p-type BG lines are conductive, and their short interval reduces the resistance of the active layer and increases the mobility.

(3) The BG-TFT is equivalent to a series of many short-channel TFTs, utilizing SCE to lower $|V_{th}|$, improve *SS*, and increase μ_{FE}. Since the voltage applied to each short-channel TFT is equal to 1/N of the entire BG-TFT's L (where N is the number of BG lines), the adverse effects of SCE, such as higher leakage current and worse HC reliability, typically observed in individual short-channel TFTs, do not occur in the BG-TFT.

Therefore, the leakage current of short-channel devices at higher drain voltage (V_{ds}) is effectively suppressed by the inherent multi-junction action of the BG structure, endowing the device with excellent on-state characteristics and low off-state leakage current.

1.3.4. Metal Oxide Thin-Film Transistors

MO TFTs are currently a focal point of research in the field of thin-film transistors, with widespread applications in both production and everyday life. They have led to the development of common structures such as the top-gate self-aligned [69], back channel *etc*hing (BCE) [70], *etc*hing stop layer (ESL) [71], and elevated-metal metal oxide (EMMO) [72]. This section will introduce these four structures:

(1) **Top-Gate Self-Aligned Structure:** In traditional TFT fabrication, precise alignment between the top gate and the channel is crucial, as misalignment can lead to performance degradation or device failure. As shown in Fig. (**1.13a**), the top-gate self-aligned structure in MO TFTs involves sequential *etc*hing to form a self-aligned insulating layer and gate electrode, followed by the deposition of source and drain electrodes on the exposed active layer at both ends. This structure eliminates the overlapping region between the gate electrode and the source/drain electrodes, reducing parasitic capacitance. Additionally, the self-alignment of the gate and insulating layer reduces the number of photomasks required, simplifying the alignment process during manufacturing. However, this structure demands high stability from the active layer material.

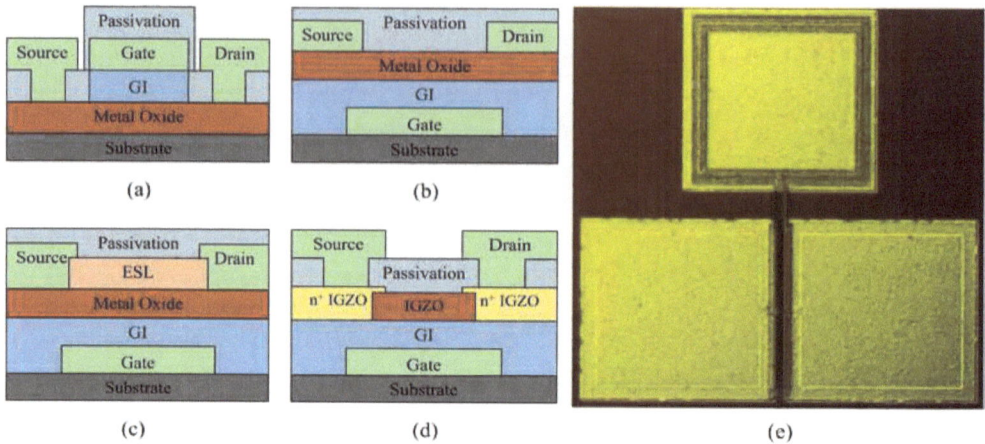

Fig. (1.13). Schematic diagrams of common IGZO TFT structures: (a) top-gate self-aligned structure, (b) BCE structure, (c) ESL structure, and (d) EMMO structure. (e) Microscopic image of an EMMO TFT.

(2) Back Channel *etch*ing Structure: Fig. (**1.13b**) illustrates the BCE structure, which is a bottom-gate top-contact configuration. After depositing the active layer on the insulating layer, source and drain electrodes are directly deposited on top of the active layer. The primary advantage of this structure is its simplicity and ease of fabrication. However, the deposition of source and drain electrodes can cause damage to the back channel of the active layer, leading to decreased device performance and stability. A passivation layer at the top is also used to protect the back channel from environmental degradation over time.

3) Etching Stop Layer Structure: In contrast to the BCE structure, the ESL structure incorporates an *etch*ing stop layer on the IGZO back channel, which effectively prevents damage to the active layer during the deposition of source and drain electrodes. However, the *etch*ing stop layer must be slightly longer than the spacing between the source and drain electrodes to provide adequate protection. This feature is not conducive to device miniaturization and conflicts with the high-resolution requirements of FPDs. Furthermore, the introduction of an *etch*ing stop layer adds an additional photomask step, increasing the overall process cost.

4) Elevated-Metal Metal Oxide Structure: As illustrated in Fig. (**1.13d and 1.13e**), the EMMO TFT design elevates the source and drain metal electrodes above the passivation layer [72]. The silicon nitride above the active layer serves a dual purpose, acting as both a passivation layer and an *etch*ing stop layer. This configuration protects the back channel during the *etc*hing process of the metal electrodes. Compared to the ESL structure, the EMMO TFT does not require an additional photomask, simplifying the fabrication process. Moreover, the source and drain regions in the EMMO TFT are formed through an annealing process, which is distinct from the traditional IGZO TFTs that rely on doping to create high-conductivity source and drain regions.

1.4. APPLICATIONS IN MODERN ELECTRONICS

Compared to metal-oxide-semiconductor field-effect transistors (MOSFETs), which are made from single-crystal silicon, TFTs composed of polycrystalline or amorphous materials exhibit a higher density of defects. This leads to a significant performance disparity between TFTs and MOSFETs of the same dimensions. For instance, the carrier mobility of a-Si TFTs is around 1 cm^2/Vs, while MO TFTs have mobilities in the tens of cm^2/Vs, and MOSFETs boast mobilities exceeding 600 cm^2/Vs. Consequently, TFTs typically operate at microampere-level currents, whereas MOSFETs can operate at milliampere-level currents. Additionally, the high defect density in the active layer of TFTs makes them susceptible to performance degradation due to voltage stress, temperature variations, or light exposure, which can lead to circuit failure. Therefore, when designing TFT circuits, compensatory circuits are necessary to mitigate these effects, limiting the application of TFTs in integrated and high-speed circuits.

However, TFTs have distinct advantages over MOSFETs in terms of both material preparation and fabrication processes. They can be made from polymers or a-Si at a low cost, and their fabrication can utilize substrates such as glass or flexible polymers. Despite their performance limitations compared to MOSFETs, TFTs are still highly useful in simple, large-area circuits, such as pixel circuits, which is why they are extensively applied in display technology. Moreover, with the evolution of TFT technology, researchers are exploring applications for TFTs beyond the display field, including flexible electronics, sensors, and artificial synapses. This chapter will delve into the various applications of TFTs.

1.4.1. Thin-Film Transistors in Active-Matrix Display

TFTs are extensively utilized in FPD technologies. The current landscape of FPDs predominantly employs active matrix display technologies, with LCD and OLED technologies being the leading contenders. LCD technology, which has been developed for a longer period, is well-established and dominates a significant portion of the mid to low-end market. However, in recent years, OLED technology has seen rapid growth, offering advantages such as a wider color gamut, lower power consumption, and the capability for flexible displays, making it increasingly mainstream. Additionally, Micro-LED technology has emerged with the potential for higher brightness, lower power consumption, and longer lifespan, making its mark in the high-end market. Emerging technologies like holographic displays, quantum dot displays, and laser projection are also being actively explored, heralding new horizons for the future of display technologies. Concurrently, advancements in flexible display technologies are opening up endless possibilities for wearable devices, curved screens, and rollable displays.

The primary distinction between AMLCD and AMOLED technologies lies in the self-emission of subpixels. As depicted in Fig. (**1.14**), AMLCD requires an additional backlight and polarizing filters to display images, resulting in higher power consumption compared to AMOLED technology, which features self-emitting subpixels. This also leads to a more complex structure in AMLCDs. However, organic light-emitting diodes are more prone to degradation over extended use compared to inorganic technologies. Despite the diversity of active matrix display technologies, the pixel circuits at their core are composed of TFTs.

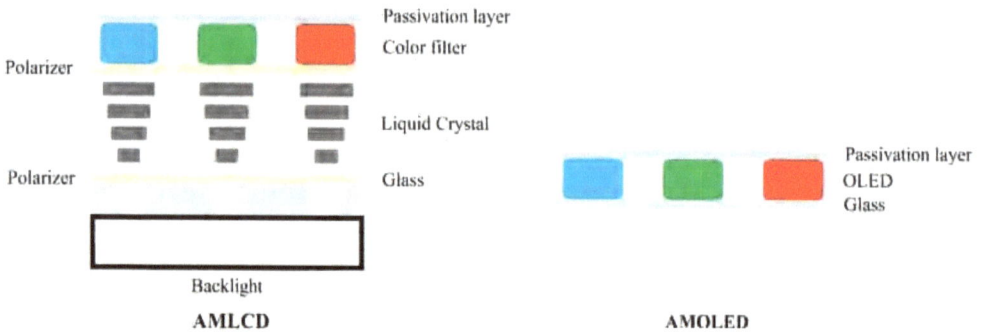

Fig. (1.14). Key differences between AMLCD and AMOLED technology.

As depicted in Fig. (**1.15**), the simplified pixel circuits for AMLCD and AMOLED display technologies exhibit distinct operational characteristics. AMLCD is characterized as a voltage-driven circuit, requiring only a single TFT within the pixel to function as a switching device, which charges and discharges the capacitor to a specific voltage level. In contrast, AMOLED is a current-driven circuit [1], necessitating the use of two TFTs per pixel: one to serve as the switching transistor and another as the driving transistor. The circuits shown in the figure are simplified representations; in practice, due to the degradation of TFTs over extended periods of operation and performance variations among devices during the pixel fabrication process, additional compensation circuits are required to ensure optimal display performance. Comparatively, since AMOLEDs are directly driven by TFTs, the compensation circuits tend to be more complex, employing configurations such as 5T1C [73] and 7T1C [74]. Consequently, the reliability of TFTs has become a critical area of focus within the industry.

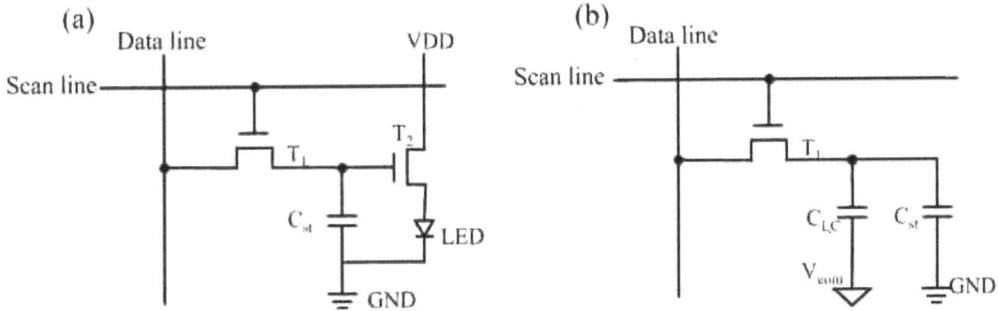

Fig. (1.15). Pixel circuit diagram: (a) AMOLED and (b) AMLCD.

With the increasing demand for higher screen resolutions, ranging from 2K to 4K and even 8K, and the rising refresh rates, the pixel density is growing exponentially. This growth implies a proportional increase in the number of drive signals and a need for faster driving speeds. Relying solely on external driving integrated circuits (ICs) for this purpose would require a significant number of pins. Therefore, the industry has shifted towards incorporating TFTs in the design of gate driver arrays (GOA), as illustrated in Fig. (**1.16**). To meet the escalating market demands, research into TFTs with higher mobility and enhanced reliability has become increasingly imperative.

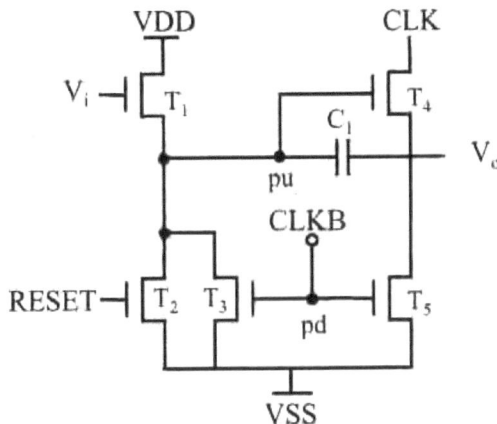

Fig. (1.16). Simplified circuit diagram of gate drive array.

1.4.2. Thin-Film Transistors in Sensors

The utilization of TFTs in the sensor domain is exhibiting a trend of increasing diversification and innovation. The range of materials available for TFT fabrication is extensive, encompassing a-Si, MO, organic materials, and even hybrid combinations of these. Within this array of materials, there are sensor-specific variants such as zinc oxide and sodium oxide, which imbue TFTs with the potential for direct application in sensor technologies by leveraging their intrinsic properties. Moreover, the fabrication process for TFTs is relatively straightforward, facilitating easier integration with a variety of sensing materials. The low-temperature processing involved allows for device fabrication on flexible substrates, significantly broadening the prospects for TFTs in the realm of flexible electronic devices.

At present, TFTs are predominantly employed in display technologies. However, many TFT materials possess photosensitive properties, as shown in Fig (**1.17**), where exposure to light within specific wavelength ranges can cause shifts in threshold voltage (V_{th}) or increases in leakage current [75-78]. Although the photoconductive (PPC) effect is common to many materials, it can be effectively negated by applying gated pulses [79]. This capability extends the application of TFTs to optoelectronic sensing, with potential uses in fingerprint recognition and X-ray detection, among others.

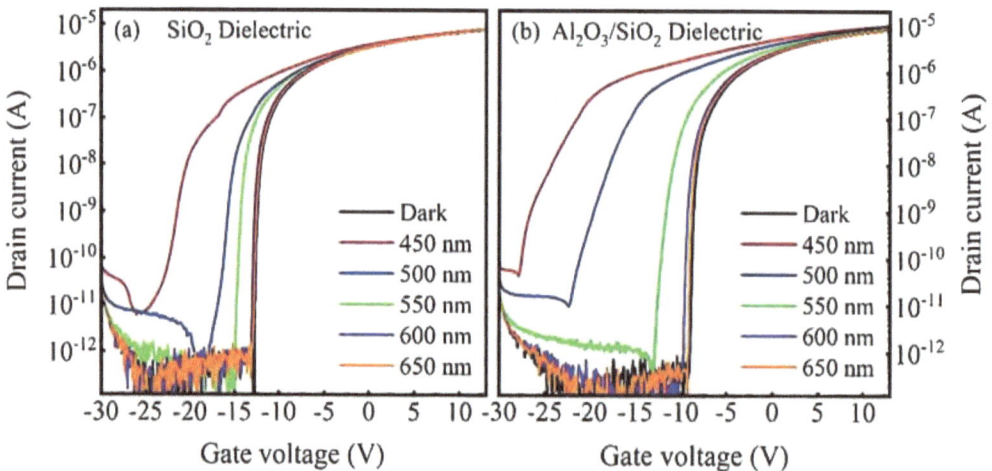

Fig. (1.17). Transfer characteristics of (a) IZO TFTs with SiO_2 dielectric and (b) IZO TFTs with Al_2O_3/SiO_2 dielectric under illuminations with different visible light λs.

In the domain of temperature sensing, while it is ideal for TFTs to have sufficient negative/positive bias illumination stress (NBIS/PBIS) reliability performance during operation, the change of TFT's performance under different temperatures can also be used for temperature sensors in some fields. With an increase in temperature, the I_{off} also rises [80], facilitating temperature detection. This is not the only method; temperature can also be detected through the relationship between V_{th} and temperature, where an increase in temperature results in a V_{th} shift [81- 83].

In terms of gas detection, traditional gas sensors typically use resistance or capacitance for detection. Although these sensors offer high sensitivity, TFT gas sensors provide better selectivity. As shown in Fig. (**1.18**), there are already TFT-based sensors for detecting gases such as methane [84], nitrogen dioxide [85], and hydrogen sulfide [86]. For pressure sensing, TFTs can be combined with piezoelectric materials to achieve touch detection [87] or pulse detection [88].

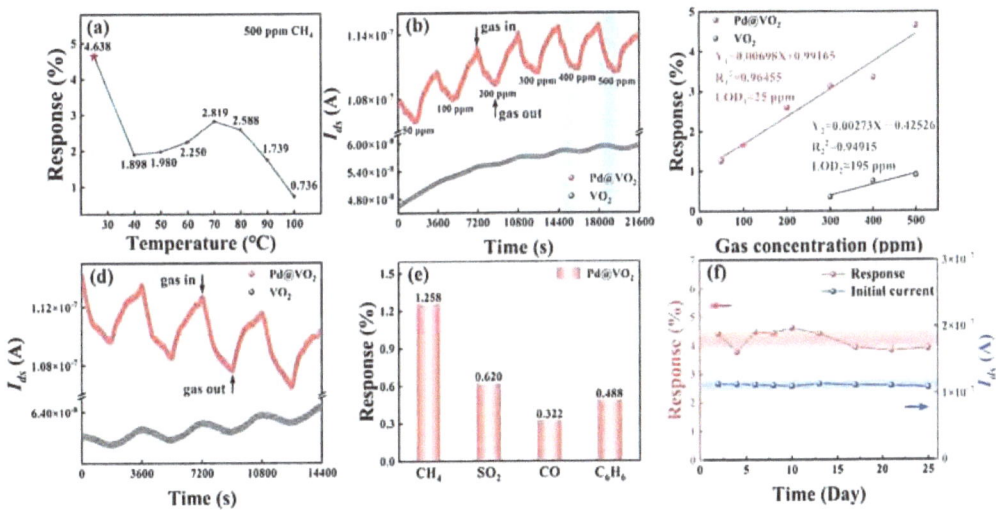

Fig. (1.18). Gas-sensing properties of TFT sensors. (a) The response values of the Pd@V$_{O2}$ TFT sensor toward 500 ppm CH$_4$ at different working temperatures. (b) The dynamic response–recovery curves, (c) the fitted curves, (d) the reproducibility test curves, and (e) the response values of the Pd@V$_{O2}$ TFT sensor toward different gases at 50 ppm. (f) The response value and base current change in the Pd@V$_{O2}$ TFT sensor measured several times within 25 days [85].

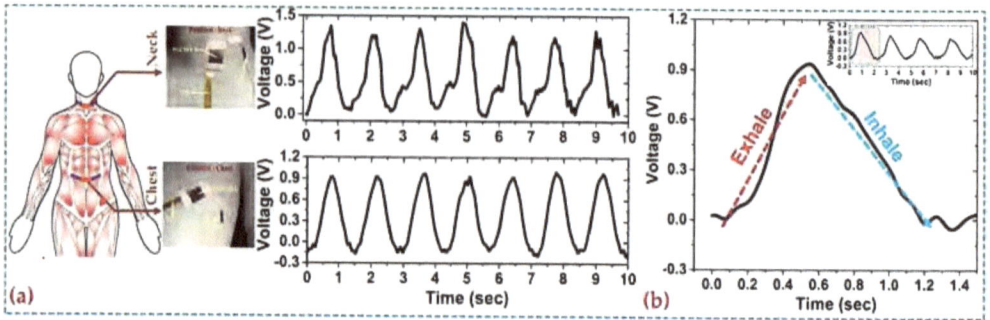

Fig. (1.19). Dynamic response of the sensor at two peripheral points of (a) neck and chest; (b) signal of one complete respiration cycle [88].

Overall, the low power consumption, flexibility (Fig. **1.19**), and high integration capabilities of TFTs open up numerous possibilities for the design of sensor devices, providing strong technical support for the intelligent and precise development of various industries. As technology continues to advance and innovate, TFT sensors will continue to play a significant role in the future.

1.4.3. Thin-Film Transistors in Other Circuits

With the advancement of TFT technology, the application of TFTs in circuits beyond display circuits has become a current research hotspot. Although TFTs may underperform compared to MOSFETs in terms of performance, they still hold significant advantages in cost-sensitive scenarios such as flexible wearable devices and edge processing circuits. Previously, researchers have implemented flexible electronic tag RFIDs based on MO TFTs [89], and some teams have even utilized these for microcontroller units (MCUs). However, due to the low mobility of TFTs and the lack of high-performance P-type TFTs, these circuits may not perform optimally. Yet, as TFT technology evolves, these issues are expected to be resolved. Additionally, TFTs can be used as front-end processors for sensors to enhance signal quality [90].

In cutting-edge research, as shown in Fig. (**1.20**), TFTs are anticipated to be applied in artificial synapses and neural networks [91, 92]. The integration of hardware-based neural networks, combined with large-area sensing arrays, can significantly reduce signal processing time and lower processing costs.

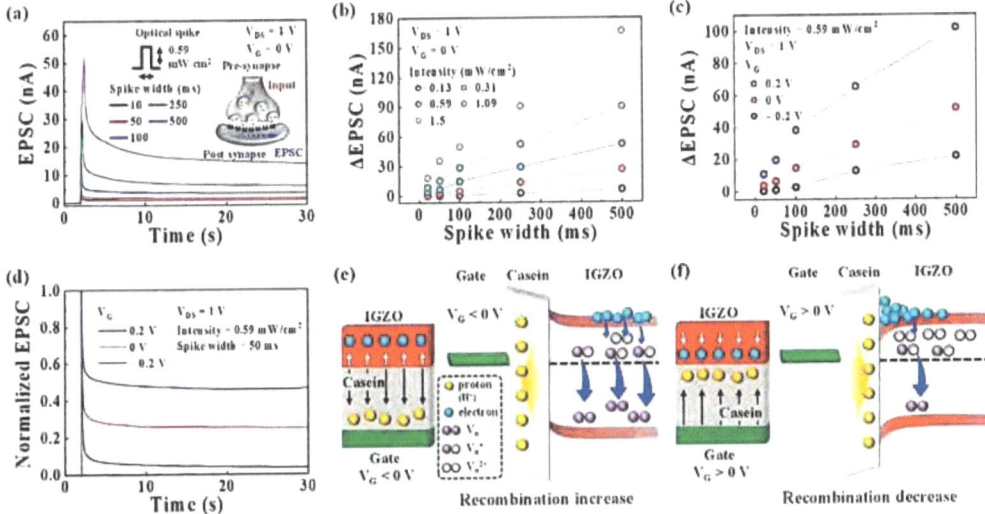

Fig. (1.20). (a) EPSC triggered by a single pre-synaptic optical spike (395 nm, 0.59 mW/cm^2) with various spike widths ranging from 20 to 500 ms at V_g of 0 V. Inset: illustration of a biological synapse. (b) ΔEPSC plotted against the light intensity ranging from 0.13 to 1.5 mW/cm^2 at V_g of 0 V for different optical spike widths. (c) ΔEPSC plotted at different VG values (−0.2 V, 0 V, and 0.2 V) with an optical spike (0.59 mW/cm^2, 50 ms). (d) Normalized EPSC induced by the optical spike (0.59 mW/cm^2, 50 ms) under different V_g values (−0.2 V, 0 V, and 0.2 V). Mechanisms of the recombination reactions that occur at (e) negative and (f) positive V_g values [91].

1.5. RELIABILITY OF THIN-FILM TRANSISTOR TECHNOLOGY

As previously discussed, TFTs are exposed to various working environments, such as moisture [93], light [34], and high temperatures [94], both in display panel designs and other applications like photodetection [81]. Although in practical applications, TFTs can often be isolated from most external environmental factors through encapsulation, these factors can still affect the device's performance to a certain extent. Additionally, TFTs operate under different voltage conditions, including direct current (DC) [95] and alternating current (AC) [96], and the application of these voltages may also lead to device degradation.

The presence of these issues means that the diverse applications of TFTs not only demand performance requirements but also set higher standards for device reliability. For example, in AMOLED displays, device degradation in pixel driving circuits can directly affect the panel's luminance and uniformity [97]; in SoP, degradation can also impact the performance of multiplexers [98], shift registers [99], and driving circuits [100]. Similarly, degradation in TFTs in other applications can also have varying degrees of impact on the overall circuit system.

1.5.1. Degradation in Active Matrix Displays

As mentioned in section 3.1, AM display circuits typically consist of addressing transistors, driving transistors, and additional compensation circuits. In the case of AMOLED pixel circuits, the addressing TFT controls the switching of the driving TFT. When the addressing TFT is turned off, the storage capacitor (C_s) continues to maintain the driving transistor in an on state, keeping the OLED lit. This means that the driving TFT is often in a state of continuous activation. Consequently, the performance and lifespan of an AMOLED are influenced by the lifespan of the TFTs and the OLEDs. On the other hand, in AMLCDs, the backlight is always on, and the TFTs primarily serve as switches.

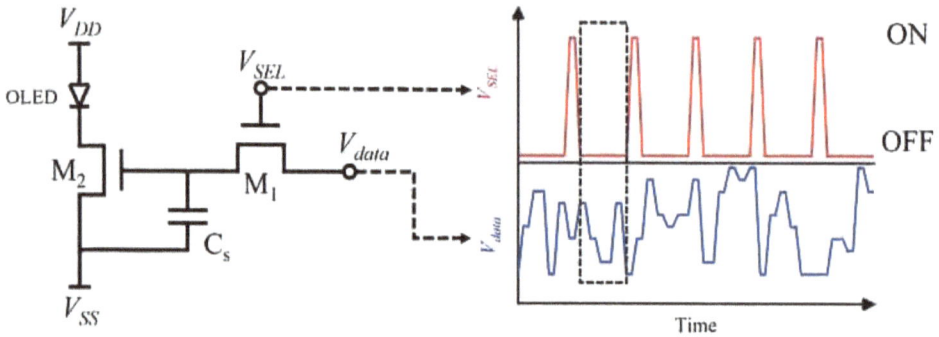

Fig. (1.21). Practical voltage pattern applied to the addressing TFT in the OFF state.

Therefore, for AM displays, particularly AMOLEDs, the reliability of the TFTs themselves is crucial. For the addressing transistors, the practical voltage patterns applied to the gate and drain electrodes of the switching TFT are shown in Fig. (**1.21**). For the select (SEL) pulse, the practical voltage pattern can be divided into two main types: the OFF state of select voltage (V_{SEL}) and the ON state of V_{SEL}. However, the ON duty ratio is quite small. For instance, in an AMOLED display with 1080 scan lines, the ON duty ratio (α) for a switching TFT in a pixel is only 1/1080 [101]. In other words, for the majority of its lifetime, the channel of the switching TFT is in the OFF state. Regarding the data pulses (V_{data}), they are irregular and do not follow a specific pattern. The situation for the switching TFT in AMLCD is similar.

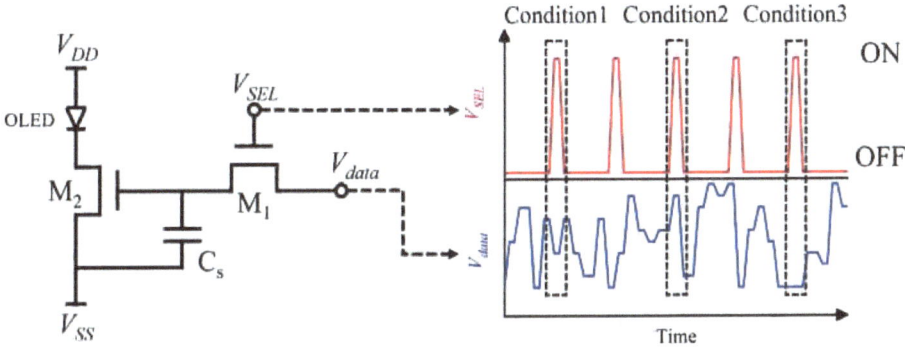

Fig. (22). Practical voltage pattern applied to the addressing TFT in the on state.

The ON state of the V_{SEL}, as shown in Fig. (**1.22**), can be represented by a V_{SEL} pulse combined with a DC V_{data} (Condition1), a forward synchronized V_{SEL} pulse and V_{data} pulse (Condition2), and a reversely synchronized V_{SEL} pulse and V_{data} pulse (Condition3). The above four types of combinations can adequately simulate the working conditions of the switching TFT [102]. For the switching TFT in AM liquid-crystal displays, the situation is similar.

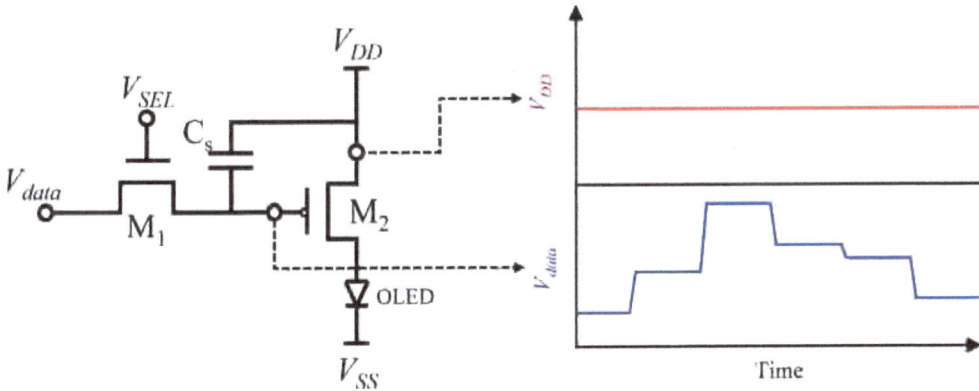

Fig. (1.23). Practical voltage pattern applied to the driving TFT.

For the driving TFT, as shown in Fig. (**1.23**), taking a p-type driving TFT as an example, the source electrode is usually connected to a positive power supply. The gate electrode suffers from irregular voltage pulses, as shown in Fig. (**1.23**). Different gate voltage values stand for different gray levels [103].

It is evident that TFT devices, under the aforementioned conditions, are typically subjected to AC, DC, and mixed AC-DC voltage signals. These voltages act as stressors within the device, leading to the generation of defects and consequently causing various degrees of degradation, such as the reduction of mobility, the degradation of V_{th}, and the deterioration of SS, among others. The primary degradation mechanisms for DC stress can be categorized into SH effect, HC effect, and positive/negative bias stress (PBS/NBS) effects. For AC stress, the main degradation is attributed to the AC HC effect, which primarily results in mobility degradation.

For a TFT operating in the saturation region, the output drain current (I_d) can be expressed as [104]:

$$I_d = \frac{1}{2}\frac{W}{L}\mu C_{ox}(V_{gs} - V_{th})^2 \tag{1}$$

Here, W and L represent the width and length of the device channel, C_{ox} is the gate insulator's unit capacitance, and V_{gs} is the gate-source voltage.

For AMOLEDs, the magnitude of I_d directly determines the intensity of the device's light emission, and a decrease in μ directly leads to a reduction in the overall display brightness. The degradation of V_{th} is also a significant indicator of device degradation, often causing a negative or positive shift in overall device performance. For example, in an n-type device, a significant positive shift in V_{th} typically requires a larger V_{gs} to control the device's reactivation. Similarly, to achieve the same brightness in the OLED, a larger V_{gs} is needed. Conversely, a negative shift in V_{th} implies that the device is more easily activated, and the original gate voltage for the off state can no longer control the device's on/off state, leading to the OLED being constantly bright at a higher level [105].

Device degradation, such as SH effects, also leads to the deterioration of the device's SS, slowing down the device's turn-on and turn-off speeds. Particularly for high-refresh-rate displays, significant SS degradation can often result in reduced refresh rates or image distortion [106].

1.5.2. Degradation in Sensors

The application of TFTs in the sensor field often capitalizes on the inherent sensitivity of their active layers to external environmental factors, such as light, temperature, and gases. Furthermore, as switching devices, TFTs can conveniently output signal variations, as discussed in section 3.2, indicating a broad prospective application for TFTs in the sensor domain.

Typically, TFTs used for sensing include MO TFTs, a-Si TFTs, and OTFTs. Compared to these, poly-Si TFTs are less frequently employed in sensor applications due to their relative stability and limited responsiveness to environmental changes. This suggests that TFTs intended for sensor use generally require heightened sensitivity to external stimuli.

However, it is important to note that for TFTs to be effective in practical applications, their reliability is also of high importance. The reliability of TFTs used in sensors can be categorized into two aspects: stability and reusability [107]. Stability refers to the ability of a TFT to maintain its performance and return to its initial state after prolonged exposure to environmental factors such as light, high temperatures, and gases, as shown in Fig. (**1.24**). For instance, MO TFTs used in photo sensors primarily function by altering the device's V_{th} in response to varying levels of external light exposure. However, under high-frequency light stress, MO TFTs can accumulate oxygen vacancies (V_os) that, upon gaining energy, are ionized as ionized oxygen vacancies (V_o^{2+}), or the active layer within the device may sustain damage due to the combined effects of light and electrical stress, leading to performance degradation [108]. Such degradation in performance can diminish the signal output of the device or even render it inoperative, resulting in reduced sensor accuracy or damage.

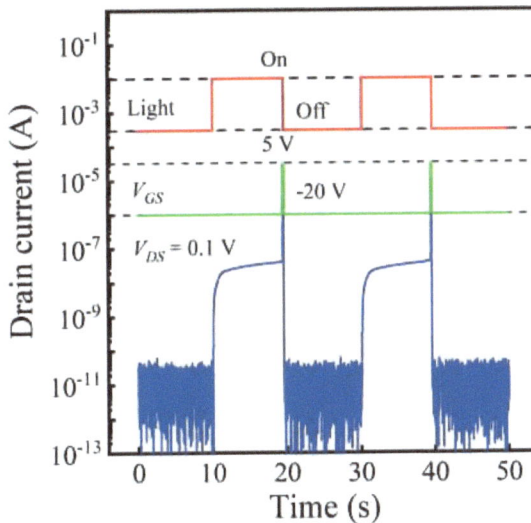

Fig. (1.24). MO TFT produces a fixed variation in drain current over multiple cycles of light exposure.

The reusability of sensors indicates that TFTs can consistently produce a defined performance change in response to the same environmental variations after multiple exposures. Specifically, a TFT is considered to have good reusability if, under several cycles of changes in a particular environment (such as light, humidity, gases, magnetic fields, *etc.*), the rate of performance change, the peak value of the change, and the speed of recovery remain almost identical [109]. If these requirements are not met, the application of TFTs in sensors will be limited to single-use sensors only.

1.5.3. Degradation in Other Applications

As mentioned in section 3.3, TFTs also have promising applications in various practical circuits, such as flexible wearable circuits, flexible electronic tags, and synaptic transistors for artificial intelligence. However, these emerging applications further increase the demands on the reliability of TFTs. For instance, in the field of flexible applications, TFTs are not only exposed to complex and diverse electrical signals but also subjected to mechanical stress [110]. For most transistors, thin-film layers are patterned into small features or narrow lines, resulting in different failure and mechanical behaviors compared to uniform layers. Additionally, in flexible and stretchable transistors, the semiconductor layer is typically very thin relative to other layers. Under these circumstances, the semiconductor is not the decisive factor that determines the overall flexibility of the transistor stacks. In most cases, the gate dielectric and passivation layers are the primary causes of failure in the total stack.

In the context of neural network applications, synaptic transistors also impose certain demands on the reliability of TFTs. Similar to their use in sensors, synaptic transistors alter the device's performance through external environmental factors or applied voltage signals [111, 112]. However, unlike sensors, synaptic transistors require that these performance changes be sustained over a period of time. Consequently, the reliability requirements for synaptic transistors encompass both the stability and reusability of TFTs.

CONCLUSION

This chapter offers a thorough introduction to TFTs, covering their basic concepts, development history, classification, and pivotal role in modern electronics, particularly in display technology. It outlines how TFTs have evolved from a-Si to more advanced materials like poly-Si and MOs, which have enabled higher performance and more versatile applications. The chapter also emphasizes the critical importance of reliability research in TFTs, highlighting how understanding and mitigating degradation mechanisms is essential for ensuring long-term performance, especially in applications like high-resolution displays and flexible electronics.

REFERENCES

[1]   S.J. Lee, S-W. Lee, K-M. Oh, S-J. Park, K-E. Lee, Y-S. Yoo, K-M. Lim, M-S. Yang, Y-S. Yang, and Y-K. Hwang, "A Novel Five-Photomask Low-Temperature Polycrystalline Silicon CMOS Structure for AMLCD Application", *IEEE Trans. Electron Dev.,* vol. 57, no. 9, pp. 2324-2329, 2010.
http://dx.doi.org/10.1109/TED.2010.2053868

[2]   Shengdong Zhang, Chunxiang Zhu, J.K.O. Sin, J.N. Li, and P.K.T. Mok, "Ultra-thin elevated channel poly-Si TFT technology for fully-integrated AMLCD system on glass", *IEEE Trans. Electron Dev.,* vol. 47, no. 3, pp. 569-575, 2000.
http://dx.doi.org/10.1109/16.824731

[3]   S. Huang, J. Jin, J. Kim, W. Wu, A. Song, and J. Zhang, "IGZO Source-Gated Transistor for AMOLED Pixel Circuit", *IEEE Trans. Electron Dev.,* vol. 70, no. 7, pp. 3637-3642, 2023.
http://dx.doi.org/10.1109/TED.2023.3274501

[4]   X. Ji, F. Wang, H. Zhou, K. Wang, L. Yin, and J. Zhang, "3400 PPI Active-Matrix Monolithic Blue and Green Micro-LED Display", *IEEE Trans. Electron Dev.,* vol. 70, no. 9, pp. 4689-4693, 2023.
http://dx.doi.org/10.1109/TED.2023.3295764

[5]   Y.Z. Lin, C. Liu, J-H. Zhang, Y-K. Yuan, W. Cai, L. Zhou, M. Xu, L. Wang, W-J. Wu, and J-B. Peng, "Active-Matrix Micro-LED Display Driven by Metal Oxide TFTs Using Digital PWM Method", *IEEE Trans. Electron Dev.,* vol. 68, no. 11, pp. 5656-5661, 2021.
http://dx.doi.org/10.1109/TED.2021.3112947

[6]   "Transistor sizing for AMLCD integrated TFT drive circuits", 2024. Available From: https://sid.onlinelibrary.wiley.com/doi/epdf/10.1889/1.1985187

[7]   B. B. Jie, "The bipolar theory of the Bipolar Field-Effect Transistor: Recent advances", *in 2008 9th International Conference on Solid-State and Integrated-Circuit Technology,* Beijing, China: IEEE, pp. 480-483, 2008.
http://dx.doi.org/10.1109/ICSICT.2008.4734585

[8]   J. Dostal, "A mathematical approximation of junction FET characteristics", *Proc. IEEE,* vol. 62, no. 6, pp. 855-856, 1974.
http://dx.doi.org/10.1109/PROC.1974.9529

[9]   Effect of Ag doping on the electrical properties of thermally deposited CdS–La2O3 TFTs", 2024.   Available   From: https://www.tandfonline.com/doi/epdf/10.1080/00207217.2011.643503?needAccess=true

[10]   H.A. Klasens, and H. Koelmans, "A tin oxide field-effect transistor", *Solid-State Electron.,* vol. 7, no. 9, pp. 701-702, 1964.
http://dx.doi.org/10.1016/0038-1101(64)90057-7

[11]   G.F. Boesen, and J.E. Jacobs, "ZnO field-effect transistor", *Proc. IEEE,* vol. 56, no. 11, pp. 2094-2095, 1968.

http://dx.doi.org/10.1109/PROC.1968.6813

[12]   A. Tsumura, H. Koezuka, and T. Ando, "Macromolecular electronic device: Field-effect transistor with a polythiophene thin film", *Appl. Phys. Lett.*, vol. 49, no. 18, pp. 1210-1212, 1986.

http://dx.doi.org/10.1063/1.97417

[13]   K. Kandpal, and N. Gupta, "Perspective of zinc oxide based thin film transistors: a comprehensive review", *Microelectron. Int.*, vol. 35, no. 1, pp. 52-63, 2018.

http://dx.doi.org/10.1108/MI-10-2016-0066

[14]   W. Zhou, Z. Meng, S. Zhao, M. Zhang, R. Chen, M. Wong, and H-S. Kwok, "Bridged-Grain Solid-Phase-Crystallized Polycrystalline-Silicon Thin-Film Transistors", *IEEE Electron Device Lett.*, vol. 33, no. 10, pp. 1414-1416, 2012.

http://dx.doi.org/10.1109/LED.2012.2210019

[15]   Meng Zhang, Wei Zhou, Rongsheng Chen, Man Wong, and Hoi-Sing Kwok, "Characterization of DC-Stress-Induced Degradation in Bridged-Grain Polycrystalline Silicon Thin-Film Transistors", *IEEE Trans. Electron Dev.*, vol. 61, no. 9, pp. 3206-3212, 2014.

http://dx.doi.org/10.1109/TED.2014.2341676

[16]   M. Zhang, W. Zhou, R. Chen, M. Wong, and H.S. Kwok, "High-performance polycrystalline silicon thin-film transistors integrating sputtered aluminum-oxide gate dielectric with bridged-grain active channel", *Semicond. Sci. Technol.*, vol. 28, no. 11, p. 115003, 2013.

http://dx.doi.org/10.1088/0268-1242/28/11/115003

[17]   D.H. Kim, S.K. Lim, H. Bae, C-K. Kim, S-W. Lee, M. Seo, S-Y. Kim, K-M. Hwang, G-B. Lee, B.H. Lee, and Y-K. Choi, "Quantitative Analysis of Deuterium Annealing Effect on Poly-Si TFTs by Low Frequency Noise and DC I – V Characterization", *IEEE Trans. Electron Dev.*, vol. 65, no. 4, pp. 1640-1644, 2018.

http://dx.doi.org/10.1109/TED.2018.2805316

[18]   L. Petti, N. Münzenrieder, C. Vogt, H. Faber, L. Büthe, G. Cantarella, F. Bottacchi, T.D. Anthopoulos, and G. Tröster, "Metal oxide semiconductor thin-film transistors for flexible electronics", *Appl. Phys. Rev.*, vol. 3, no. 2, p. 021303, 2016.

http://dx.doi.org/10.1063/1.4953034

[19]   R.L. Hoffman, B.J. Norris, and J.F. Wager, "ZnO-based transparent thin-film transistors", *Appl. Phys. Lett.*, vol. 82, no. 5, pp. 733-735, 2003.

http://dx.doi.org/10.1063/1.1542677

[20]   R.L. Hoffman, "ZnO-channel thin-film transistors: Channel mobility", *J. Appl. Phys.*, vol. 95, no. 10, pp. 5813-5819, 2004.

http://dx.doi.org/10.1063/1.1712015

[21]   E.M.C. Fortunato, P.M.C. Barquinha, A.C.M.B.G. Pimentel, A.M.F. Gonçalves, A.J.S. Marques, R.F.P. Martins, and L.M.N. Pereira, "Wide-bandgap high-mobility ZnO thin-film transistors produced at room temperature", *Appl. Phys. Lett.*, vol. 85, no. 13, pp. 2541-2543, 2004.

http://dx.doi.org/10.1063/1.1790587

[22] T. Kamiya, K. Nomura, and H. Hosono, "Present status of amorphous In–Ga–Zn–O thin-film transistors", *Sci. Technol. Adv. Mater.*, vol. 11, no. 4, p. 044305, 2010.
http://dx.doi.org/10.1088/1468-6996/11/4/044305 PMID: 27877346

[23] G.L. Olson, and J.A. Roth, "Kinetics of solid phase crystallization in amorphous silicon", *Mater. Sci. Rep.*, vol. 3, no. 1, pp. 1-77, 1988.
http://dx.doi.org/10.1016/S0920-2307(88)80005-7

[24] R.E. Presley, D. Hong, H.Q. Chiang, C.M. Hung, R.L. Hoffman, and J.F. Wager, "Transparent ring oscillator based on indium gallium oxide thin-film transistors", *Solid-State Electron.*, vol. 50, no. 3, pp. 500-503, 2006.
http://dx.doi.org/10.1016/j.sse.2006.02.004

[25] N. Matsuo, A. Heya, and H. Hamada, "Review—Technology Trends of Poly-Si TFTs from the Viewpoints of Crystallization and Device Performance", *ECS J. Solid State Sci. Technol.*, vol. 8, no. 4, pp. P239-P252, 2019.
http://dx.doi.org/10.1149/2.0211903jss

[26] Interface Engineering of Metal Oxide Semiconductors for Biosensing Applications", 2024. Available From: https://onlinelibrary.wiley.com/doi/epdf/10.1002/admi.201700020

[27] X.Z. Bo, N. Yao, S.R. Shieh, T.S. Duffy, and J.C. Sturm, "Large-grain polycrystalline silicon films with low intragranular defect density by low-temperature solid-phase crystallization without underlying oxide", *J. Appl. Phys.*, vol. 91, no. 5, pp. 2910-2915, 2002.
http://dx.doi.org/10.1063/1.1448395

[28] G.J. Qi, S. Zhang, T.T. Tang, J.F. Li, X.W. Sun, and X.T. Zeng, "Experimental study of aluminum-induced crystallization of amorphous silicon thin films", *Surf. Coat. Tech.*, vol. 198, no. 1-3, pp. 300-303, 2005.
http://dx.doi.org/10.1016/j.surfcoat.2004.10.092

[29] G.M. Wu, C.N. Chen, W.S. Feng, and H.C. Lu, "Improved AMOLED with aligned poly-Si thin-film transistors by laser annealing and chemical solution treatments", *Physica B,* vol. 404, no. 23-24, pp. 4649-4652, 2009.
http://dx.doi.org/10.1016/j.physb.2009.08.147

[30] S. Jeon, S. Park, I. Song, J.H. Hur, J. Park, H. Kim, S. Kim, S. Kim, H. Yin, U.I. Chung, E. Lee, and C. Kim, "Nanometer-scale oxide thin film transistor with potential for high-density image sensor applications", *ACS Appl. Mater. Interfaces,* vol. 3, no. 1, pp. 1-6, 2011.
http://dx.doi.org/10.1021/am1009088 PMID: 21171647

[31] L. Lu, and M. Wong, "The Resistivity of Zinc Oxide Under Different Annealing Configurations and Its Impact on the Leakage Characteristics of Zinc Oxide Thin-Film Transistors", *IEEE Trans. Electron Dev.*, vol. 61, no. 4, pp. 1077-1084, 2014.
http://dx.doi.org/10.1109/TED.2014.2302431

[32] R.E. Presley, C.L. Munsee, C-H. Park, D. Hong, J.F. Wager, and D.A. Keszler, "Tin oxide transparent thin-film transistors", *J. Phys. D Appl. Phys.*, vol. 37, no. 20, pp. 2810-2813, 2004.
http://dx.doi.org/10.1088/0022-3727/37/20/006

[33]   H.Q. Chiang, J.F. Wager, R.L. Hoffman, J. Jeong, and D.A. Keszler, "High mobility transparent thin-film transistors with amorphous zinc tin oxide channel layer", *Appl. Phys. Lett.,* vol. 86, no. 1, p. 013503, 2005.

http://dx.doi.org/10.1063/1.1843286

[34]   P. Barquinha, A. Pimentel, A. Marques, L. Pereira, R. Martins, and E. Fortunato, "Effect of UV and visible light radiation on the electrical performances of transparent TFTs based on amorphous indium zinc oxide", *Journal of Non-Crystalline Solids,* vol. 352, no. 9–20, pp. 1756-1760, 2006.

http://dx.doi.org/10.1016/j.jnoncrysol.2006.01.068

[35]   Y. Hanyu, K. Domen, K. Nomura, H. Hiramatsu, H. Kumomi, H. Hosono, and T. Kamiya, "Hydrogen passivation of electron trap in amorphous In-Ga-Zn-O thin-film transistors", *Appl. Phys. Lett.,* vol. 103, no. 20, p. 202114, 2013.

http://dx.doi.org/10.1063/1.4832076

[36]   H. Hosono, "How we made the IGZO transistor", *Nat Electron,* vol. 1, pp. 428, 2018.

http://dx.doi.org/10.1038/s41928-018-0106-0

[37]   D.-G. Yang, J.-H. Kim, H.-D. Kim, and H.-S. Kim, "Flexible Amorphous Indium-Tin-Zinc Oxide (a-ITZO) Thin-Film Transistors on Polyimide Substrate",

[38]   Y. Shi, Y.S. Shiah, K. Sim, M. Sasase, J. Kim, and H. Hosono, "High-performance a-ITZO TFTs with high bias stability enabled by self-aligned passivation using a-GaOx", *Appl. Phys. Lett.,* vol. 121, no. 21, p. 212101, 2022.

http://dx.doi.org/10.1063/5.0123253

[39]   A. Ding, R. You, S. Luo, J. Gong, S. Song, K. Wang, B. Dai, and H. Sun, "Influence of annealing temperature on the optoelectronic properties of ITZO thin films", *Nanotechnology,* vol. 32, no. 40, p. 405701, 2021.

http://dx.doi.org/10.1088/1361-6528/ac0dda PMID: 34161926

[40]   Y. Zhou, F. Liu, H. Yang, X. Zhou, G. Li, M. Zhang, R. Chen, S. Zhang, and L. Lu, "Competition between heating and cooling during dynamic self-heating degradation of amorphous InGaZnO thin-film transistors", *Solid-State Electron.,* vol. 195, p. 108393, 2022.

http://dx.doi.org/10.1016/j.sse.2022.108393

[41]   Z. Chen, M. Zhang, S. Deng, Z. Jiang, Y.Yan, S. Han, Y. Zhou, M.Wong, and H-S. Kwok, "Effect of Moisture Exchange Caused by Low-Temperature Annealing on Device Characteristics and Instability in InSnZnO Thin-Film Transistors", *Adv. Mater. Interfaces,* vol. 9, pp. 2102584, 2022.

http://dx.doi.org/10.1002/admi.202102584

[42]   G. Cantarella, J. Costa, T. Meister *et al.,* "Review of recent trends in flexible metal oxide thin-film transistors for analog applications", *Flexible and Printed Electronics,* vol. 5, no. 3, 2020.

http://dx.doi.org/10.1088/2058-8585/aba79a

[43]   W.E. Wei, H-Y. Li, C-Y. Han, J.C-M. Li, J-J. Huang, I-C. Cheng, C-N. Liu, and Y-H. Yeh, "A Flexible TFT Circuit Yield Optimizer Considering Process Variation, Aging, and Bending Effects", *J. Disp. Technol.,* vol. 10, no. 12, pp. 1055-1063, 2014.

http://dx.doi.org/10.1109/JDT.2014.2340892

[44]   S.B. Qian, Y. Shao, W.J. Liu, D.W. Zhang, and S.J. Ding, "Erasing-Modes Dependent Performance of a-IGZO TFT Memory With Atomic-Layer-Deposited Ni Nanocrystal Charge Storage Layer", *IEEE Trans. Electron Dev.*, vol. 64, no. 7, pp. 3023-3027, 2017.
http://dx.doi.org/10.1109/TED.2017.2702702

[45]   T.H. Yang, T.Y. Chen, N.T. Wu, Y.T. Chen, and J.J. Huang, "IGZO-TFT Biosensors for Epstein–Barr Virus Protein Detection", *IEEE Trans. Electron Dev.*, vol. 64, no. 3, pp. 1294-1299, 2017.
http://dx.doi.org/10.1109/TED.2016.2646379

[46]   K. Imamura, T. Sakai, H. Yakushiji *et al.*, "Organic photoconductive film–stacked active pixel sensor pixel circuits using indium–tin–zinc-oxide thin-film transistors", *Japanese Journal of Applied Physics*, vol. 61, no. 7, 2022.
http://dx.doi.org/10.35848/1347-4065/ac73cc

[47]   S.H. Kim, E.B. Kim, J.H. Oh, J.H. Hur, and J. Jang, "A 2-in. a-Si:H TFT-LCD with embedded backlight control TFT sensors with various channel widths", *J. Soc. Inf. Disp.*, vol. 16, no. 3, pp. 415-419, 2008.
http://dx.doi.org/10.1889/1.2896318

[48]   S.G. Park, J.H. Lee, W.K. Lee, and M.K. Han, "Investigation of the hysteresis phenomenon of an a-Si:H TFT at an elevated temperature for AMOLED displays", *J. Soc. Inf. Disp.*, vol. 15, no. 12, pp. 1145-1149, 2007.
http://dx.doi.org/10.1889/1.2825105

[49]   H. Hosono, "How we made the IGZO transistor", *Nat. Electron.*, vol. 1, no. 7, pp. 428-428, 2018.
http://dx.doi.org/10.1038/s41928-018-0106-0

[50]   D. Kong, "P-5: New Capacitive Active-Pixel Sensor Based on LTPS-TFT Backplane Technology for Fingerprint Recognition", *SID Symposium Digest of Technical Papers,* vol. vol. 52, pp. 1074-1077, 2021.
http://dx.doi.org/10.1002/sdtp.14878

[51]   Y.M. Kim, H.B. Kang, G.H. Kim, C.S. Hwang, and S.M. Yoon, "Improvement in Device Performance of Vertical Thin-Film Transistors Using Atomic Layer Deposited IGZO Channel and Polyimide Spacer", *IEEE Electron Device Lett.*, vol. 38, no. 10, pp. 1387-1389, 2017.
http://dx.doi.org/10.1109/LED.2017.2736000

[52]   H.R. Kim, M. Furuta, and S.M. Yoon, "Highly Robust Flexible Vertical-Channel Thin-Film Transistors Using Atomic-Layer-Deposited Oxide Channels and Zeocoat Spacers on Ultrathin Polyimide Substrates", *ACS Appl. Electron. Mater.*, vol. 1, no. 11, pp. 2363-2370, 2019.
http://dx.doi.org/10.1021/acsaelm.9b00544

[53]   K. Beom, P. Yang, D. Park, H.J. Kim, H.H. Lee, C.J. Kang, and T.S. Yoon, "Single- and double-gate synaptic transistor with TaO x gate insulator and IGZO channel layer", *Nanotechnology,* vol. 30, no. 2, p. 025203, 2019.
http://dx.doi.org/10.1088/1361-6528/aae8d2 PMID: 30387440

[54] T.C. Chen, Y. Kuo, T.C. Chang, M.C. Chen, and H.M. Chen, "Stability of double gate amorphous In–Ga–Zn–O thin-film transistors with various top gate designs", *Jpn. J. Appl. Phys.,* vol. 56, no. 12, p. 120303, 2017.
http://dx.doi.org/10.7567/JJAP.56.120303

[55] W.J. Cho, and M.J. Ahn, "Bias stress instability of double-gate a-IGZO TFTs on polyimide substrate", *J. Korean Phys. Soc.,* vol. 71, no. 6, pp. 325-328, 2017.
http://dx.doi.org/10.3938/jkps.71.325

[56] X. He, L. Wang, X. Xiao, W. Deng, L. Zhang, M. Chan, and S. Zhang, "Implementation of Fully Self-Aligned Homojunction Double-Gate a-IGZO TFTs", *IEEE Electron Device Lett.,* vol. 35, no. 9, pp. 927-929, 2014.
http://dx.doi.org/10.1109/LED.2014.2336232

[57] G.L. Olson, J.A. Roth, L.D. Hess, and J. Narayan, *Kinetics of solid-phase crystallization in ion-implanted and deposited amorphous-silicon films.* US/Japan Seminar on Solid Phase Epitaxy and Interface Kinetics, 1983, pp. 20-24.

[58] N. Van Hieu, "Formation of source and drain of a-Si:H TFT by ion implantation through metal technique", *Physica B,* vol. 392, no. 1-2, pp. 38-42, 2007.
http://dx.doi.org/10.1016/j.physb.2006.10.037

[59] S. Tomita, S. Jurichich, and K.C. Saraswat, "Transistor sizing for AMLCD integrated TFT drive circuits", *J. Soc. Inf. Disp.,* vol. 5, no. 4, pp. 399-404, 1997.
http://dx.doi.org/10.1889/1.1985187

[60] Y. Tang, R. Zhang, H. Gao, K. Liu, G. Zhao, X. Yang, Q. Li, Y. Liang, N. Ye, H. Liu, and S. Liu, "Partially light-controlled imaging system based on High Temperature Poly-Silicon Thin Film Transistor-Liquid Crystal Display", *Opt. Express,* vol. 18, no. 10, pp. 10616-10626, 2010.
http://dx.doi.org/10.1364/OE.18.010616 PMID: 20588914

[61] M. Mirshojaeian Hosseini and R. Nawrocki, "A Review of the Progress of Thin-Film Transistors and Their Technologies for Flexible Electronics", *Micromachines*, vol. 12, no. 6, 2021.
http://dx.doi.org/10.3390/mi12060655

[62] D. Ding, J.-P. Raskin, G. Lumbeeck, D. Schryvers, and H. Idrissi, "TEM investigation of the role of the polycrystalline-silicon film/substrate interface in high quality radio frequency silicon substrates," *Materials Characterization*, vol. 161, 2020.
http://dx.doi.org/10.1007/s12633-019-00129-1

[63] F. Kezzoula, M. Kechouane, T. Mohammed-Brahim, and H. Menari, "Crystallization of P Type Amorphous Silicon (a-Si: H) by AIC Method: Effect of Aluminum Thickness," Silicon, vol. 12, no. 2, pp. 405-411, 2019.
http://dx.doi.org/10.1007/s12633-019-00129-1

[64] M. G. Kang, K.H. Cho, S.M. Oh *et al.,* "Low-temperature crystallization and electrical properties of BST thin films using excimer laser annealing," *Current Applied Physics,* vol. 11, no. 3, pp. S66-S69, 2011.
http://dx.doi.org/10.1007/s12633-019-00129-1

[65] L. Xu, R. Liu, B. Yuan, Z. Du, Y. Liu, and J. Li, "6.4: The Effect of LDD Structure on the Characteristics of TFT Devices", *Dig. Tech. Pap.,* vol. 50, no. S1, pp. 55-56, 2019.
http://dx.doi.org/10.1002/sdtp.13385

[66] M. Mativenga, M.H. Choi, W. Choi, J.W. Choi, J. Jang, R. Mruthyunjaya, T.J. Tredwell, E. Mozdy, and C. Kosik-Williams, "Reduction of Hot Carrier Effects in Silicon-on-Glass TFTs", *J. Electrochem. Soc.,* vol. 158, no. 6, pp. J169-J174, 2011.
http://dx.doi.org/10.1149/1.3573769

[67] R. Chen, "Bridged-Grain Metal-Induced Crystallization Poly-Si TFTs with Silicon Self-Implantation",

[68] Z. Xia, M. Zhang, W. Zhou, R. Chen, M. Wong, and H.-S. Kwok, "Dynamic Reliability of Bridged-Grain Poly-Si Thin Film Transistors",

[69] R. Chen, W. Zhou, M. Zhang, M. Wong, and H.S. Kwok, "Self-aligned top-gate InGaZnO thin film transistors using SiO2/Al2O3 stack gate dielectric", *Thin Solid Films,* vol. 548, pp. 572-575, 2013.
http://dx.doi.org/10.1016/j.tsf.2013.09.020

[70] Y.C. Park, J.G. Um, M. Mativenga, and J. Jang, "Enhanced Operation of Back-Channel-Etched a-IGZO TFTs by Fluorine Treatment during Source/Drain Wet-Etching", *ECS J. Solid State Sci. Technol.,* vol. 6, no. 5, pp. P300-P303, 2017.
http://dx.doi.org/10.1149/2.0201705jss

[71] C.C. Pan, S-B. Yang, L-L. Chen, J-F. Shi, X. Sun, X-F. Li, and J-H. Zhang, "Improvement in Bias Stability of IGZO TFT With Etching Stop Structure by UV Irradiation Treatment of Active Layer Island", *IEEE J. Electron Devices Soc.,* vol. 8, pp. 524-529, 2020.
http://dx.doi.org/10.1109/JEDS.2020.2983251

[72] Y. Zhang, Z. Wang, M. Wang, D. Zhang, H. Wang, and M. Wong, "A Unified Degradation Model of Elevated-Metal Metal Oxide (EMMO) TFTs Under Positive Gate Bias With or Without an Illumination", *IEEE Trans. Electron Dev.,* vol. 68, no. 3, pp. 1081-1087, 2021.
http://dx.doi.org/10.1109/TED.2021.3053915

[73] Y. Zeng, H. Chu, N. Wei, Y. Li, X. Wang, and H. He, "A 5T1C pixel circuit compensating mobility and threshold voltage variation", *Microelectronics J.,* vol. 117, p. 105283, 2021.
http://dx.doi.org/10.1016/j.mejo.2021.105283

[74] E. Song, and H. Nam, "Novel voltage programming n-channel TFT pixel circuit for low power and high performance AMOLED displays", *Displays,* vol. 35, no. 3, pp. 118-125, 2014.
http://dx.doi.org/10.1016/j.displa.2014.04.002

[75] J. Huang, "Enhanced Visible Light Response of Amorphous InZnO Thin-Film Transistors by Hydrogen Doping via Al2O3/SiO2 Gate Dielectric",
http://dx.doi.org/10.1002/sdtp.16955

[76] T. Zou, B. Xiang, Y. Xu, Y. Wang, C. Liu, J. Chen, K. Wang, Q. Dai, S. Zhang, Y-Y. Noh, and H. Zhou, "Pixellated Perovskite Photodiode on IGZO Thin Film Transistor Backplane for Low Dose Indirect X-Ray Detection", *IEEE J. Electron Devices Soc.,* vol. 9, pp. 96-101, 2021.
http://dx.doi.org/10.1109/JEDS.2020.3040771

[77]   H.Y. Rho, A. Bala, A. Sen, U. Jeong, J. Shim, J. Oh, Y. Ju, M. Naqi, and S. Kim, "Plasma-Engineered Amorphous Metal Oxide Nanostructure-Based Low-Power Highly Responsive Phototransistor Array for Next-Generation Optoelectronics", *ACS Appl. Nano Mater.*, vol. 6, no. 17, pp. 15990-15999, 2023.
http://dx.doi.org/10.1021/acsanm.3c02973

[78]   H. Liu, X. Zhou, C. Fan, J. Chen, L. Lu, H. Zhou, and S. Zhang, "Thorough Elimination of Persistent Photoconduction in Amorphous InZnO Thin-Film Transistor via Dual-Gate Pulses", *IEEE Electron Device Lett.*, vol. 43, no. 8, pp. 1247-1250, 2022.
http://dx.doi.org/10.1109/LED.2022.3183840

[79]   S. Jeon, S.E. Ahn, I. Song, C.J. Kim, U.I. Chung, E. Lee, I. Yoo, A. Nathan, S. Lee, K. Ghaffarzadeh, J. Robertson, and K. Kim, "Gated three-terminal device architecture to eliminate persistent photoconductivity in oxide semiconductor photosensor arrays", *Nat. Mater.*, vol. 11, no. 4, pp. 301-305, 2012.
http://dx.doi.org/10.1038/nmat3256 PMID: 22367002

[80]   A. Nakashima, Y. Sagawa, and M. Kimura, "Temperature Sensor Using Thin-Film Transistor", *IEEE Sens. J.*, vol. 11, no. 4, pp. 995-998, 2011.
http://dx.doi.org/10.1109/JSEN.2010.2060720

[81]   H. Jeong, C.S. Kong, S.W. Chang, K.S. Park, S.G. Lee, Y.M. Ha, and J. Jang, "Temperature Sensor Made of Amorphous Indium–Gallium–Zinc Oxide TFTs", *IEEE Electron Device Lett.*, vol. 34, no. 12, pp. 1569-1571, 2013.
http://dx.doi.org/10.1109/LED.2013.2286824

[82]   F. Haque, S. Lim, S. Lee, Y. Park, and M. Mativenga, "Highly Sensitive and Ambient Air-Processed Hybrid Perovskite TFT Temperature Sensor", *IEEE Electron Device Lett.*, vol. 41, no. 7, pp. 1086-1089, 2020.
http://dx.doi.org/10.1109/LED.2020.2995086

[83]   E. Yu, S.G. Kim, S.J. Kang, H.S. Lee, S.J. Moon, J.M. Lee, S.B. An, and B.S. Bae, "Serially connected tantalum and amorphous indium tin oxide for sensing the temperature increase in IGZO thin-film transistor backplanes", *J. Inf. Disp.*, vol. 24, no. 3, pp. 205-213, 2023.
http://dx.doi.org/10.1080/15980316.2023.2185563

[84]   F. Ji, J. Hu, and Y. Zhang, "Functionalized Carbon-Nanotubes-Based Thin-Film Transistor Sensor for Highly Selective Detection of Methane at Room Temperature", *Chemosensors (Basel)*, vol. 11, no. 7, p. 365, 2023.
http://dx.doi.org/10.3390/chemosensors11070365

[85]   M.T. Vijjapu, S.G. Surya, S. Yuvaraja, X. Zhang, H.N. Alshareef, and K.N. Salama, "Fully Integrated Indium Gallium Zinc Oxide NO 2 Gas Detector", *ACS Sens.*, vol. 5, no. 4, pp. 984-993, 2020.
http://dx.doi.org/10.1021/acssensors.9b02318 PMID: 32091191

[86]   X. Li, Y. Jiang, G. Xie, H. Tai, P. Sun, and B. Zhang, "Copper phthalocyanine thin film transistors for hydrogen sulfide detection", *Sens. Actuators B Chem.*, vol. 176, pp. 1191-1196, 2013.
http://dx.doi.org/10.1016/j.snb.2012.09.084

[87] X. Wang, X. ling, Y. Hu, X. Hu, Q. Zhang, K. Sun, and Y. Xiang, "Electronic skin based on PLLA/TFT/PVDF-TrFE array for Multi-Functional tactile sensing and visualized restoring", *Chem. Eng. J.,* vol. 434, p. 134735, 2022.
http://dx.doi.org/10.1016/j.cej.2022.134735

[88] A. Rasheed, E. Iranmanesh, W. Li, Y. Xu, Q. Zhou, H. Ou, and K. Wang, "An Active Self-Driven Piezoelectric Sensor Enabling Real-Time Respiration Monitoring", *Sensors (Basel),* vol. 19, no. 14, p. 3241, 2019.
http://dx.doi.org/10.3390/s19143241 PMID: 31340564

[89] K. Myny, "15.2 A flexible ISO14443-A compliant 7.5mW 128b metal-oxide NFC barcode tag with direct clock division circuit from 13.56MHz carrier", *2017 IEEE International Solid-State Circuits Conference (ISSCC),* pp. 258-259, 2017.
http://dx.doi.org/10.1109/ISSCC.2017.7870359

[90] T. Moy, L. Huang, W. Rieutort-Louis, C. Wu, P. Cuff, S. Wagner, J.C. Sturm, and N. Verma, "An EEG Acquisition and Biomarker-Extraction System Using Low-Noise-Amplifier and Compressive-Sensing Circuits Based on Flexible, Thin-Film Electronics", *IEEE J. Solid-State Circuits,* vol. 52, no. 1, pp. 309-321, 2017.
http://dx.doi.org/10.1109/JSSC.2016.2598295

[91] H.S. Kim, H. Park, and W.J. Cho, "Light-Stimulated IGZO Transistors with Tunable Synaptic Plasticity Based on Casein Electrolyte Electric Double Layer for Neuromorphic Systems", *Biomimetics (Basel),* vol. 8, no. 7, p. 532, 2023.
http://dx.doi.org/10.3390/biomimetics8070532 PMID: 37999173

[92] M. Kumar, D.K. Ban, and J. Kim, "Photo-induced pyroelectric spikes for neuromorphic sensors", *Mater. Lett.,* vol. 225, pp. 46-49, 2018.
http://dx.doi.org/10.1016/j.matlet.2018.04.106

[93] Z. Chen, M. Zhang, S. Deng, Z. Jiang, Y. Yan, S. Han, Y. Zhou, M. Wong, and H-S. Kwok, "Effect of Moisture Exchange Caused by Low-Temperature Annealing on Device Characteristics and Instability in InSnZnO Thin-Film Transistors", *Adv. Mater. Interfaces,* vol. 9, no. 14, p. 2102584, 2022.
http://dx.doi.org/10.1002/admi.202102584

[94] C.S. Huang, and P.T. Liu, "Impact of Negative-Bias-Temperature-Instability on Channel Bulk of Polysilicon TFT by Gated PIN Diode Analysis", *Electrochem. Solid-State Lett.,* vol. 14, no. 5, p. H194, 2011.
http://dx.doi.org/10.1149/1.3551463

[95] N. Yoshida, J.P. Bermundo, Y. Ishikawa, T. Nonaka, K. Taniguchi, and Y. Uraoka, "Low temperature cured poly-siloxane passivation for highly reliable a -InGaZnO thin-film transistors", *Appl. Phys. Lett.,* vol. 112, no. 21, p. 213503, 2018.
http://dx.doi.org/10.1063/1.5029521

[96] Y.H. Tai, S.C. Huang, and C.K. Chen, "Analysis of Poly-Si TFT Degradation Under Gate Pulse Stress Using the Slicing Model", *IEEE Electron Device Lett.,* vol. 27, no. 12, pp. 981-983, 2006.
http://dx.doi.org/10.1109/LED.2006.886416

[97]   C.H. Jeon, K.J. Kwon, S.K. Hong, Y.M. Ha, H. Kim, and J. Jang, "Study on Residual Image in Low-Temperature Poly-Si Oxide TFT-Based OLED Display on Polyimide Substrate", *IEEE Trans. Electron Dev.,* vol. 69, no. 9, pp. 4958-4961, 2022.

http://dx.doi.org/10.1109/TED.2022.3188954

[98]   S.M. Sadati Faramarzi, B. Luo, J. Poortmans, J. Genoe, and K. Myny, "Thin-Film Transistor-Based Sensor Interface Circuits Enabling Distributed Local In-Module Solar Cell Temperature Monitoring", *IEEE J. Solid-State Circuits,* vol. 59, no. 1, pp. 307-315, 2024.

http://dx.doi.org/10.1109/JSSC.2023.3281253

[99]   N.D. Young, A. Gill, and M.J. Edwards, "Hot carrier degradation in low temperature processed polycrystalline silicon thin film transistors", *Semicond. Sci. Technol.,* vol. 7, no. 9, pp. 1183-1188, 1992.

http://dx.doi.org/10.1088/0268-1242/7/9/007

[100]   M. Fujii, "Thermal Analysis of Degradation in Ga_2O_3–In_2O_3–ZnO Thin-Film Transistors", *JJAP,* vol. 47, no. 8R, p. 6236, 2008.

http://dx.doi.org/10.1143/JJAP.47.6236

[101]   M. Zhang, Y. Yan, G. Li, S. Deng, W. Zhou, R. Chen, M. Wong, and H-S. Kwok, "OFF-State-Stress-Induced Instability in Switching Polycrystalline Silicon Thin-Film Transistors and Its Improvement by a Bridged-Grain Structure", *IEEE Electron Device Lett.,* vol. 39, no. 11, pp. 1684-1687, 2018.

http://dx.doi.org/10.1109/LED.2018.2872350

[102]   M. Zhang, X. Ma, S. Deng, W. Zhou, Y. Yan, M. Wong, and H-S. Kwok, "Degradation Induced by Forward Synchronized Stress in Poly-Si TFTs and Its Reduction by a Bridged-Grain Structure", *IEEE Electron Device Lett.,* vol. 40, no. 9, pp. 1467-1470, 2019.

http://dx.doi.org/10.1109/LED.2019.2931007

[103]   M. Zhang, W. Zhou, R. Chen, M. Wong, and H.S. Kwok, ""Driving"-Stress-Induced Degradation in Polycrystalline Silicon Thin-Film Transistors and Its Suppression by a Bridged-Grain Structure", *IEEE Electron Device Lett.,* vol. 38, no. 1, pp. 52-55, 2017.

http://dx.doi.org/10.1109/LED.2016.2626481

[104]   J.J. Lih, C.F. Sung, C.H. Li, T.H. Hsiao, and H.H. Lee, "Comparison of a-Si and Poly-Si for AMOLED displays", *J. Soc. Inf. Disp.,* vol. 12, no. 4, pp. 367-371, 2004.

http://dx.doi.org/10.1889/1.1847734

[105]   A. Rjoub, B. Tarawneh, and R. Alghsoon, "Active matrix organic light emitting diode displays (AMOLED) new pixel design", *Microelectron. Eng.,* vol. 212, pp. 42-52, 2019.

http://dx.doi.org/10.1016/j.mee.2019.04.001

[106]   J.S. Park, W.J. Maeng, H.S. Kim, and J.S. Park, "Review of recent developments in amorphous oxide semiconductor thin-film transistor devices", *Thin Solid Films,* vol. 520, no. 6, pp. 1679-1693, 2012.

http://dx.doi.org/10.1016/j.tsf.2011.07.018

[107]   J. Zhu, X. Wang, and H. Wang, "Highly sensitive organic thin-film transistors based sensor for putrescine detection", *Appl. Phys., A Mater. Sci. Process.,* vol. 126, no. 6, p. 463, 2020.

http://dx.doi.org/10.1007/s00339-020-03627-x

[108]   S. Park, K. Park, H. Kim, H.W. Park, K.B. Chung, and J.Y. Kwon, "Light-induced bias stability of crystalline indium-tin-zinc-oxide thin film transistors", *Appl. Surf. Sci.,* vol. 526, p. 146655, 2020.

http://dx.doi.org/10.1016/j.apsusc.2020.146655

[109]   Y. Zhang, W. Jiang, D. Feng, C. Wang, Y. Xu, Y. Shan, J. Wang, Z. Yin, H. Deng, X. Mi, and N. Dai, "Bio-Separated and Gate-Free 2D MoS2 Biosensor Array for Ultrasensitive Detection of BRCA1", *Nanomaterials (Basel),* vol. 11, no. 2, p. 545, 2021.

http://dx.doi.org/10.3390/nano11020545 PMID: 33669986

[110]   M.Z. Li, L-C. Guo, G. Liu, Y. Yan, M. Zhang, S-T. Han, and Y. Zhou, "Manipulating Strain in Transistors: From Mechanically Sensitive to Insensitive", *Adv. Electron. Mater.,* vol. 8, no. 7, p. 2101288, 2022.

http://dx.doi.org/10.1002/aelm.202101288

[111]   S. Hong, H. Cho, B.H. Kang, K. Park, D. Akinwande, H.J. Kim, and S. Kim, "Neuromorphic Active Pixel Image Sensor Array for Visual Memory", *ACS Nano,* vol. 15, no. 9, pp. 15362-15370, 2021.

http://dx.doi.org/10.1021/acsnano.1c06758 PMID: 34463475

[112]   Y. Jang, J. Park, J. Kang, and S.Y. Lee, "Amorphous InGaZnO (a-IGZO) Synaptic Transistor for Neuromorphic Computing", *ACS Appl. Electron. Mater.,* vol. 4, no. 4, pp. 1427-1448, 2022.

http://dx.doi.org/10.1021/acsaelm.1c01088

CHAPTER 2

Overview of Reliability Issues in Thin-Film Transistors

Abstract. This chapter provides an in-depth analysis of the reliability issues associated with thin-film transistors (TFTs), focusing on the common defects found in silicon-based and metal-oxide TFTs. It discusses the various types of defect states and the typical degradation mechanisms that impact the performance of TFTs. The chapter emphasizes the need for comprehensive understanding and strategies to mitigate these reliability challenges in TFT applications.

Keywords: Defect states, Degradation mechanisms, Metal oxide TFTs, Reliability, Silicon-based TFTs.

2.1. INTRODUCTION

The reliability of thin-film transistors (TFTs) is a critical determinant of their practical application. Thus, an analysis of TFT reliability is imperative. Unlike metal-oxide-semiconductor (MOS) devices, TFTs contain a significant number of defects. The reliability issues of TFTs often stem from the various defect states within the device. Therefore, understanding the defect state density within the device and its impact on performance is fundamental for subsequent reliability analysis.

In this chapter, we delineate common defect state densities found in silicon-based and metal-oxide (MO) TFTs, along with their effects on device performance. Building on this foundation, we also introduce several common classification methods for defect states and discuss the impact of different types of defect states on device performance. Lastly, we briefly outline several typical degradation mechanisms in TFTs.

For TFT to meet the requirements of actual production, characterizing the fundamental performance of the devices is a crucial initial step. The *I-V* (current-voltage) characteristics of TFTs can primarily be categorized into output characteristics and transfer characteristics, as shown in Fig. (**2.1**). Typically, the output characteristics of a device are measured by fixing the gate voltage and varying the drain voltage, while the transfer characteristics are obtained by fixing the drain voltage and varying the gate voltage. Commonly, the drain voltage is set

Meng Zhang & Mingxiang Wang

to 0.1 V and 5 V (or their negative equivalents) depending on the polarity of the TFT's active layer to represent the device's performance in linear and saturation conditions, respectively. When testing the organic TFTs, the drain voltage can be set to a larger level.

2.2. BASIC PERFORMANCE CHARACTERIZATION OF THIN-FILM TRANSISTORS

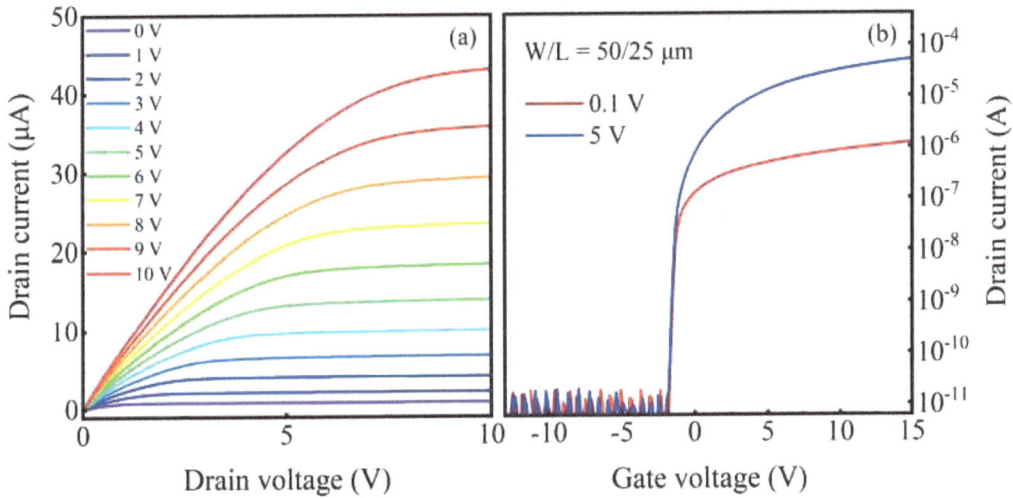

Fig. (2.1). (a) Output characteristic curve and (b) transfer characteristic curve of a typical TFT.

By analyzing the IV curves of TFTs, numerous key fundamental performance characteristics of the devices can be extracted:

(1) Mobility (μ)

Mobility is a key indicator of device performance, defined as the average drift velocity of charge carriers under the influence of a unit electric field. It reflects the ease with which charge carriers drift in response to an external electric field, with units of m^2/Vs or cm^2/Vs. The magnitude of mobility is primarily related to scattering within the device, such as lattice scattering and ionized impurity scattering. Unlike MOS devices, TFTs have a significant number of defect states within the bulk of their active layers, which can capture charge carriers during their motion. Therefore, the density of defect states in the TFT's active layer directly affects its mobility [1, 2]. Mobility can be simply derived from the linear region of the TFT's transfer characteristic curve using the following formula [3]:

$$\mu = g_m \cdot \frac{L}{W \cdot C_{ox} \cdot V_{ds}} \tag{2.1}$$

Where g_m is the maximum transconductance of the TFT, which can be obtained from the slope of the transfer characteristic curve. Besides, the saturation mobility can be simply derived from the saturation region of the TFT's transfer characteristic curve using the following formula:

$$\mu_{sat} = \frac{\left(\frac{d\sqrt{I_D}}{dV_G} \right)^2}{\frac{1}{2} C_{OX} \frac{W}{L}} \tag{2.2}$$

Generally, increasing temperature can enhance carrier thermal motion [4], thus increasing mobility, but excessively high temperatures may also lead to an increase in lattice defects, which can reduce mobility [5-7]. Electric fields also affect carrier mobility; for example, devices can experience mobility degradation under negative/positive bias temperature instability (PBTI/NBTI) stress [8, 9] and negative/positive bias illumination stress (NBIS/PBIS) conditions [10].

(2) Threshold Voltage (V_{th})

The V_{th} in metal-oxide-semiconductor field-effect transistors (MOSFETs) typically refers to the gate voltage required for the device to just enter strong inversion. TFTs, on the other hand, usually operate in the accumulation region. Thus, the V_{th} for a TFT indicates the gate voltage at which charge carriers begin to accumulate in the channel. There are two common methods for extracting V_{th}. The first method is based on the fundamental definition of V_{th}, which identifies it as the gate voltage corresponding to a certain drain current level, typically defined as [11]:

$$V_{th} = V_{gs}(I_{th} = 1nA, V_{ds} = 0.1V) \tag{2.3}$$

$$V_{th} = V_{gs}(I_{th} = 10nA, V_{ds} = 5V) \tag{2.4}$$

The second method involves extracting the on-state region of the transfer characteristic curve in linear coordinates [12], extending the linear portion backward to intersect the horizontal axis, and defining the corresponding voltage value as V_{th}. Factors such as semiconductor material properties [13], insulator thickness [14], gate structure [13], temperature [5, 15, 16], light exposure [17-20], and applied voltage can all influence the V_{th} of a TFT. In gate bias reliability, positive bias stress (PBS) stress typically causes a positive shift in device V_{th}, while negative gate bias (NBS) stress leads to a negative shift [21]. The degradation principle mainly involves charge carriers accumulating at the interface and being captured by defect states at the interface, and some carriers are being injected into the gate insulator (GI) under a strong electric field. The charged defect states and fixed charges can shield part of the gate electric field, leading to the V_{th}_shift. In PBTI/NBTI stress, carriers are more easily injected into the GI, resulting in a more severe V_{th} shift [22].

(3) Subthreshold Swing (*SS*)

In TFTs, the subthreshold region indicates a zone where the current sharply increases from the off-state to the on-state. The steepness of this transition is quantified by the *SS*, which can be expressed as [23]:

$$SS = \left(\frac{dlog\left(I_{ds}\right)}{dV_{gs}} \right)^{-1} \tag{2.5}$$

A larger *SS* implies a more gradual curve in this region, while a smaller *SS* corresponds to a steeper curve; hence, a smaller *SS* is generally preferred [23]. *SS* can be optimized by adjusting the device structure, process parameters, and materials. The number of traps at the gate oxide/active layer interface in TFT devices also affects the magnitude of *SS*, which can be used to assess the extent of interface defect states, as expressed by the formula [24]:

$$N_{tot} = \frac{C_{ox}}{q} \left(\frac{qSS}{kTln\,10} - 1 \right) \tag{2.6}$$

Here, C_{ox} represents the gate oxide capacitance per unit area, q is the elementary charge, T is the absolute temperature, and k is the Boltzmann constant. Poor interface quality can lead to a larger *SS*. However, unlike MOSFETs, where defects are primarily concentrated at the interface, TFTs have a significant number of defect states within the active layer itself. The distribution of these defect states within the bandgap also affects the TFT's *SS*, with deep-level defects often being a more dominant factor in influencing *SS* [24].

(4) Contact Resistance (Rc)

Semiconductor devices typically utilize metals as electrodes, and forming a good ohmic contact between the metal and semiconductor is crucial for maintaining device performance. The method of extracting the over-threshold linear region curve from the transmission characteristic curve set of a TFT, once expanded, is widely known as the transmission line method (TLM) [25-27]. The TLM takes into account the contact resistance between the source/drain electrodes and the active layer, making it an accurate calculation method. Assuming the semiconductor material is uniform and the resistivity of the conductive channel is constant, the total resistance can be expressed as the sum of the source and drain contact resistances, the channel resistance, and the sheet resistance of the channel, as shown below [27]:

$$R_{tot} = R_C + R_s = 2R_{S/D} + R_s \qquad (2.7)$$

Further expressed as [28, 29]:

$$R_{tot} = 2R_{S/D} + \frac{R_{sh}}{W}L = 2\frac{R_{sh}}{W}L_T + \frac{R_{sh}}{W}L \qquad (2.8)$$

where R_{tot} is the measured total resistance, $R_{S/D}$ is the source/drain_contact resistance, R_s is the channel resistance, R_{sh} is the sheet resistance of the channel, W is the device channel width, and L is the device channel length. From equation 2.10, the fitted linear equation is $Y = B + kX$, as shown in Fig. (**2.2**). Here, B $= 2R_{S/D} = 2$ $R_{sh}L_T/W$ and k $= R_{sh}/W$ thus allowing for the calculation of the contact resistance R_c and the channel resistance R_s [27].

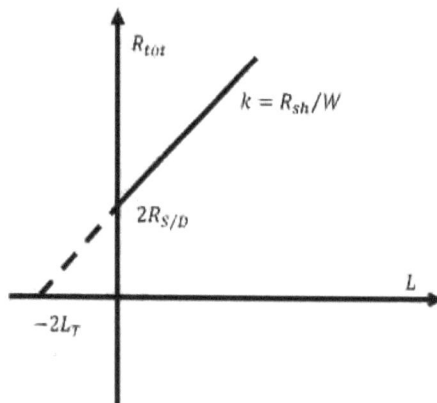

Fig. (2.2). Fitted plot of the relationship between R_{tot} and L.

(5) Off-State Current (I_{off})

The I_{off} of a TFT refers to the magnitude of the current that flows through the device when the channel is depleted. Ideally, when a TFT is off, the conduction current should be zero. However, in practice, due to structural features, material properties, and fabrication process imperfections of the TFT, a small number of charge carriers pass through the channel, resulting in I_{off}. The value of I_{off} may be related to two operational regions: reverse subthreshold and Poole-Frenkel emission [30-32]. When the device operates in these regions, the Fermi level of the active layer is located within deep energy levels, and thus, the density and distribution characteristics of these deep states significantly affect the I_{off} [31].

When a TFT operates in the reverse subthreshold region, the conductive channel is in the back channel, and the state of the active layer/GI interface influences the magnitude of I_{off}. In the Poole-Frenkel emission region, the conductive channel is in the front channel, and the state of the active layer/GI interface affects I_{off} [30, 32] Additionally, Poole-Frenkel emission is related to the longitudinal electric field in the G/D overlap region [33], so factors such as the GI material, active layer thickness, and applied voltage can all influence the I_{off} of a TFT. For instance, the hot carrier (HC) mechanism generally causes an increase in I_{off} when the gate-source voltage (V_{gs}) is low and the V_{ds} is high. Over time, under a strong electric field, trapped states in the grain boundaries can release electrons and holes with high kinetic energy that enter the gate oxide and shield V_{gs}, leading to the degradation of the device's electrical parameters [34]. Conversely, the self-heating (SH) mechanism, which occurs when both V_{ds} and V_{gs} are high, leads to a cumulative heating effect that raises the device temperature and causes a degradation in electrical characteristics, resulting in a decrease in I_{off} [35].

(6) Current Switching Ratio (I_{on}/I_{off})

The I_{on}/I_{off} describes the on-off characteristics of a TFT, representing the ratio of the on-state current (I_{on}) to the I_{off}. A larger I_{on}/I_{off} ratio indicates better switching performance. A higher I_{on} allows for faster charging of the signal storage capacitor, while a lower I_{off} means lower leakage current, leading to lower power consumption when the device is idle or off. Additionally, a lower I_{off} helps retain the charge in the storage capacitor for a longer period, enabling longer signal retention. A higher switching ratio also results in more pronounced contrast, with the device's brightness levels being more distinct, leading to more saturated colors and improved display quality on external screens [36].

2.3. COMMON DEFECTS IN SILICON-BASED THIN-FILM TRANSISTORS

The performance of a crystal is closely related to the internal arrangement of its atoms; even minor issues can significantly alter the electrical properties of silicon-based TFTs. The periodic structure of a crystal is formed by the arrangement of atoms in a pattern that repeats throughout the solid, meaning that the fundamental arrangement of certain atoms is consistently replicated across the entire crystal [37]. As a result, once this basic periodicity is identified, the behavior of the crystal at one point is identical to its behavior at other equivalent points.

The periodicity of a crystal is defined by the symmetric arrangement of lattice points in space [38, 39]. At each lattice point, atoms can be added in a pattern known as a basis, which may consist of a single atom or a group of atoms with the same spatial arrangement, thus forming a complete crystal. In every case, the lattice contains a unit that represents the entire lattice and repeats regularly throughout the crystal [39].

In crystals, even extremely minute defect states can have a decisive impact on the physical and chemical properties of semiconductor materials. Defects within a crystal that arise from disruptions in periodicity can be classified as point defects, line defects, and planar defects. Point defects are primarily characterized by vacancies and interstitial atoms, line defects are mainly dislocations, and planar defects are typically stacking faults [39-43].

For Si-based TFTs, the active layers are primarily composed of amorphous silicon (a-Si) and polycrystalline silicon (poly-Si). In the case of a-Si, the structure itself is made up of silicon atoms arranged in a disordered manner, resulting in a significant number of structural defects within the a-Si. For poly-Si thin films, the structure is mainly composed of individual monocrystalline Si grains, with defects primarily arising from grain boundaries, which are planar defects.

2.3.1. Common Defect in Amorphous Silicon Thin Film

The a-Si commonly exhibits a variety of defect structures that differ from the typical diamond lattice structure of crystalline Si, as depicted in Fig. (**2.3**). In crystalline Si, the Si-Si bonds are uniform with a fixed bond length of approximately 1.55×10^{-10} meters and a fixed bond angle of 109°28' [44]. However, in a-Si devices, structural instability leads to variations in the bond energy, bond length, and bond angles after Si atoms form covalent bonds within the a-Si thin film. The bond length typically varies by less than 1%, while the bond angle can deviate by about 10%. The further the bond angle strays from 109°28', the weaker the corresponding bond energy, with the weaker Si-Si bonds often referred to as weak bonds [44, 45].

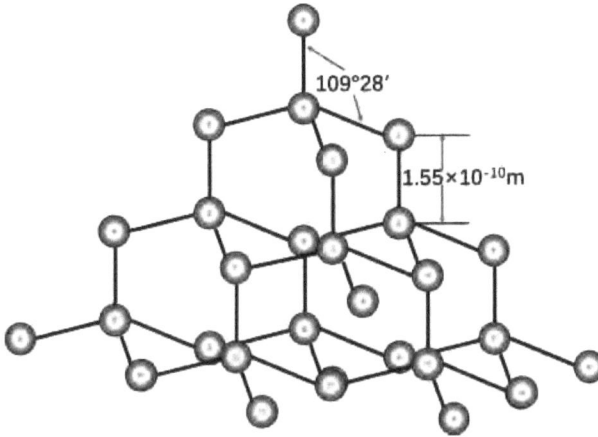

Fig. (2.3). Alignment of silicon atoms.

A significant difference between a-Si and single-crystal Si is that while single-crystal Si atoms typically bond with four neighboring Si atoms, a-Si thin films may not follow this pattern, resulting in a large number of Si atoms that are bonded to only three other Si atoms, leaving the last bond in a dangling state. These valence bonds are commonly known as dangling bonds. Similarly, at the interface of a-Si devices, if Si atoms do not form a strong bond with the GI layer, a significant number of Si atoms with dangling bonds can also be present. These dangling bonds within the a-Si thin film create a multitude of localized states that can significantly trap nearby electrons. Fig. (**2.4**) shows the schematic of weak bonds and dangling bonds. Therefore, the presence of dangling bonds within a-Si is a key factor affecting the conductivity of a-Si [46].

To address the issue of dangling bonds in a-Si, hydrogen (H) atoms can be incorporated into the a-Si thin film to repair the defect states, resulting in hydrogenated amorphous silicon (a-Si:H) thin films [47]. However, it is important to note that Si-H bonds are also not very stable and can easily break under various types of stress, especially under SH stress, leading to device performance degradation [48-50].

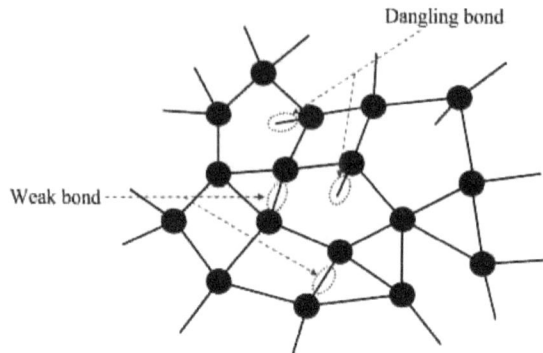

Fig. (2.4). Schematic of weak bonds between silicon atoms and dangling bonds.

The weak Si-Si bonds, Si-H bonds, and dangling bonds present in a-Si thin films typically manifest as numerous localized defect states within the bandgap [44-47]. Depending on the position of these defect states within the bandgap, they often act as traps, capturing a large number of free charge carriers, thereby significantly affecting the device's performance in terms of I_{on}, V_{th}, and SS.

2.3.2. Common Defects in Polycrystalline Silicon Thin Film

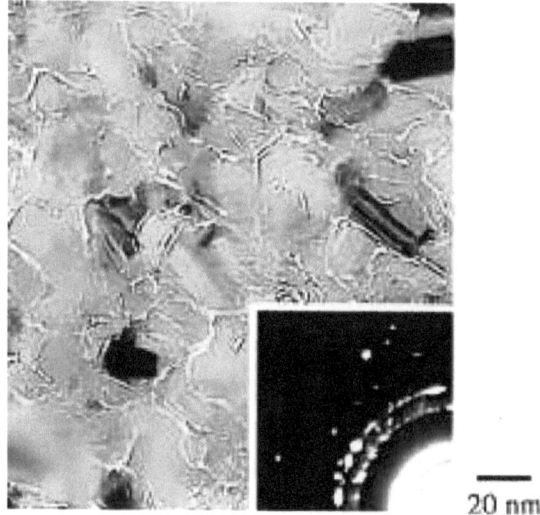

20 nm

Fig. (2.5). Crystalline grains [54].

Poly-Si differs significantly from single-crystal silicon and a-Si primarily due to the presence of grain boundaries within the poly-Si thin film [51], as shown in Fig. (**2.5**). While the crystal structure of poly-Si is composed of periodically arranged Si atoms, similar to single-crystal silicon, the grain boundaries are where a significant number of weak bonds, dangling bonds, and other defects are found [52, 53]. The quality of a poly-Si thin film can be assessed based on several factors:

Grain boundary inclination angle: The angle and orientation (Fig. **2.6**) at which adjacent grains meet play a crucial role. High-angle grain boundaries, which have a larger orientation misorientation angle compared to low-angle grain boundaries, typically possess higher energy and greater electrical resistance. The misorientation angle affects the extent of defects and related electrical characteristics [55-57].

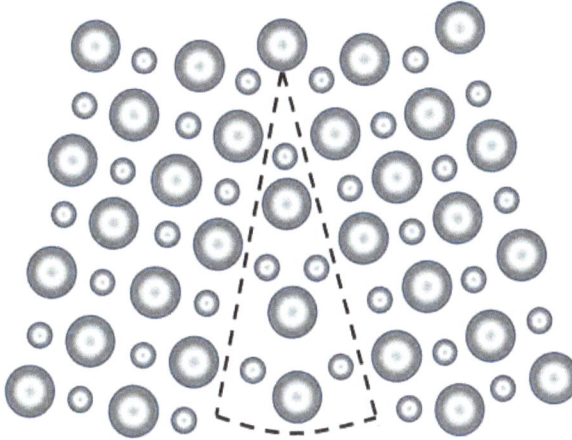

Fig. (2.6). Grain boundary inclination.

Grain boundary type: Different types of grain boundaries exhibit distinct properties. For instance, 'tilted' grain boundaries involve displacement along a single crystal axis, while 'twist' grain boundaries involve rotation around a single crystal axis, as shown in Fig. (**2.7**). Their distinct structures influence the electronic and mechanical properties of the grain boundaries [58-60].

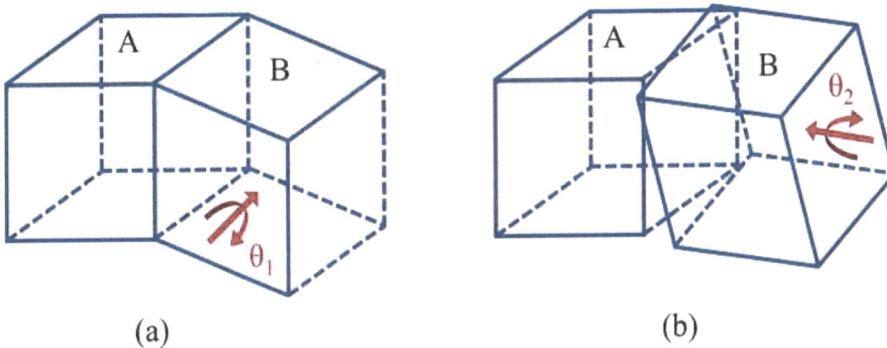

(a)   (b)

Fig. (2.7). Tilted and twisted grain boundaries.

Grain size: The size of individual grains within the material impacts the grain boundary density, as shown in Fig. (**2.8**). Smaller grain sizes lead to higher grain boundary densities and, consequently, higher defect densities. However, when grain sizes are too large, such as greater than 1 μm, the channel of the entire TFT may only span a few grain boundaries, and the difference in the number of grain boundaries can significantly affect device performance, leading to reduced uniformity in TFTs. Therefore, it is necessary to control the grain size within an appropriate range to ensure optimal device performance [61-63].

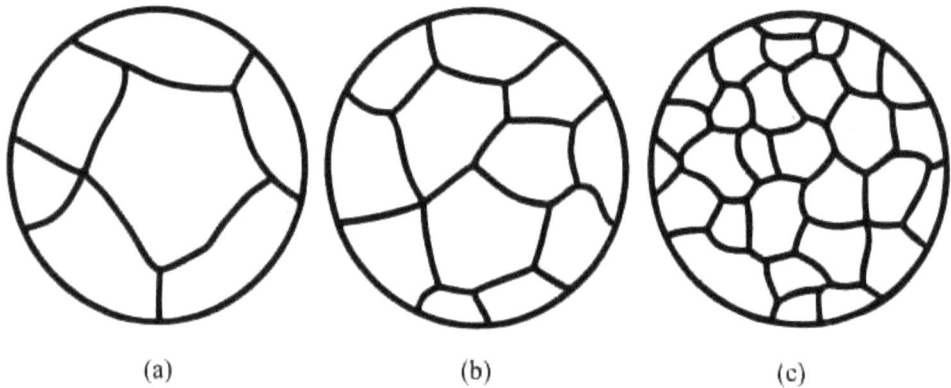

(a)   (b)   (c)

Fig. (2.8) Different grain sizes in poly-Si thin film.

2.4. COMMON DEFECTS IN METAL OXIDE THIN FILM

Poly-Si's semi-ordered polycrystalline structure endows it with weak bond and interface defect structures similar to those found in single crystals, as well as unique grain boundary structures that are inherent to polycrystalline materials. In contrast, metal oxides typically exist in an amorphous form. Although weak bonds and interface defects are also present in metal oxides, their formation mechanisms differ from those in poly-Si. Additionally, due to the amorphous structure of metal oxides, the grain boundary structures found in poly-Si are absent in metal oxides. Apart from weak bonds and interface defects, another significant factor affecting the performance of MO TFT devices is the presence of oxygen vacancies (V_os) in the metal oxide.

2.4.1. Impact of Weaken Bonds

Bonds in metal oxides primarily exist in the form of metal-oxygen and metal-metal bonds. The introduction of external impurities, such as H, can weaken the bond strength of the coordinated chemical bonds [64], leading to the formation of weak bonds. For example, in MO TFTs, the introduction of H impurities during the film fabrication process can result in the presence of -OH bonds within the metal oxide, which weakens the chemical bonds of metal ions, making the metal elements more prone to desorption [64].

The presence of weak bonds increases the instability of MO TFTs and degrades the reliability of the devices. When there is a high concentration of weak bonds in the active layer of a MO TFT, the lattice vibrations in the active layer become intense at high temperatures, which may cause these weak bonds to break. The broken weak bonds can form uncoordinated chemical bonds at the breakage sites, which can capture electrons or holes in the active layer, affecting the number of free charge carriers and thus reducing the performance of the TFT [64].

2.4.2. Impact of Interface Traps

Similarly, during the manufacturing process of MO TFTs, when different materials come into contact, imperfect coordination between chemical elements often leaves a certain amount of uncoordinated dangling bonds at the contact interface, also known as interface defects. Like poly-Si TFTs, interface defects in MO TFTs are classified as donor-like and acceptor-like. Since MO TFTs conduct electrons, acceptor-like interface defects are the primary interface defects affecting the performance of metal oxide devices.

Interface defects located at the interface between the GI layer and the active layer have the most significant impact on device performance. In MO TFTs, as the device transitions from the off-state to the on-state, i.e., when the gate voltage shifts from negative to positive, charge carriers gradually accumulate at the interface between the active layer and the GI layer. At this time, interface defects can capture charge carriers in the active layer, hindering their free movement and affecting the formation of a conductive channel at the interface, leading to a degradation in the *SS* of the device's transfer characteristic curve [65-68]. Interface defects in metal oxides can also cause hysteresis effects similar to those observed in poly-Si TFTs [69].

Furthermore, interface defects can affect the reliability of TFTs after stress application. When a positive gate bias stress is applied to the device, electrons in the active layer accumulate at the interface between the GI layer and the active layer. These electrons can be captured by interface defects, forming negatively charged fixed charges at the interface, making the device harder to turn on and manifesting as a positive shift in the device's transfer characteristic curve and an increase in V_{th} [65-68]. Conversely, when a negative gate bias stress is applied, holes in the active layer (which are less abundant in metal oxides and generally originate from photogenerated electron-hole pairs or impact ionization) accumulate at the interface, where they can be captured by interface defects, forming positively charged fixed charges at the interface, making the device easier to turn on and manifesting as a negative shift in the device's transfer characteristic curve and a decrease in V_{th}.

2.4.3. Impact of Oxygen Vacancies

The V_os, unique defects in MO TFTs, significantly influence the V_{th} and mobility of these devices [70-78]. With oxygen atoms having six valence electrons, the removal of an oxygen atom from the metal oxide leaves behind two electrons within the material. Whether these two electrons are confined to the V_o site or one or both become free electrons largely depends on the formation energy of the oxygen defect. When both electrons are confined to the oxygen vacancy, the V_o state is referred to as the neutral oxygen vacancy; when one electron is confined and the other becomes a free electron, the V_o^+ state is known as the singly charged oxygen vacancy; and when both electrons are free, the ionized oxygen vacancies (V_o^{2+}) state is called the doubly charged oxygen vacancy [79-81]. These defect states exist in three charge states: V_o, V_o^+, and V_o^{2+}

$$E_f(V_o,q) = E_{tot}(V_o,q) - E_{tot}(IGZO) + \mu_o + q\mu_e \qquad (2.9)$$

As shown in a typical density of states (DOS) distribution diagram for IGZO (Fig. **2.9**), the bandgap of IGZO is approximately 3.5 eV, and the width of the deep states is about 1.5 eV. The deep states near the valence band edge, composed of neutral V_os, are primarily formed during the deposition of the IGZO thin film. The formation energy of V_os can be defined as the energy required to remove an oxygen atom from the metal oxide and create a vacancy. This energy is a critical factor in determining the electrical properties and overall performance of MO TFTs, as it affects the concentration and distribution of V_os within the material.

The mobility and V_{th} of MO TFTs are directly related to the number of V_os in the active layer, acting as shallow donor-like defects. Generally, the more V_os present in the active layer of a TFT, the easier it is to turn on the device, the lower the V_{th}, and the higher the mobility. Conversely, the fewer the V_os, the harder it is to turn on the device, the higher the V_{th}, and the lower the mobility. Additionally, since temperature and light can affect the ionization state of V_os, their quantity also determines the sensitivity of TFTs to temperature and light exposure. V_os is more likely to ionize under high temperatures and light exposure, which refers to the transition from a neutral V_o state, releasing two electrons, to a positively charged V_o^{2+} state. This implies that the electrical properties of the device can change after exposure to light or high temperatures, primarily manifesting as increased mobility, negative shifts in the transfer characteristic curve, and the negative shift of V_{th}.

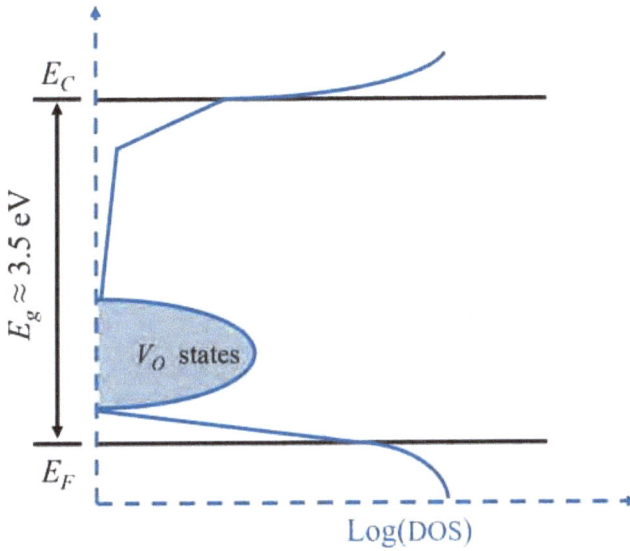

Fig. (2.9). DOS distribution of a typical IGZO TFT.

Furthermore, V_o^{2+} is closely related to the carrier concentration in the active layer. When the carrier concentration is low, V_os is more likely to ionize, transitioning from a neutral V_o state to a charged V_o^{2+} state by releasing two electrons. When a positive gate bias stress is applied to a TFT, electrons in the active layer move toward the interface, reducing the number of electrons within the active layer and causing V_os to ionize. This phenomenon is known as oxygen vacancy redistribution [82-84] The V_o^{2+} can lower the potential barrier between the active layer and the source/drain electrodes, making the device easier to turn on and negatively shifting the V_{th}.

2.5. TYPES OF DEFECT STATES IN THIN-FILM TRANSISTORS

One of the key differences between TFTs and MOSFETs is that the active layer of a TFT typically contains a significant number of defects, and poly-Si TFTs are no exception, with the primary defects stemming from the grain boundaries within the poly-Si [85-87]. The presence of defects not only degrades the performance of the device but is also the source of many reliability issues. Therefore, understanding the nature of these defect states is an essential theoretical foundation for subsequent analysis of TFT reliability.

Defect states within a semiconductor are usually caused by the non-periodicity of the lattice. In poly-Si, most defect states originate from the grain boundaries within the poly-Si. These defect states are typically distributed within the bandgap and exist as energy levels within the band structure. Unlike impurity levels, which provide free carriers, defect levels in semiconductors usually serve to trap free carriers.

Based on the location of the defect states within the device, they can be classified as bulk defects and interface defects. Depending on the ionization state of the defect states, they can be classified as donor-like defects and acceptor-like defects. According to their position within the band structure, defect states can be classified as shallow-level defects and deep-level defects.

2.5.1. Donor-Like States and Acceptor-Like States

Defect states in semiconductor devices, based on their ionization states, can be categorized into donor-like and acceptor-like defects. Donor-like defects, when ionized, capture electrons and become neutral, while in their unionized state, they exhibit a positive charge. Conversely, acceptor-like defects, when ionized, exhibit a negative charge after capturing an electron and are neutral when un-ionized [88]. As depicted in Fig. (**2.10**), donor-like defects are typically located near the valence band, while acceptor-like defects are found near the conduction band. In thermal equilibrium, donor-like defects are filled with electrons, and when free holes are present in the semiconductor, these holes near the valence band are easily captured by the donor-like defects, leading to an increase in Coulomb scattering and the formation of a potential barrier, which can degrade device performance [89]. Similarly, acceptor-like defects, when ionized, increase Coulomb scattering and reduce free carriers, enhancing scattering within the device.

Due to their positions in the band structure and their ionization states, donor-like defects primarily affect hole transport in the channel, while acceptor-like defects mainly influence electron transport. Typically, for p-type TFTs, donor-like defects are considered the primary type of defect affecting device performance, while for n-type TFTs, acceptor-like defects are the main concern [87]. However, for MO TFTs, which often contain a significant number of V_os (primarily considered donor-like defects), the ionization of these vacancies not only provides a large number of electrons but also inevitably affects the performance of n-type MO TFTs [90- 92].

The positive charges from V_o^{2+} can significantly reduce the potential barrier between the source and drain, leading to a severe negative shift in the device's V_{th}.

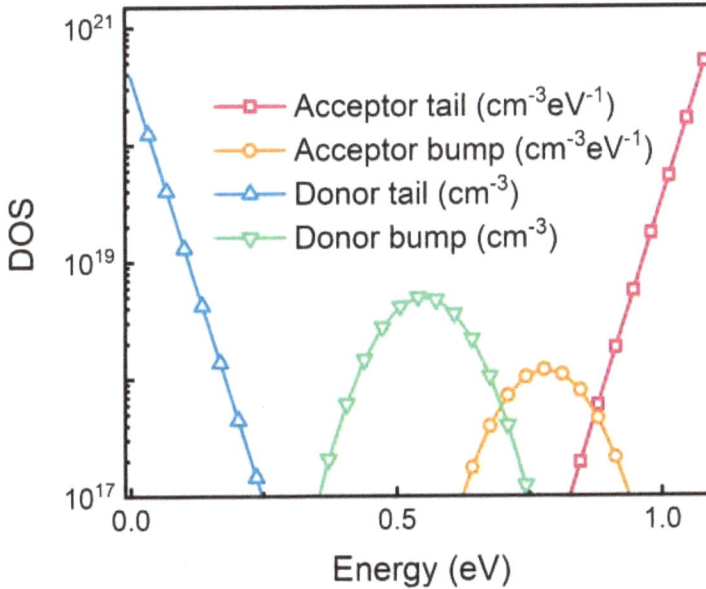

Fig. (2.10). DOS distribution in a poly-Si active layer.

2.5.2. Deep-Level Defects and Shallow-Level Defects

Defect states can also be classified based on their position in the band structure as deep-level defects and shallow-level defects or tail states, such as weak bonds in a-Si. Deep-level defects are typically found in the middle of the bandgap with a low density of states, while tail states are located near the band edges with a higher density of states.

As an example, for donor-like defects, the tail states are usually distributed at the edge of the valence band, while deep levels are found in the middle of the valence band. When a p-type device is in thermal equilibrium, the Fermi level near the interface is typically positioned in the middle of the bandgap, with most of the tail states occupied by electrons and some deep levels unoccupied. As the device transitions from thermal equilibrium to the on-state, the Fermi level shifts towards the valence band, and the deep levels are progressively occupied by holes. When the device is in the on-state, the Fermi level is near the valence band, and many tail states are occupied. Notably, the occupation of defects leads to a reduction in free holes, and based on the state of the device when the defects are occupied, it can be inferred that deep-level defects primarily affect the subthreshold region of the device, while tail states mainly influence the on-state region [93].

From this analysis, it is clear that defect states at different positions have varying impacts on the device's performance depending on the device's operating state. During the transition from the off-state to the on-state, *i.e.,* the subthreshold region, a small number of carriers participate in conduction as free carriers. The presence of deep-level defects captures some of these carriers, leading to delays in device turn-on or degradation of the subthreshold slope. Since there are fewer free carriers in the subthreshold region, even a low density of deep-level states can significantly affect the device's subthreshold performance. Once the device is in the on-state, the Fermi level is already near the conduction or valence band, and the deep-level states are almost fully occupied Due to their low density of states, they have difficulty significantly affecting the on-state performance. Tail states, which are typically several orders of magnitude denser than deep-level states and are close to the bands [87], can significantly impact the on-state performance, with a large number of tail states leading to a noticeable reduction in I_{on}.

2.5.3. Interface Defects and Bulk Defects

Defect states in semiconductor devices can be categorized based on their distribution within the device structure into interface state defects and bulk defects. For poly-Si TFTs, defects at the interface are often due to the breaking of Si-Si bonds at the interface of poly-Si, where a significant number of dangling bonds at the interface inhibit the flow of charge carriers [94]. Bulk defects typically originate from grain boundaries within poly-Si, where the periodic arrangement of Si atoms in each grain is interrupted at the grain boundaries, introducing defect states [85].

Normally, the density of bulk defects in a TFT is much higher than that of interface state defects [95]. However, because a large number of charge carriers gather near the interface and move laterally to produce current during actual operation, more charge carriers are susceptible to being captured by interface states. Therefore, a relatively low density of interface states can often lead to significant degradation in device performance. In the case of bulk defects in poly-Si thin films, they manifest as potential barriers in the flow of charge carriers, requiring carriers to have sufficient energy to overcome these barriers, which results in reduced mobility.

2.6. TYPICAL DEGRADATION MECHANISMS IN THIN-FILM TRANSISTORS

TFTs are one of the core components in active matrix (AM) displays, and their performance directly impacts various aspects of the display. As TFTs primarily operate in environments with complex electrical signals, research on their reliability focuses on the electrical reliability of TFTs. This research explores the impact of different environments, such as light exposure [96, 97], humidity [98, 99], and high temperatures [100, 101], on device performance based on various electrical reliability characteristics. The electrical reliability of TFTs can be mainly categorized into direct current (DC) reliability [102-110] and alternating current (AC) reliability [111, 112]. The primary degradation mechanisms for these different types of reliability can be explained by several common degradation mechanisms:

2.6.1. Hot Carrier Effect

The HC typically occurs when the V_{ds} is large and the V_{gs} is slightly above the V_{th} [103, 109]. A large V_{ds} significantly screens the longitudinal electric field near the drain, reducing the carriers near the drain and causing the device to 'pinch off' At this point, the electric field generated by the larger V_{ds} mainly falls in the pinch-off region, creating an electric field greater than 10^5 V/cm. Carriers near the drain gain significant energy and are accelerated into HCs [103].

HCs are usually those with energy several times greater than the thermal energy (kT) above the Fermi level. These carriers are not in thermal equilibrium with the lattice. When their energy reaches or exceeds the Si/SiO_2 interface barrier (3.2 eV for electron injection and 4.5 eV for hole injection), they can be injected into the oxide layer, creating interface states, oxide defects, or being captured by traps, increasing the charge in the oxide layer [109]. Typically, carriers injected into the GI or at the interface can cause a shift in the device's V_{th}, as shown in Fig. (**2.11**).

Similarly, these HCs can also collide with the lattice within the semiconductor, especially in TFTs with a higher concentration of defect states within the active layer, making them more susceptible to the effects of HC injection. This type of degradation typically leads to an increase in the potential barrier near the source and drain, resulting in a reduction in the device's I_{on}.

2.6.2. Self-Heating Effect

SH stress is typically induced by applying a high voltage simultaneously to the source and gate electrodes. Operating the device under these conditions leads to a significant increase in the working current. Additionally, due to the application of drain stress, the carrier concentration distribution across the channel becomes non-uniform, with fewer carriers near the drain end, resulting in a higher generation of Joule heat at the drain [111, 112]. Figs. (**12** and **13**) show the typical SH degradation behaviors in poly-Si TFTs and MO TFTs.

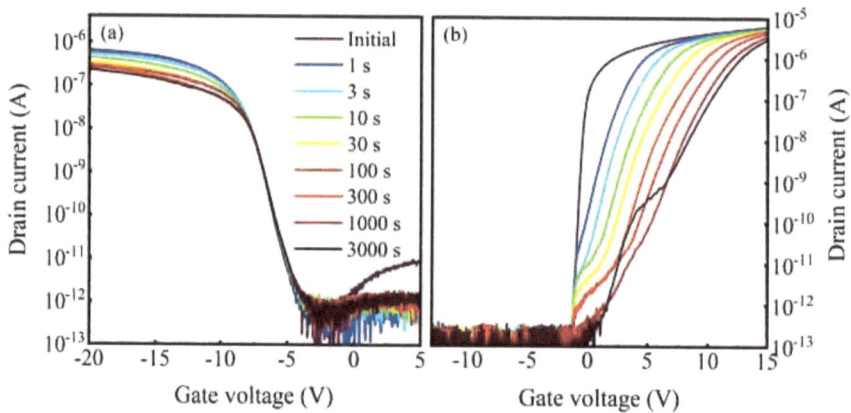

Fig. (2.11). Typical degradation of (a) poly-Si TFT device and (b) MO TFT device under HC stresses.

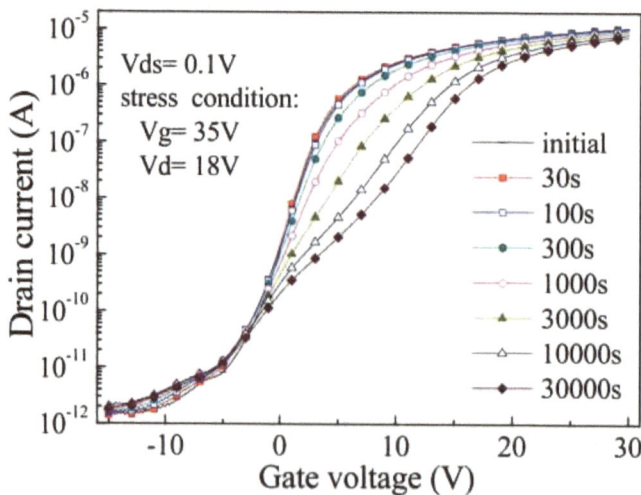

Fig. (2.12). Degradation of a typical poly-Si TFT device under SH stresses [104].

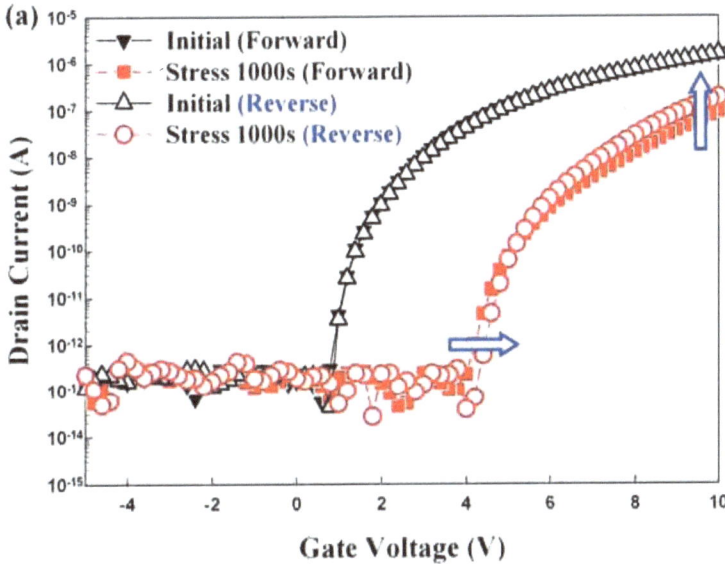

Fig. (2.13). Degradation of a typical MO TFT device under SH stresses [108].

The accumulation of Joule heat within the active layer can raise the internal temperature of the device to over 200°C. Such high temperatures can further cause weak bonds within the active layer film, such as Si-H bonds at grain boundaries, to break, generating a large number of dangling bonds [104]. These dangling bonds can act as deep-level defects within the active layer, leading to a shift in the V_{th} and degradation of the SS, as shown in Fig. (**2.12**) [104].

SH stress also leads to an increase in the device's I_{off}. Under SH stress, the I_{off} initially exhibits a significant and rapid degradation over a short period of stress because many electrons quickly become injected into the GI layer, forming fixed charges that shield the electric field and weaken the tunneling effect. However, as SH stress continues, I_{off} recovers. The main reason for this recovery is that defects generated at the interface and within the channel can assist in tunneling. Even though fixed charges shield a significant portion of the electric field, the large number of defects produced can lead to a recovery in I_{off}.

2.6.3. Gate Bias Effect and Impact of Environment

PBS and NBS are applied to TFTs by subjecting the gate to a substantial positive or negative voltage, respectively [102, 105-107, 110]. PBS stress typically results in a positive shift in the devices's V_{th}, whereas NBS stress causes a negative shift. The primary cause of this degradation is the accumulation of a large number of charge carriers at the interface due to the applied gate voltage. Some of these carriers are captured by defect states at the interface, while others are injected into the GI under the influence of a strong electric field. The charged defect states and fixed charges can shield part of the gate electric field, leading to a V_{th} shift. Fig. (**2.14**) shows the typical PBS and NBS degradation behaviors in MO TFTs and poly-Si TFTs.

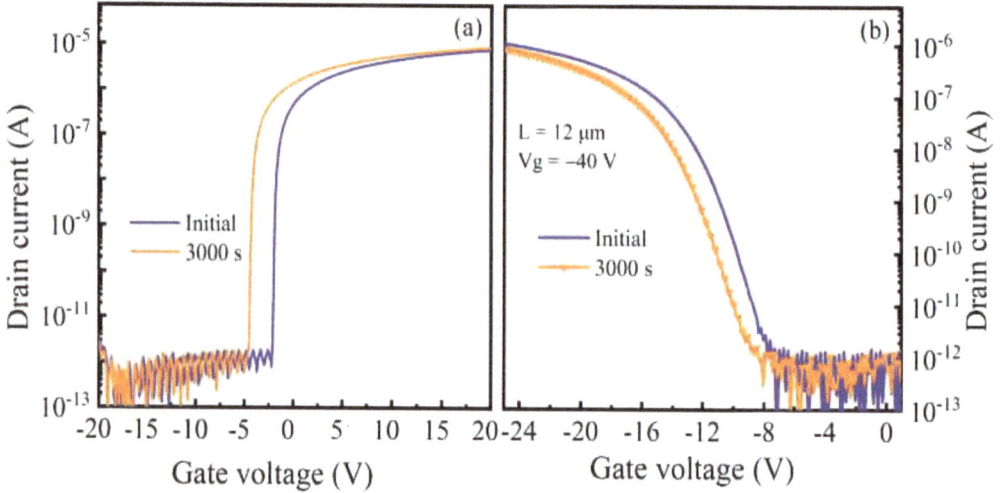

Fig. (2.14). (a) Typical MO TFT device stress degradation at PBS. (b) Typical poly-Si TFT device stress degradation at NBS.

Environmental factors can further exacerbate the degradation of TFTs. In PBTI/NBTI stress conditions, higher temperatures can provide charge carriers with additional energy, making them more likely to be injected into the GI, which can result in a more severe V_{th} shift [102, 106, 110]. Additionally, some poly-Si TFTs may experience a simultaneous degradation of I_{on} and mobility during PBTI/NBTI stress. This degradation can be attributed to damage to the poly-Si grain boundaries caused by high temperatures, which trap more charge carriers, reducing the device's I_{on} and further degrading the device's V_{th}. Similarly, in PBIS/NBIS stress conditions, the impact of light exposure can intensify the degradation of devices. This is mainly because charge carriers absorb photon energy, making them more likely to be injected into the GI or trapped by grain boundaries, leading to further device degradation.

CONCLUSION

This chapter comprehensively examines the fundamental aspects and performance characteristics of TFTs, focusing on various types of defects that can arise in silicon-based and metal-oxide thin films. It begins with an introduction to basic performance metrics essential for evaluating TFT functionality, followed by a detailed analysis of common defects specific to amorphous and polycrystalline silicon films. The chapter highlights critical issues such as weakened bonds, interface traps, and oxygen vacancies in metal oxide thin films, which can adversely affect device performance. Furthermore, it categorizes defect states into donor-like, acceptor-like, deep-level, shallow-level, interface, and bulk defects, providing insights into their contributions to TFT behavior. The discussion on typical degradation mechanisms, including the hot-carrier effect, self-heating effect, and the influence of gate bias and environmental factors, underscores the complex interplay of these phenomena in determining the reliability and longevity of TFTs. By identifying and understanding these various defects and degradation mechanisms, this chapter lays a solid foundation for future research aimed at enhancing TFT performance and reliability in practical applications.

REFERENCES

[1]   L. Li, N. Lu, and M. Liu, "Field Effect Mobility Model in Oxide Semiconductor Thin Film Transistors With Arbitrary Energy Distribution of Traps", *IEEE Electron Device Lett.,* vol. 35, no. 2, pp. 226-228, 2014.
        http://dx.doi.org/10.1109/LED.2013.2291782

[2]   J.Y. Hwang, C.C. Kuo, L.C. Chen, and K.H. Chen, "Correlating defect density with carrier mobility in large-scaled graphene films: Raman spectral signatures for the estimation of defect density", *Nanotechnology,* vol. 21, no. 46, p. 465705, 2010.
        http://dx.doi.org/10.1088/0957-4484/21/46/465705 PMID: 20972312

[3]   Y. Yang, M. Zhang, L. Lu, M. Wong, and H.S. Kwok, "Low-Frequency Noise in Bridged-Grain Polycrystalline Silicon Thin-Film Transistors", *IEEE Trans. Electron Dev.,* vol. 69, no. 4, pp. 1984-1988, 2022.
        http://dx.doi.org/10.1109/TED.2022.3148697

[4]   S. Lee, "Temperature dependent electron transport in amorphous oxide semiconductor thin film transistors", *2011 International Electron Devices Meeting,* p. 6, 2011.
        http://dx.doi.org/10.1109/IEDM.2011.6131554

[5]   K. Takechi, M. Nakata, T. Eguchi, H. Yamaguchi, and S. Kaneko, "Temperature-Dependent Transfer Characteristics of Amorphous InGaZnO 4 Thin-Film Transistors", *Jpn. J. Appl. Phys.,* vol. 48, no. 1R, p. 011301, 2009.
        http://dx.doi.org/10.1143/JJAP.48.011301

[6]   H.H. Hsu, C.H. Cheng, P. Chiou, Y.C. Chiu, S.S. Yen, C.H. Tung, and C.Y. Chang, "Temperature-Dependent Transfer Characteristics of Low Turn-On Voltage InGaZnO Metal-Oxide Devices With Thin Titanium Oxide Capping Layers", *J. Disp. Technol.,* vol. 11, no. 6, pp. 512-517, 2015.

http://dx.doi.org/10.1109/JDT.2014.2355876

[7]   H. Godo, D. Kawae, S. Yoshitomi, T. Sasaki, S. Ito, H. Ohara, H. Kishida, M. Takahashi, A. Miyanaga, and S. Yamazaki, "Temperature Dependence of Transistor Characteristics and Electronic Structure for Amorphous In–Ga–Zn-Oxide Thin Film Transistor", *Jpn. J. Appl. Phys.,* vol. 49, no. 3S, p. 03CB04, 2010.

http://dx.doi.org/10.1143/JJAP.49.03CB04

[8]   J. Lee, Y. Lee, T. Kang, H. Chu, and J. Kwag, "Alleviation of abnormal NBTI phenomenon in LTPS TFTs on polyimide substrate for flexible AMOLED", *J. Soc. Inf. Disp.,* vol. 28, no. 4, pp. 333-341, 2020.

http://dx.doi.org/10.1002/jsid.883

[9]   A. Chasin, "Understanding and modelling the PBTI reliability of thin-film IGZO transistors", *2021 IEEE International Electron Devices Meeting (IEDM),* p. 31.1, 2021.

http://dx.doi.org/10.1109/IEDM19574.2021.9720666

[10]   J. Lee, D. Kim, S. Lee, J. Cho, H. Park, and J. Jang, "High Field Effect Mobility, Amorphous In-Ga-Sn-O Thin-Film Transistor With No Effect of Negative Bias Illumination Stress", *IEEE Electron Device Lett.,* vol. 40, no. 9, pp. 1443-1446, 2019.

http://dx.doi.org/10.1109/LED.2019.2931089

[11]   T. Kawamura, M. Matsumura, T. Kaitoh, T. Noda, M. Hatano, T. Miyazawa, and M. Ohkura, "A Model for Predicting On-Current Degradation Caused by Drain-Avalanche Hot Carriers in Low-Temperature Polysilicon Thin-Film Transistors", *IEEE Trans. Electron Dev.,* vol. 56, no. 1, pp. 109-115, 2009.

http://dx.doi.org/10.1109/TED.2008.2008376

[12]   D. Hong, G. Yerubandi, H.Q. Chiang, M.C. Spiegelberg, and J.F. Wager, "Electrical Modeling of Thin-Film Transistors", *Crit. Rev. Solid State Mater. Sci.,* vol. 33, no. 2, pp. 101-132, 2008.

http://dx.doi.org/10.1080/10408430701384808

[13]   J. Lee, J.S. Park, Y.S. Pyo, D.B. Lee, E.H. Kim, D. Stryakhilev, T.W. Kim, D.U. Jin, and Y.G. Mo, "The influence of the gate dielectrics on threshold voltage instability in amorphous indium-gallium-zinc oxide thin film transistors", *Appl. Phys. Lett.,* vol. 95, no. 12, p. 123502, 2009.

http://dx.doi.org/10.1063/1.3232179

[14]   D.H. Kim, M.J. Park, and H.I. Kwon, "Effect of Channel Layer Thickness on Electrical and Thermal Stabilities of High-Mobility Zinc Oxynitride Thin-Film Transistors", *ECS J. Solid State Sci. Technol.,* vol. 6, no. 9, pp. Q109-Q113, 2017.

http://dx.doi.org/10.1149/2.0101709jss

[15]   Y. Liu, W.J. Wu, L. Qiang, L. Wang, Y.F. En, and B. Li, "Temperature-Dependent Drain Current Characteristics and Low Frequency Noises in Indium Zinc Oxide Thin Film Transistors", *Chin. Phys. Lett.,* vol. 32, no. 8, p. 088506, 2015.

http://dx.doi.org/10.1088/0256-307X/32/8/088506

[16]   K. Zhong, Y. Liu, S.T. Cai, and X.M. Xiong, "Temperature dependence of conduction and low frequency noise characteristics in hydrogenated amorphous silicon thin film transistors", *Mod. Phys. Lett. B,* vol. 33, no. 2, p. 1950009, 2019.
http://dx.doi.org/10.1142/S021798491950009X

[17]   K.Y. Lee, Y.K. Fang, C.W. Chen, K.C. Hwang, M.S. Liang, and S.G. Wuu, "To suppress UV damage on the subthreshold characteristic of TFT during hydrogenation for high density TFT SRAM", *IEEE Electron Device Lett.,* vol. 18, no. 1, pp. 4-6, 1997.
http://dx.doi.org/10.1109/55.553058

[18]   A. Koyama, H. Miyoshi, K. Shimazoe, Y. Otaka, M. Nitta, F. Nishikido, T. Yamaya, and H. Takahashi, "Radiation stability of an InGaZnO thin-film transistor in heavy ion radiotherapy", *Biomed. Phys. Eng. Express,* vol. 3, no. 4, p. 045009, 2017.
http://dx.doi.org/10.1088/2057-1976/aa78ae

[19]   Y.J. Jeong, T.K. An, D.J. Yun, L.H. Kim, S. Park, Y. Kim, S. Nam, K.H. Lee, S.H. Kim, J. Jang, and C.E. Park, "Photo-Patternable ZnO Thin Films Based on Cross-Linked Zinc Acrylate for Organic/Inorganic Hybrid Complementary Inverters", *ACS Appl. Mater. Interfaces,* vol. 8, no. 8, pp. 5499-5508, 2016.
http://dx.doi.org/10.1021/acsami.6b00259 PMID: 26840992

[20]   B. Kim, E. Chong, D. Hyung Kim, Y. Woo Jeon, D. Hwan Kim, and S. Yeol Lee, "Origin of threshold voltage shift by interfacial trap density in amorphous InGaZnO thin film transistor under temperature induced stress", *Appl. Phys. Lett.,* vol. 99, no. 6, p. 062108, 2011.
http://dx.doi.org/10.1063/1.3615304

[21]   Seonghyun Jin, Tae-Woong Kim, Young-Gug Seol, M. Mativenga, and Jin Jang, "Reduction of Positive-Bias-Stress Effects in Bulk-Accumulation Amorphous-InGaZnO TFTs", *IEEE Electron Device Lett.,* vol. 35, no. 5, pp. 560-562, 2014.
http://dx.doi.org/10.1109/LED.2014.2311172

[22]   Z. Lin, J. Zhao, X. Li, L. Kang, J. Li, Y. Wu, J. Xu, and M. Si, "Universal PBTI Relaxation on the Negative V TH Shift in Oxide Semiconductor Transistors and New Insights", *IEEE Electron Device Lett.,* vol. 44, no. 7, pp. 1136-1139, 2023.
http://dx.doi.org/10.1109/LED.2023.3274771

[23]   D. Palumbo, S. Masala, P. Tassini, A. Rubino, and D. della Sala, "Electrical Stress Degradation of Small-Grain Polysilicon Thin-Film Transistors", *IEEE Trans. Electron Dev.,* vol. 54, no. 3, pp. 476-482, 2007.
http://dx.doi.org/10.1109/TED.2006.890377

[24]   C.L. Fan, F.P. Tseng, B.J. Li, Y.Z. Lin, S.J. Wang, W.D. Lee, and B.R. Huang, "Improvement in reliability of amorphous indium–gallium–zinc oxide thin-film transistors with Teflon/SiO 2 bilayer passivation under gate bias stress", *Jpn. J. Appl. Phys.,* vol. 55, no. 2S, p. 02BC17, 2016.
http://dx.doi.org/10.7567/JJAP.55.02BC17

[25]   E.N. Cho, J.H. Kang, and I. Yun, "Contact resistance dependent scaling-down behavior of amorphous InGaZnO thin-film transistors", *Curr. Appl. Phys.,* vol. 11, no. 4, pp. 1015-1019, 2011.
http://dx.doi.org/10.1016/j.cap.2011.01.017

[26]   M. Zhang, H. Lin, S. Deng, R. Chen, G. Li, S.T. Han, Y. Zhou, Y. Yan, W. Zhou, M. Wong, and H.S. Kwok, "High-Performance Polycrystalline Silicon Thin-Film Transistors without Source/Drain Doping by Utilizing Anisotropic Conductivity of Bridged-Grain Lines", *Adv. Electron. Mater.,* vol. 6, no. 2, p. 1900961, 2020.
http://dx.doi.org/10.1002/aelm.201900961

[27]   Y. Liu, S. Deng, R. Chen, B. Li, Y.F. En, and Y. Chen, "Low-Frequency Noise in Hybrid-Phase- Microstructure ITO-Stabilized ZnO Thin-Film Transistors", *IEEE Electron Device Lett.,* vol. 39, no. 2, pp. 200-203, 2018.
http://dx.doi.org/10.1109/LED.2017.2784844

[28]   Y. Yan, X.J. She, H. Zhu, and S.D. Wang, "Origin of bias stress induced instability of contact resistance in organic thin film transistors", *Org. Electron.,* vol. 12, no. 5, pp. 823-826, 2011.
http://dx.doi.org/10.1016/j.orgel.2011.02.019

[29]   W.S. Kim, Y.K. Moon, K.T. Kim, J.H. Lee, B. Ahn, and J.W. Park, "An investigation of contact resistance between metal electrodes and amorphous gallium–indium–zinc oxide (a-GIZO) thin-film transistors", *Thin Solid Films,* vol. 518, no. 22, pp. 6357-6360, 2010.
http://dx.doi.org/10.1016/j.tsf.2010.02.044

[30]   W.J. Wu, R.H. Yao, S.H. Li, Y.F. Hu, W.L. Deng, and X.R. Zheng, "A Compact Model for Polysilicon TFTs Leakage Current Including the Poole–Frenkel Effect", *IEEE Trans. Electron Dev.,* vol. 54, no. 11, pp. 2975-2983, 2007.
http://dx.doi.org/10.1109/TED.2007.906968

[31]   G.F. Chen, H.C. Chen, T.C. Chang, S.P. Huang, H.M. Chen, P.Y. Liao, J.J. Chen, C.W. Kuo, W.C. Lai, A.K. Chu, S.C. Lin, C.Y. Yeh, C.S. Chang, C.M. Tsai, M.C. Yu, and S. Zhang, "An Energy-Band Model for Dual-Gate-Voltage Sweeping in Hydrogenated Amorphous Silicon Thin-Film Transistors", *IEEE Trans. Electron Dev.,* vol. 66, no. 6, pp. 2614-2619, 2019.
http://dx.doi.org/10.1109/TED.2019.2908859

[32]   R. Swain, K. Jena, and T.R. Lenka, "Modeling of Forward Gate Leakage Current in MOSHEMT Using Trap-Assisted Tunneling and Poole–Frenkel Emission", *IEEE Trans. Electron Dev.,* vol. 63, no. 6, pp. 2346-2352, 2016.
http://dx.doi.org/10.1109/TED.2016.2555851

[33]   B.H. Hamadani, C.A. Richter, D.J. Gundlach, R.J. Kline, I. McCulloch, and M. Heeney, "Influence of source-drain electric field on mobility and charge transport in organic field-effect transistors", *J. Appl. Phys.,* vol. 102, no. 4, p. 044503, 2007.
http://dx.doi.org/10.1063/1.2769782

[34]   N. On, B.K. Kim, S. Lee, E.H. Kim, J.H. Lim, and J.K. Jeong, "Hot Carrier Effect in Self-Aligned In–Ga–Zn–O Thin-Film Transistors With Short Channel Length", *IEEE Trans. Electron Dev.,* vol. 67, no. 12, pp. 5544-5551, 2020.
http://dx.doi.org/10.1109/TED.2020.3032383

[35]   K. Takechi, M. Nakata, H. Kanoh, S. Otsuki, and S. Kaneko, "Dependence of self-heating effects on operation conditions and device structures for polycrystalline silicon TFTs", *IEEE Trans. Electron Dev.,* vol. 53, no. 2, pp. 251-257, 2006.
http://dx.doi.org/10.1109/TED.2005.861729

[36]   C.Y. Liang, F.Y. Gan, P.T. Liu, F.S. Yeh, S.H.L. Chen, and T.C. Chang, "A Novel Self-Aligned Etch-Stopper Structure With Lower Photo Leakage for AMLCD and Sensor Applications", *IEEE Electron Device Lett.,* vol. 27, no. 12, pp. 978-980, 2006.
http://dx.doi.org/10.1109/LED.2006.886418

[37]   M. Bouzerdoum, and B. Birouk, "Characterization by SEM and FTIR of B-LPCVD polysilicon films after thermal oxidation", *in 2012 24th International Conference on Microelectronics (ICM),* pp. 1-4, 2012.
http://dx.doi.org/10.1109/ICM.2012.6471438

[38]   S.S. Kubatur, and M.L. Comer, "Simulation of grain growth in polycrystalline materials: A signal processing perspective", *2015 IEEE Global Conference on Signal and Information Processing (GlobalSIP),* IEEE: Orlando, FL, USA, pp. 853-857, 2015.
http://dx.doi.org/10.1109/GlobalSIP.2015.7418318

[39]   N. Banno, T. Takeuchi, H. Kitaguchi, K. Tsuchiya, and K. Kurushima, "Three-dimensional structural analysis for crystal defects in phase-transformed Nb$_3$Al", *IEEE Trans. Appl. Supercond.,* pp. 1-1, 2015.
http://dx.doi.org/10.1109/TASC.2015.2509779

[40]   D. Lu, Q. Jiang, X. Ma, Q. Zhang, X. Fu, and L. Fan, "Defect-Related Etch Pits on Crystals and Their Utilization", *Crystals (Basel),* vol. 12, no. 11, p. 1549, 2022.
http://dx.doi.org/10.3390/cryst12111549

[41]   H. Föll, U. Gösele, and B.O. Kolbesen, "Microdefects in silicon and their relation to point defects", *J. Cryst. Growth,* vol. 52, pp. 907-916, 1981.
http://dx.doi.org/10.1016/0022-0248(81)90397-3

[42]   G.P. Parry, "Rotational Symmetries of Crystals with Defects", *J. Elast.,* vol. 94, no. 2, pp. 147-166, 2009.
http://dx.doi.org/10.1007/s10659-008-9188-7

[43]   M. Li, C. Zhang, M. Li, F. Liu, L. Zhou, Z. Gao, J. Sun, D. Han, and J. Gong, "Growth defects of organic crystals: A review", *Chem. Eng. J.,* vol. 429, p. 132450, 2022.
http://dx.doi.org/10.1016/j.cej.2021.132450

[44]   J. Robertson, and M.J. Powell, "Deposition, defect and weak bond formation processes in a-Si:H", *Thin Solid Films,* vol. 337, no. 1-2, pp. 32-36, 1999.
http://dx.doi.org/10.1016/S0040-6090(98)01171-7

[45]   E. Tarnow, and R.A. Street, "Bonding of hydrogen to weak Si-Si bonds", *Phys. Rev. B Condens. Matter,* vol. 45, no. 7, pp. 3366-3371, 1992.
http://dx.doi.org/10.1103/PhysRevB.45.3366 PMID: 10001910

[46]   N. Ishii, and T. Shimizu, "Cluster-model calculations of hyperfine coupling constants of dangling bond and weak bond in a-Si : H", *Solid State Commun.,* vol. 102, no. 9, pp. 647-651, 1997.
http://dx.doi.org/10.1016/S0038-1098(97)80074-9

[47]   H. Oheda, "Density of Si–H bonds responsible for structural flexibility in a-Si:H", *J. Non-Cryst. Solids,* vol. 227-230, pp. 120-123, 1998.
http://dx.doi.org/10.1016/S0022-3093(98)00027-1

[48]   Ling Wang, T.A. Fjeldly, B. Iniguez, H.C. Slade, and M. Shur, "Self-heating and kink effects in a-Si:H thin film transistors", *IEEE Trans. Electron Dev.*, vol. 47, no. 2, pp. 387-397, 2000.
http://dx.doi.org/10.1109/16.822285

[49]   S.C. Kao, H.W. Zan, J.J. Huang, and B.C. Kung, "Self-Heating Effect on Bias-Stressed Reliability for Low-Temperature a-Si:H TFT on Flexible Substrate", *IEEE Trans. Electron Dev.*, vol. 57, no. 3, pp. 588-593, 2010.
http://dx.doi.org/10.1109/TED.2009.2039261

[50]   H.L. Chen, W.J. Chen, P.Y. Liu, K.H. Cheng, M.S. Lai, C.W. Wang, and C.T. Liu, "Universal Bias Dependence of Excess Current Induced by Self-Heating Effect for a-Si:H TFTs", *IEEE Trans. Electron Dev.*, vol. 54, no. 5, pp. 1238-1243, 2007.
http://dx.doi.org/10.1109/TED.2007.893191

[51]   Tien-Fu Chen, Ching-Fa Yeh, and Jen-Chung Lou, "Investigation of grain boundary control in the drain junction on laser-crystalized poly-Si thin film transistors", *IEEE Electron Device Lett.*, vol. 24, no. 7, pp. 457-459, 2003.
http://dx.doi.org/10.1109/LED.2003.814007

[52]   M. Kimura, and C.A. Dimitriadis, "Characteristic Degradation of Poly-Si Thin-Film Transistors With Large Grains From the Viewpoint of Grain Boundary Location", *IEEE Trans. Electron Dev.*, vol. 58, no. 6, pp. 1748-1751, 2011.
http://dx.doi.org/10.1109/TED.2011.2135356

[53]   Y. Uraoka, K. Kitajima, H. Kirimura, H. Yano, T. Hatayama, and T. Fuyuki, "Degradation in Low-Temperature Poly-Si Thin Film Transistors Depending on Grain Boundaries", *Jpn. J. Appl. Phys.*, vol. 44, no. 5R, p. 2895, 2005.
http://dx.doi.org/10.1143/JJAP.44.2895

[54]   A. Heya, A.Q. He, N. Otsuka, and H. Matsumura, "Anomalous grain boundary and carrier transport in cat-CVD poly-Si films", *J. Non-Cryst. Solids,* vol. 227-230, pp. 1016-1020, 1998.
http://dx.doi.org/10.1016/S0022-3093(98)00256-7

[55]   J.E. Brandenburg, L.A. Barrales-Mora, D.A. Molodov, and G. Gottstein, "Effect of inclination dependence of grain boundary energy on the mobility of tilt and non-tilt low-angle grain boundaries", *Scr. Mater.*, vol. 68, no. 12, pp. 980-983, 2013.
http://dx.doi.org/10.1016/j.scriptamat.2013.02.054

[56]   J.E. Brandenburg, L.A. Barrales-Mora, and D.A. Molodov, "On migration and faceting of low-angle grain boundaries: Experimental and computational study", *Acta Mater.*, vol. 77, pp. 294-309, 2014.
http://dx.doi.org/10.1016/j.actamat.2014.06.006

[57]   D.M. Kirch, E. Jannot, L.A. Barrales-Mora, D.A. Molodov, and G. Gottstein, "Inclination dependence of grain boundary energy and its impact on the faceting and kinetics of tilt grain boundaries in aluminum", *Acta Mater.*, vol. 56, no. 18, pp. 4998-5011, 2008.
http://dx.doi.org/10.1016/j.actamat.2008.06.017

[58]   B. Schönfelder, G. Gottstein, and L.S. Shvindlerman, "Comparative study of grain-boundary migration and grain-boundary self-diffusion of [0 0 1] twist-grain boundaries in copper by atomistic simulations", *Acta Mater.*, vol. 53, no. 6, pp. 1597-1609, 2005.

http://dx.doi.org/10.1016/j.actamat.2004.12.010

[59]   M. Winning, "Stress-induced migration of tilt and twist grain boundaries", *Z. Metallk.,* vol. 95, no. 4, pp. 233-238, 2004.

http://dx.doi.org/10.3139/146.017940

[60]   Z.H. Liu, Y.X. Feng, and J.X. Shang, "Characterizing twist grain boundaries in BCC Nb by molecular simulation: Structure and shear deformation", *Appl. Surf. Sci.,* vol. 370, pp. 19-24, 2016.

http://dx.doi.org/10.1016/j.apsusc.2016.02.097

[61]   K. Ishida, "Effect of grain size on grain boundary segregation", *J. Alloys Compd.,* vol. 235, no. 2, pp. 244-249, 1996.

http://dx.doi.org/10.1016/0925-8388(95)02094-2

[62]   K. Shirai, F. Oshiro, T. Noguchi, H. Koo, and H. Choi, "Influence of Grain Size Deviation on the Characteristics of Poly-Si Thin Film Transistor", *J. Korean Phys. Soc.,* vol. 59, no. 2, pp. 298-303, 2011.

http://dx.doi.org/10.3938/jkps.59.298

[63]   H. Kuriyama, S. Kiyama, S. Noguchi, T. Kuwahara, S. Ishida, T. Nohda, K. Sano, H. Iwata, H. Kawata, M. Osumi, S. Tsuda, S. Nakano, and Y. Kuwano, "Enlargement of Poly-Si Film Grain Size by Excimer Laser Annealing and Its Application to High-Performance Poly-Si Thin Film Transistor", *Jpn. J. Appl. Phys.,* vol. 30, no. 12S, p. 3700, 1991.

http://dx.doi.org/10.1143/JJAP.30.3700

[64]   T. Miyase, K. Watanabe, I. Sakaguchi, N. Ohashi, K. Domen, K. Nomura, H. Hiramatsu, H. Kumomi, H. Hosono, and T. Kamiya, "Roles of Hydrogen in Amorphous Oxide Semiconductor In-Ga-Zn-O: Comparison of Conventional and Ultra-High-Vacuum Sputtering", *ECS J. Solid State Sci. Technol.,* vol. 3, no. 9, pp. Q3085-Q3090, 2014.

http://dx.doi.org/10.1149/2.015409jss

[65]   J.M. Lee, I.T. Cho, J.H. Lee, and H.I. Kwon, "Bias-stress-induced stretched-exponential time dependence of threshold voltage shift in InGaZnO thin film transistors", *Appl. Phys. Lett.,* vol. 93, no. 9, p. 093504, 2008.

http://dx.doi.org/10.1063/1.2977865

[66]   F.H. Chen, T.M. Pan, C.H. Chen, J.H. Liu, W.H. Lin, and P.H. Chen, "Two-step Electrical Degradation Behavior in α-InGaZnO Thin-film Transistor Under Gate-bias Stress", *IEEE Electron Device Lett.,* vol. 34, no. 5, pp. 635-637, 2013.

http://dx.doi.org/10.1109/LED.2013.2248115

[67]   J.F. Conley, "Instabilities in Amorphous Oxide Semiconductor Thin-Film Transistors", *IEEE Trans. Device Mater. Reliab.,* vol. 10, no. 4, pp. 460-475, 2010.

http://dx.doi.org/10.1109/TDMR.2010.2069561

[68]   I.T. Cho, J.M. Lee, J.H. Lee, and H.I. Kwon, "Charge trapping and detrapping characteristics in amorphous InGaZnO TFTs under static and dynamic stresses", *Semicond. Sci. Technol.,* vol. 24, no. 1, p. 015013, 2009.

http://dx.doi.org/10.1088/0268-1242/24/1/015013

[69]   M. Kimura, T. Nakanishi, K. Nomura, T. Kamiya, and H. Hosono, "Trap densities in amorphous-InGaZnO4 thin-film transistors", *Appl. Phys. Lett.,* vol. 92, no. 13, p. 133512, 2008.
http://dx.doi.org/10.1063/1.2904704

[70]   A. Janotti, and C.G. Van de Walle, "Oxygen vacancies in ZnO", *Appl. Phys. Lett.,* vol. 87, no. 12, p. 122102, 2005.
http://dx.doi.org/10.1063/1.2053360

[71]   A. de Jamblinne de Meux, A. Bhoolokam, G. Pourtois, J. Genoe, and P. Heremans, "Oxygen vacancies effects in a-IGZO: Formation mechanisms, hysteresis, and negative bias stress effects", *Phys. Status Solidi., A Appl. Mater. Sci.,* vol. 214, no. 6, p. 1600889, 2017. [a].
http://dx.doi.org/10.1002/pssa.201600889

[72]   T. Kamiya, K. Nomura, and H. Hosono, "Origins of High Mobility and Low Operation Voltage of Amorphous Oxide TFTs: Electronic Structure, Electron Transport, Defects and Doping", *J. Disp. Technol.,* vol. 5, no. 7, pp. 273-288, 2009.
http://dx.doi.org/10.1109/JDT.2009.2021582

[73]   W. Körner, D.F. Urban, and C. Elsässer, "Origin of subgap states in amorphous In-Ga-Zn-O", *J. Appl. Phys.,* vol. 114, no. 16, p. 163704, 2013.
http://dx.doi.org/10.1063/1.4826895

[74]   M. Nakashima, M. Oota, N. Ishihara, Y. Nonaka, T. Hirohashi, M. Takahashi, S. Yamazaki, T. Obonai, Y. Hosaka, and J. Koezuka, "Origin of major donor states in In–Ga–Zn oxide", *J. Appl. Phys.,* vol. 116, no. 21, p. 213703, 2014.
http://dx.doi.org/10.1063/1.4902859

[75]   N. On, Y. Kang, A. Song, B.D. Ahn, H.D. Kim, J.H. Lim, K.B. Chung, S. Han, and J.K. Jeong, "Origin of Electrical Instabilities in Self-Aligned Amorphous In–Ga–Zn–O Thin-Film Transistors", *IEEE Trans. Electron Dev.,* vol. 64, no. 12, pp. 4965-4973, 2017.
http://dx.doi.org/10.1109/TED.2017.2766148

[76]   A. Janotti, and C.G. Van de Walle, "Native point defects in ZnO", *Phys. Rev. B Condens. Matter Mater. Phys.,* vol. 76, no. 16, p. 165202, 2007.
http://dx.doi.org/10.1103/PhysRevB.76.165202

[77]   Y. Nonaka, Y. Kurosawa, Y. Komatsu, N. Ishihara, M. Oota, M. Nakashima, T. Hirohashi, M. Takahashi, S. Yamazaki, T. Obonai, Y. Hosaka, J. Koezuka, and J. Yamauchi, "Investigation of defects in In–Ga–Zn oxide thin film using electron spin resonance signals", *J. Appl. Phys.,* vol. 115, no. 16, p. 163707, 2014.
http://dx.doi.org/10.1063/1.4873638

[78]   Y. Yue, H. Zhu, X. Liu, Y. Song, and X. Zuo, "First-principles study on non-radiative carrier captures of point defects associated with proton generation in silica", *AIP Adv.,* vol. 11, no. 1, p. 015214, 2021.
http://dx.doi.org/10.1063/5.0033421

[79]   Jianke Yao, Ningsheng Xu, Shaozhi Deng, Jun Chen, Juncong She, H.D. Shieh, Po-Tsun Liu, and Yi-Pai Huang, "Electrical and Photosensitive Characteristics of a-IGZO TFTs Related to Oxygen Vacancy", *IEEE Trans. Electron Dev.,* vol. 58, no. 4, pp. 1121-1126, 2011.

http://dx.doi.org/10.1109/TED.2011.2105879

[80] K.C. Chang, K. Liu, L. Hu, L. Li, X. Lin, S. Zhang, R. Zhang, H.J. Liu, and T.P. Kuo, "Supercritical Ammoniation-Enabled Interfacial Polarization for Function-Mode Transformation and Overall Optimization of Thin-Film Transistors", *ACS Appl. Mater. Interfaces,* vol. 13, no. 33, pp. 40053-40061, 2021.

http://dx.doi.org/10.1021/acsami.1c09673 PMID: 34392676

[81] J.H. Park, Y. Kim, S. Yoon, S. Hong, and H.J. Kim, "Simple method to enhance positive bias stress stability of In-Ga-Zn-O thin-film transistors using a vertically graded oxygen-vacancy active layer", *ACS Appl. Mater. Interfaces,* vol. 6, no. 23, pp. 21363-21368, 2014.

http://dx.doi.org/10.1021/am5063212 PMID: 25402628

[82] G. Zhu, M. Zhang, Z. Jiang, J. Huang, Y. Huang, S. Deng, L. Lu, M. Wong, and H.S. Kwok, "Significant Degradation Reduction in Metal Oxide Thin-Film Transistors via the Interaction of Ionized Oxygen Vacancy Redistribution, Self-Heating Effect, and Hot Carrier Effect", *IEEE Trans. Electron Dev.,* vol. 70, no. 8, pp. 4198-4205, 2023.

http://dx.doi.org/10.1109/TED.2023.3283940

[83] Y. Zhang, Z. Wang, M. Wang, D. Zhang, H. Wang, and M. Wong, "A Unified Degradation Model of Elevated-Metal Metal Oxide (EMMO) TFTs Under Positive Gate Bias With or Without an Illumination", *IEEE Trans. Electron Dev.,* vol. 68, no. 3, pp. 1081-1087, 2021.

http://dx.doi.org/10.1109/TED.2021.3053915

[84] S. Li, M. Wang, D. Zhang, H. Wang, and Q. Shan, "A Unified Degradation Model of a-InGaZnO TFTs Under Negative Gate Bias With or Without an Illumination", *IEEE J. Electron Devices Soc.,* vol. 7, pp. 1063-1071, 2019.

http://dx.doi.org/10.1109/JEDS.2019.2946383

[85] C.-P. Chou, Y.-X. Lin, K.-Y. Hsieh, and Y.-H. Wu, "Poly-GeSn junctionless P-TFTs featuring a record high ION/IOFF ratio and hole mobility by defect engineering", *Journal of Materials Chemistry C*, vol. 7, no. 17, pp. 5201-5208, 2019.

http://dx.doi.org/10.1039/c8tc04972f

[86] S.C. Lee, and M.J. Lee, "Effects of multi-energetic grain-boundary trapping states on the electrical characteristics of poly-CdSe thin film transistors", *J. Appl. Phys.,* vol. 88, no. 4, pp. 1999-2004, 2000.

http://dx.doi.org/10.1063/1.1305908

[87] P.M. Walker, S. Uno, and H. Mizuta, "Simulation Study of the Dependence of Submicron Polysilicon Thin-Film Transistor Output Characteristics on Grain Boundary Position", *Jpn. J. Appl. Phys.,* vol. 44, no. 12R, p. 8322, 2005.

http://dx.doi.org/10.1143/JJAP.44.8322

[88] N. Jankovic, "Numerical simulations of N-type CdSe poly-TFT electrical characteristics with trap density models of Atlas/Silvaco", *Microelectron. Reliab.,* vol. 52, no. 11, pp. 2537-2541, 2012.

http://dx.doi.org/10.1016/j.microrel.2012.03.031

[89] J. Levinson, F.R. Shepherd, P.J. Scanlon, W.D. Westwood, G. Este, and M. Rider, "Conductivity behavior in polycrystalline semiconductor thin film transistors", *J. Appl. Phys.,* vol. 53, no. 2, pp. 1193-1202, 1982.

http://dx.doi.org/10.1063/1.330583

[90] H.C. Chen, G.F. Chen, P.H. Chen, S.P. Huang, J.J. Chen, K.J. Zhou, C.W. Kuo, Y.C. Tsao, A.K. Chu, H.C. Huang, W.C. Lai, and T.C. Chang, "A Novel Heat Dissipation Structure for Inhibiting Hydrogen Diffusion in Top-Gate a-InGaZnO TFTs", *IEEE Electron Device Lett.,* vol. 40, no. 9, pp. 1447-1450, 2019.
http://dx.doi.org/10.1109/LED.2019.2927422

[91] H. Im, H. Song, J. Jeong, Y. Hong, and Y. Hong, "Effects of defect creation on bidirectional behavior with hump characteristics of InGaZnO TFTs under bias and thermal stress", *Jpn. J. Appl. Phys.,* vol. 54, no. 3S, p. 03CB03, 2015.
http://dx.doi.org/10.7567/JJAP.54.03CB03

[92] P.R. Xu, and R.H. Yao, "A model for threshold voltage shift under negative gate bias stress in amorphous InGaZnO thin film transistors", *Eur. Phys. J. Appl. Phys.,* vol. 72, no. 3, p. 30102, 2015.
http://dx.doi.org/10.1051/epjap/2015150375

[93] M. Koyanagi, Y. Baba, K. Hata, I.W. Wu, A.G. Lewis, M. Fuse, and R. Bruce, "The charge-pumping technique for grain boundary trap evaluation in polysilicon TFTs", *IEEE Electron Device Lett.,* vol. 13, no. 3, pp. 152-154, 1992.
http://dx.doi.org/10.1109/55.144994

[94] R.B. Wehrspohn, M.J. Powell, S.C. Deane, I.D. French, and P. Roca i Cabarrocas, "Dangling-bond defect state creation in microcrystalline silicon thin-film transistors", *Appl. Phys. Lett.,* vol. 77, no. 5, pp. 750-752, 2000.
http://dx.doi.org/10.1063/1.127107

[95] W. Lee, J. Oh, J.H. Chu, S. Choi, T. Kang, H. Chu, and H. Kim, "Comparative study of C–V-based extraction methods of interface state density for a low-temperature polysilicon thin film", *Mater. Res. Express,* vol. 8, no. 8, p. 085902, 2021.
http://dx.doi.org/10.1088/2053-1591/ac1aa6

[96] M.D.H. Chowdhury, P. Migliorato, and J. Jang, "Light induced instabilities in amorphous indium–gallium–zinc–oxide thin-film transistors", *Appl. Phys. Lett.,* vol. 97, no. 17, p. 173506, 2010.
http://dx.doi.org/10.1063/1.3503971

[97] K. Park, H.W. Park, H.S. Shin, J. Bae, K.S. Park, I. Kang, K.B. Chung, and J.Y. Kwon, "Reliability of Crystalline Indium–Gallium–Zinc-Oxide Thin-Film Transistors Under Bias Stress With Light Illumination", *IEEE Trans. Electron Dev.,* vol. 62, no. 9, pp. 2900-2905, 2015.
http://dx.doi.org/10.1109/TED.2015.2458987

[98] D. Li, E.J. Borkent, R. Nortrup, H. Moon, H. Katz, and Z. Bao, "Humidity effect on electrical performance of organic thin-film transistors", *Appl. Phys. Lett.,* vol. 86, no. 4, p. 042105, 2005.
http://dx.doi.org/10.1063/1.1852708

[99] K. Hoshino, B. Yeh, and J.F. Wager, "Impact of humidity on the electrical performance of amorphous oxide semiconductor thin-film transistors", *J. Soc. Inf. Disp.,* vol. 21, no. 7, pp. 310-316, 2013.

http://dx.doi.org/10.1002/jsid.184

[100] Y.C. Tsao, T.C. Chang, S.P. Huang, Y.L. Tsai, Y.C. Chien, M.C. Tai, H.Y. Tu, and J.W. Huang, "Reliability Test Integrating Electrical and Mechanical Stress at High Temperature for a-InGaZnO Thin Film Transistors", *IEEE Trans. Device Mater. Reliab.,* vol. 19, no. 2, pp. 433-436, 2019.

http://dx.doi.org/10.1109/TDMR.2019.2915807

[101] Y.P. Chen, M. Si, B.K. Mahajan, Z. Lin, P.D. Ye, and M.A. Alam, "Positive Bias Temperature Instability and Hot Carrier Degradation of Back-End-of-Line, nm-Thick, In 2 O 3 Thin-Film Transistors", *IEEE Electron Device Lett.,* vol. 43, no. 2, pp. 232-235, 2022.

http://dx.doi.org/10.1109/LED.2021.3134902

[102] M. Zhang, "P-15: Gate-Bias-Stress-Induced Instability in Hybrid-Phase Microstructural ITO-Stabilized ZnO TFTs", *SID Symposium Digest of Technical Papers,* vol. vol. 50, pp. 1267-1270, 2019.

http://dx.doi.org/10.1002/sdtp.13164

[103] Meng Zhang, Wei Zhou, Rongsheng Chen, Man Wong, and Hoi-Sing Kwok, "Characterization of DC-Stress-Induced Degradation in Bridged-Grain Polycrystalline Silicon Thin-Film Transistors", *IEEE Trans. Electron Dev.,* vol. 61, no. 9, pp. 3206-3212, 2014.

http://dx.doi.org/10.1109/TED.2014.2341676

[104] M. Wang, "Degradation Mechanisms of Low Temperature Poly-Si Thin-Film Transistors under Circuit Applications", *Proceedings of China Display/Asia Display 2011,* B. Wang, Ed., pp. 62-65, 2011.

[105] Z. Jiang, *Device Performance Improvement Caused by Negative Gate Voltage Bias Stress in Indium-Tin-Zinc Oxide Thin-Film Transistors.* 2018.

[106] Q.J. Sun, J. Wu, M. Zhang, Y. Yuan, X. Gao, S.D. Wang, Z. Tang, C.C. Kuo, and Y. Yan, "Enhanced Electrical Performance and Bias-Stress Stability of Solution-Processed Bilayer Metal Oxide Thin-Film Transistors", *Phys. Status Solidi., A Appl. Mater. Sci.,* vol. 219, no. 22, p. 2200311, 2022. [a].

http://dx.doi.org/10.1002/pssa.202200311

[107] Y. Yan, Z. Huang, X. Ma, Z. Jiang, and M. Zhang, "Gate-Voltage-Stress-Induced Instability in C 8 -BTBT Thin-Film Transistors with Aluminium Oxide as Gate Dielectric", *in 2019 IEEE 26th International Symposium on Physical and Failure Analysis of Integrated Circuits (IPFA),* Hangzhou, China: IEEE, pp. 1-4, 2019.

http://dx.doi.org/10.1109/IPFA47161.2019.8984911

[108] T.Y. Hsieh, T.C. Chang, T.C. Chen, M.Y. Tsai, Y.T. Chen, Y.C. Chung, H.C. Ting, and C.Y. Chen, "Origin of self-heating effect induced asymmetrical degradation behavior in InGaZnO thin-film transistors", *Appl. Phys. Lett.,* vol. 100, no. 23, p. 232101, 2012.

http://dx.doi.org/10.1063/1.4723573

[109] M. Zhang, W. Zhou, R. Chen, S. Zhao, M. Wong, and H.S. Kwok, "P-6: Static Reliability of Bridged-Grain Poly-Si TFTs", *Dig. Tech. Pap.,* vol. 45, no. 1, pp. 964-967, 2014.

http://dx.doi.org/10.1002/j.2168-0159.2014.tb00250.x

[110]　M. Zhang, W. Zhou, R. Chen, M. Wong, and H.S. Kwok, "Water-enhanced negative bias temperature instability in p-type low temperature polycrystalline silicon thin film transistors", *Microelectron. Reliab.,* vol. 54, no. 1, pp. 30-32, 2014.

http://dx.doi.org/10.1016/j.microrel.2013.07.082

[111]　F. Liu, Y. Zhou, H. Yang, X. Zhou, X. Zhang, G. Li, M. Zhang, S. Zhang, and L. Lu, "Roles of Hot Carriers in Dynamic Self-Heating Degradation of a-InGaZnO Thin-Film Transistors", *IEEE Electron Device Lett.,* vol. 43, no. 1, pp. 40-43, 2022.

http://dx.doi.org/10.1109/LED.2021.3133011

[112]　J.H. Stathis, and S. Zafar, "The negative bias temperature instability in MOS devices: A review", *Microelectron. Reliab.,* vol. 46, no. 2-4, pp. 270-286, 2006.

http://dx.doi.org/10.1016/j.microrel.2005.08.001

CHAPTER 3

Reliability Analysis Methods for Thin-Film Transistors

Abstract. This chapter explores the various methods used for reliability analysis of Thin-Film Transistors (TFTs). It covers transfer curve degradation analysis techniques, capacitance-voltage curve analysis, low-frequency noise analysis, and thin-film quality assessment. Additionally, the chapter discusses simulation analysis methods, including TCAD simulation and thermal simulation, which are crucial for evaluating and enhancing the reliability of TFTs in electronic applications.

Keywords: CV curve analysis, Low-frequency noise analysis, Reliability analysis methods, Simulation analysis.

3.1. INTRODUCTION

Thin-film transistors (TFTs) can be characterized by a variety of performance measurements, each with its strengths and weaknesses due to different emphases. For reliability analysis of TFTs, different characterizations must be employed based on the specific degradation phenomena and mechanisms.

This chapter presents the measurement and analysis methods for basic transfer characteristic curves, output characteristic curves, and capacitance-voltage characteristic curves. It then delves into the analysis methods for low-frequency noise and several quality assessment techniques for devices. Finally, it introduces two commonly used simulation approaches for semiconductor devices: device simulation and thermal simulation.

3.2. BASIC TRANSFER CURVE DEGRADATION ANALYSIS

3.2.1. Degradation in On-State Region

The on-state region of a thin-film transistor (TFT) device typically begins after the device's threshold voltage (V_{th}) [1]. In logarithmic coordinates, it is usually located in the region where the current growth flattens for large gate voltages, as shown in Fig. (**3.1**). As stated in section 2.1, the current at a fixed gate voltage can generally be considered the on-state current (I_{on}).

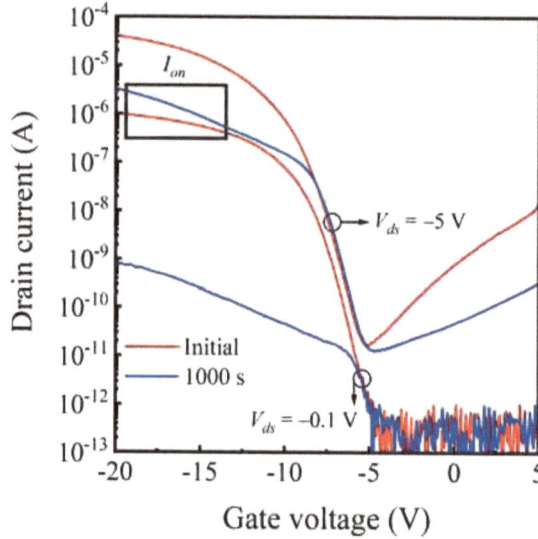

Fig. (3.1). The on-state region of a TFT device and its degradation.

Various stresses typically affect the device's I_{on}, which often experiences significant degradation, such as that caused by hot carrier (HC) stress. Defects that cause I_{on} degradation are usually tail states, which is due to their higher density and closer proximity to the conduction or valence bands, making them more likely to capture free charge carriers [2]. In reliability analysis, the change rate of I_{on} (ΔI_{on}) is commonly used to assess the reliability of the device in the on-state, expressed by the following equation [3]:

$$\Delta I_{on} = \frac{I_{on}\left(stress\right) - I_{on}\left(initial\right)}{I_{on}\left(initial\right)} \tag{3.1}$$

However, there can be some issues in determining the I_{on} using the fixed voltage method. For instance, when threshold voltage (V_{th}) shifts significantly, the I_{on} will correspondingly increase or decrease. However, this type of I_{on} degradation is usually not caused by tail-state defects. To express the I_{on} more accurately, the following formula can be used [4]:

$$I_{on} = I_d\left(V_{th} + 6\ V\right) \tag{3.2}$$

This method can effectively eliminate the impact of V_{th} shift on I_{on}. However, the I_{on} only provides a rough indication of the overall device performance. The true cause of device degradation requires further analysis using other methods.

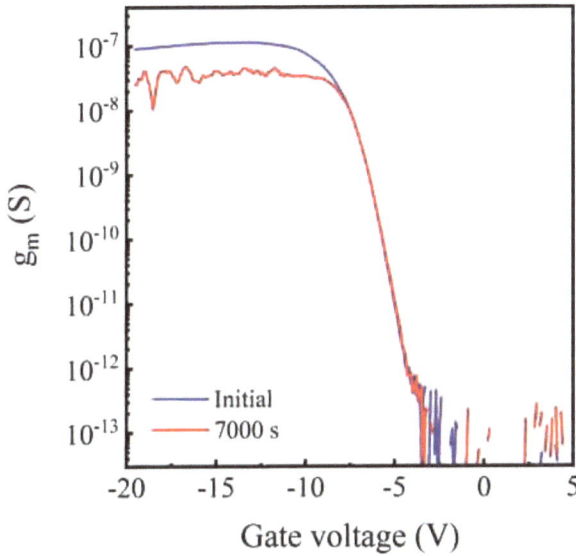

Fig. (3.2). Variation of g_m with V_g voltage for TFT device.

In addition, the mobility of a device can be extracted in the on-state region, as shown in Fig. (**3.2**). The magnitude of the device's mobility is directly related to the scattering within the active layer of the device. It can be represented by the following formula [5]:

$$\frac{1}{\mu} = \frac{1}{\mu_i} + \frac{1}{\mu_s} + \frac{1}{\mu_0}$$

(**3.3**)

Here, μ_i represents acoustic phonon scattering, μ_s represents ionized impurity scattering, and μ_0 represents optical phonon scattering [6]. Typically, during various types of degradation, the increase in various types of defect states can significantly enhance ionized impurity scattering or other types of scattering. This leads to a noticeable decrease in the device's mobility, which in turn contributes to further degradation in the on-state region of the device.

3.2.2. Degradation in Subthreshold Region

The subthreshold region of a device is a critical area affecting its operational performance, typically representing when the device turns on and how quickly it does so. Two main factors are generally studied in this region: the V_{th} and the subthreshold swing (*SS*), both of which degrade differently from phenomena observed in the on-state region [7, 8].

V_{th} is one of the important electrical characteristics of a TFT. For metal-oxide-semiconductor field-effect transistors (MOSFETs), V_{th} corresponds to the gate voltage at which the device operates in strong inversion. TFTs, on the other hand, often operate in the accumulation region, and V_{th} is commonly defined as the gate voltage at which carriers begin to accumulate in the channel [9-11]. V_{th} is sensitive to fixed charges within the device, such as trapped charges in the gate insulator (GI) or positive charges like H^+ in the active layer. These charged particles can enhance or shield the gate electric field, leading to an increase or decrease in V_{th}.

Similarly, *SS* reflects the speed of the device's switching and is also determined in the subthreshold region. The degradation of *SS* is typically due to an increase in interface state defects. Changes in *SS* can be used to extract the density of interface state defects in the device, as shown in the following formula [8]:

$$N_{it} = \left[\frac{SS \log(e)}{kT/q} - 1 \right] \frac{C_{ox}}{q} \tag{3.4}$$

In the equation provided, N_{it} represents the interface state defect density, k is the Boltzmann constant, q is the elementary charge, e is the base of the natural logarithm, C_{ox} is the charge per unit area in the GI layer, T is the temperature, and *SS* is the subthreshold slope.

In terms of energy bands, the degradation of V_{th} and *SS* can generally be explained by the increase of deep-level defects. For example, in p-type TFT devices, the Fermi level is positioned in the middle of the bandgap during thermal equilibrium. As the device gradually transitions to the on-state, the Fermi level moves closer to the valence band. During this process, deep levels positioned in the center of the bandgap can have a significant impact on the subthreshold region. Fig. (**3.3**) illustrates the typical range of V_{th} for polycrystalline silicon (poly-Si) TFT devices. In terms of V_{th} reliability, it can shift with stress time as stress is applied. V_{th} can change under stress conditions such as negative bias temperature instability (NBTI) or negative gate bias (NBS) [7-9, 12], which may result in either a negative or positive shift, depending on the type of stress applied. The change in V_{th} is typically measured by the amount of V_{th} shift, and generally, a positive gate voltage stress will increase the device's V_{th}, while a negative gate voltage stress will decrease it.

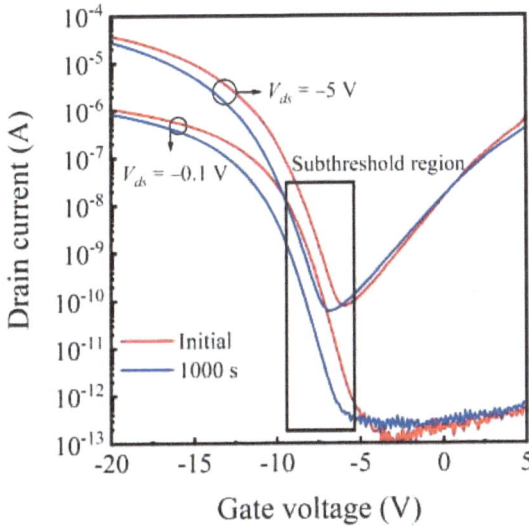

Fig. (3.3). The subthreshold region of a TFT device and its degradation.

3.2.3. Degradation in Off-State Region

As shown in Fig. (**3.4**), the black region represents the off-state region. The mechanisms for off-state current (I_{off}) generation primarily include thermionic emission, tunneling, and thermionic field effect emission [13, 14]. Thermionic emission mainly occurs as carriers are excited thermally to become free carriers, thereby generating the I_{off}. The magnitude of the current from thermionic emission is mainly related to the temperature of the device, and typically, higher temperatures result in larger I_{off}.

Tunneling is mainly due to the significant narrowing of the potential barrier between the electrode and the active layer under a strong electric field, which greatly increases the probability of carriers tunneling through the bandgap, thus generating I_{off} [13]. Consequently, the larger the drain voltage, the larger the I_{off} in a certain range. Thermionic field effect emission is related to defect states within the device and, under the influence of an electric field, these defect states can assist carrier tunneling, further shortening the tunneling barrier, and carriers tunnel through the bandgap in two or more steps, generating the I_{off}. The I_{off} generated by thermionic field effect emission is not only related to the electric field but also to the density of defect states within the device.

It can be seen that when the drain voltage is small (*e.g.*, 0.1 V), the electric field is small, and the potential barrier width between the electrode and the active layer is wide, so the I_{off} is dominated by thermionic emission. However, when the drain voltage is large (*e.g.*, 5 V), the electric field is significant, and the I_{off} will not only come from thermionic emission but also from tunneling and thermionic field effect emission [15]. Typically, the I_{off} generated by thermionic emission is smaller, while that from tunneling and thermionic field effect emission is larger. Therefore, the variation of I_{off} at higher drain voltages is usually the main focus of the study.

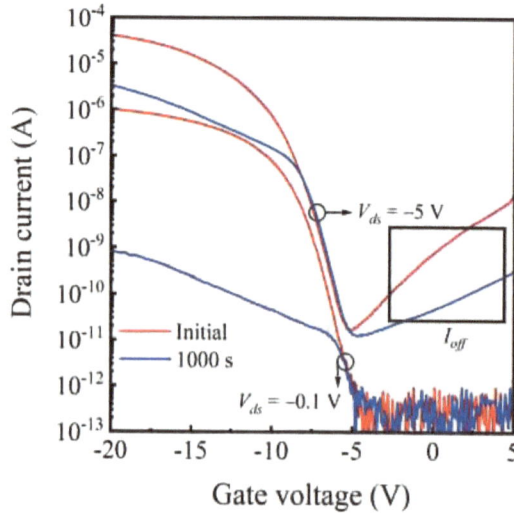

Fig. (3.4). Off-state region of TFT device and its degradation phenomenon.

The size of the I_{off} generated by thermionic emission is mainly related to the bandgap width of the active layer and temperature and thus does not usually exhibit significant changes [16]. The I_{off} generated by the tunneling and thermionic field effect is highly sensitive to the magnitude of the electric field, and a partial reduction in the electric field can significantly decrease the device's I_{off}. As shown in Fig. (**3.4**), a significant degradation of I_{off} can be observed, which may be due to the HC effect, positive bias stress (PBS) or NBS effects, or simply a large voltage applied to the drain, which can significantly reduce the device's I_{off}. This is because the aforementioned effects can generate fixed charges near the drain, which, if present, will shield part of the drain field, and the device's I_{off} will be significantly reduced. However, I_{off} will not degrade indefinitely; when the amount of fixed charges reaches a certain level, the I_{off} generated by tunneling and thermionic field effect will almost completely disappear, and at this point, the device's I_{off} will be dominated by thermionic emission [13].

In summary, when analyzing TFT degradation, observing how the transfer curve changes with stress time provides valuable insights into the device's reliability.

(1) Changes in the transfer curve reveal shifts in the threshold voltage, which is a critical indicator of TFT degradation. Over time, V_{th} may shift either positively or negatively, depending on the stress conditions and material properties. This shift is typically caused by charge trapping or defect generation in the dielectric or interface, providing insight into issues like charge injection and stability under prolonged operation. (2) Changes in the slope of the transfer curve reflect variations in the subthreshold swing. An increase in SS indicates a higher density of interface traps or defects, suggesting degradation in the gate insulator or channel interface. Poor SS leads to degraded switching performance and higher power consumption, which directly impacts the TFT's efficiency and reliability over time. (3) The shape of the transfer curve can also reveal degradation in carrier mobility. Stress-induced mobility degradation occurs due to increased scattering at the channel, often as a result of defect generation or charge trapping. Lower mobility reduces the device's current drive capability, affecting performance, especially in high-frequency or high-current applications. (4) A reduction in I_{on} observed from the transfer curve suggests degradation in the TFT's ability to conduct under bias. This can result from mobility degradation, charge trapping, or increased contact resistance. In display applications, reduced I_{on} affects the ability to_drive pixels effectively. (5) An increase in I_{off} indicates worsened leakage characteristics, often due to trap generation, defect accumulation, or stress-induced breakdown in the gate dielectric. Higher I_{off} negatively impacts the power consumption and signal integrity of the device, reducing overall reliability. This information is crucial for improving device reliability, optimizing materials, and refining manufacturing processes.

3.3. CAPACITANCE-VOLTAGE CURVE ANALYSIS

3.3.1. Capacitance-Voltage Measurements in Thin-Film Transistors

The CV characteristic curve measurement method for TFTs is similar to that for MOS transistors. This technique uses a small signal superimposed on the voltage to measure capacitance. Typically, capacitance changes are observed at different frequencies and gate voltages. In CV measurements, the alternating current (AC) small signal is transmitted through a distributed network composed of internal capacitances and resistances [17]. For such a circuit, the common measurement method is to measure the impedance Z of the total circuit from outside the equivalent circuit. Semiconductor impedance measurement is a common method for extracting capacitance values. For an RC network, the impedance Z can be expressed as [18]:

$$Z = R - jX_c$$

(3.5)

Here, the real part R of the impedance Z represents the total resistance in the combined circuit, and the imaginary part X_c represents the capacitive reactance in the combined circuit. The capacitance value of the Device Under Test (DUT) can be obtained from the capacitive reactance. The most common method for measuring device capacitance values in the 10 MHz frequency range is the automatic bridge measurement principle. The device's capacitance is determined by applying an AC voltage, measuring the AC and phase, and simultaneously applying or scanning a direct current (DC) voltage in the device. Fig. (**3.5**) shows the CV curve obtained by measuring poly-Si TFT at different frequencies.

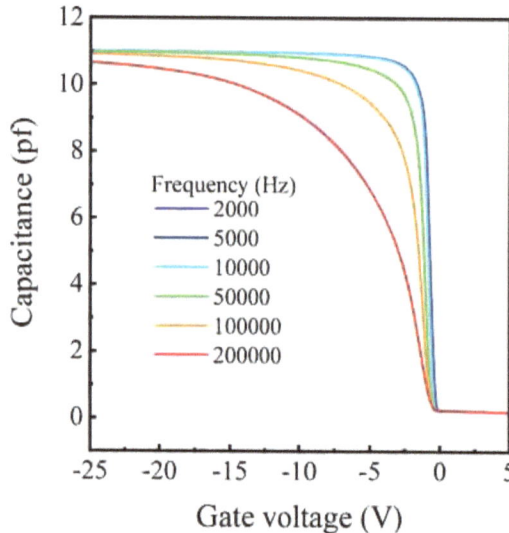

Fig. (3.5). CV characteristic curve of a typical TFT.

In addition to the typical method of measuring C_{gsd} capacitance with the gate connected to the anode and the source and drain shorted to the cathode [19], one can also observe the trend of capacitance changes at different biasing positions by measuring the gate-drain capacitance C_{gd} or the gate-source capacitance C_{gs} [20]. This approach can effectively analyze the distribution of defect states within the active layer, as well as the distribution of charge, as shown in Fig. (**3.6**). For some asymmetric TFT degradation behaviors, such as the HC effect, the internal defect state distribution can be analyzed by comparing the measurement results of C_{gd} and C_{gs}.

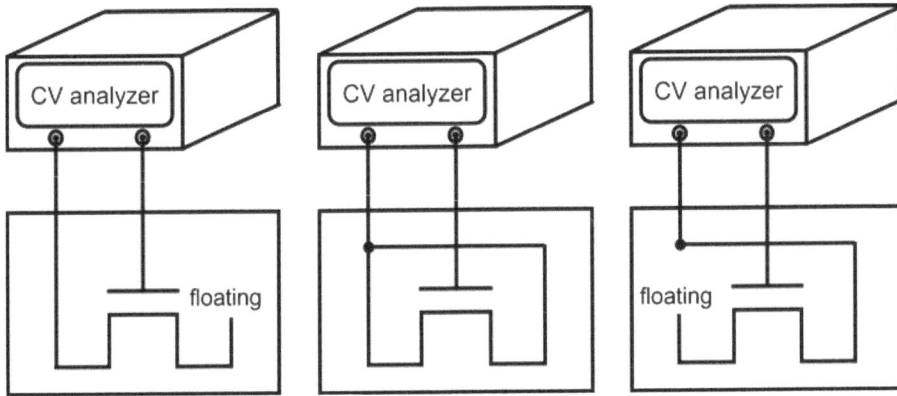

Fig. (3.6). Different connections model of TFTs for CV testing.

3.3.2. Density of States Calculation

The density of defect states is a critical factor that directly affects various aspects of device performance, and the extraction of these defect states is essential for characterizing device degradation. In the analysis of AC reliability for poly-Si, the AC HC effect is one of the most significant degradation mechanisms. It not only leads to an increase in bulk defects in poly-Si TFT devices but also results in an increase in interface state defects. Monitoring the changes in defect state density is advantageous for further analysis of AC reliability.

3.3.2.1. Extraction Methods for Bulk Defects

The density of defect states is a critical factor directly affecting various aspects of device performance. The extraction of defect states plays a vital role in characterizing device degradation. There are currently two main methods for extracting bulk defect states: Multi-Frequency Capacitance–Voltage (MFM) [19] and Differential Ideality Factor Technique (DIFT) [21]. The MFM method involves converting frequency-dependent CV characteristics into frequency-independent gate capacitance (C_G) and extracting the defect distribution within the active layer using parameters such as C_G. On the other hand, the DIFT method is based on the transfer characteristic curve of the TFT and extracts the variation relationship between the Gate-Voltage-Dependent Ideality Factor ($\eta(V_{gs})$) and V_g to determine the defect distribution within the active layer.

(1) Multi-Frequency Capacitance Voltage Method [19]

Sangwon Lee from Korea National University proposed a method for extracting the density of defect states within TFTs in 2010. By measuring the CV (Capacitance-Voltage) and RV (Resistance-Voltage) curves and using an equivalent circuit model, the dual RC equivalent model of the device is separated into the C_{ox}, contact resistance (R_s), and the channel's equivalent capacitance (C_{CH}) and equivalent resistance (R_{CH}). These are then transformed into capacitances related to localized charge (C_{LOC}) and free charge (C_{free}), as well as an equivalent resistance (R_L) that are independent of frequency and related to defect states, as shown in Fig. (**3.7**).

Based on the above mechanism and the following formulae, the defect state density of the device can be extracted:

$$R_{CH} = \sqrt{\frac{C_M(1 + D_M^2) - C_{OX}}{\omega^2 C_{CH}^2 C_{OX} - \omega^2 C_{CH} C_M (1 + D_M^2)(C_{CH} + C_{OX})}} \tag{3.6}$$

$$C_{CH} = \frac{bC_{OX}^2 - b^2 C_{OX}}{[(abw)^2 + 1]C_{OX} - 2bC_{OX} + b^2} \tag{3.7}$$

$$a = \frac{R_M}{wC_M(1 + D_M^2)} - R_S \tag{3.8}$$

$$b = C_M(1 + D_M^2) \tag{3.9}$$

$$DM = \frac{1}{wC_M R_M} \tag{3.10}$$

$$R_L = \sqrt{\frac{w^2 C_{CH} R_{CH}^2 (C_{LOC} + C_{free})(C_{LOC} + C_{free} - C_{CH}) - (C_{LOC} + C_{free})}{w^2 C_{LOC}^2 C_{free}[1 + w^2 C_{CH} R_{CH}^2 (C_{CH} - C_{free})]}} \tag{3.11}$$

$$C'_{LOC} = \frac{[C_{LOC}(V_{GS1}) - C_{LOC}(V_{GS2})]}{W \cdot L \cdot T} \tag{3.12}$$

$$g(E) = \frac{C'_{LOC}}{q^2} \tag{3.13}$$

$$\phi_s = \int_{V_{FB}}^{V_{GS}} \left(1 - \frac{C_G}{C_{OX}}\right) dV_{GS} \tag{3.14}$$

In the process, CM and RM represent the initial capacitance and resistance measured from the CV and RV curves, respectively. Based on Fig. (**3.7**), the equivalent capacitance (C_{CH}) and equivalent resistance (R_{CH}) of the device's active layer can be extracted. Further extraction can determine the capacitance (C_{LOC}) representing fixed defect states within the active layer and the capacitance (C_{free}) representing free charge, along with the corresponding equivalent resistance (R_L). Since R_L is independent of frequency, C_{LOC} and C_{free} can be calculated separately based on $R_L(w_1) = R_L(w_2) = R_L(w_3)$. Subsequently, the distribution of defect states as a function of gate voltage (V_{gs}) can be calculated using equations (3.12) and (3.13). Finally, by converting V_{gs} to energy using equation (3.14), the distribution of defect states within the bandgap of the device can be obtained.

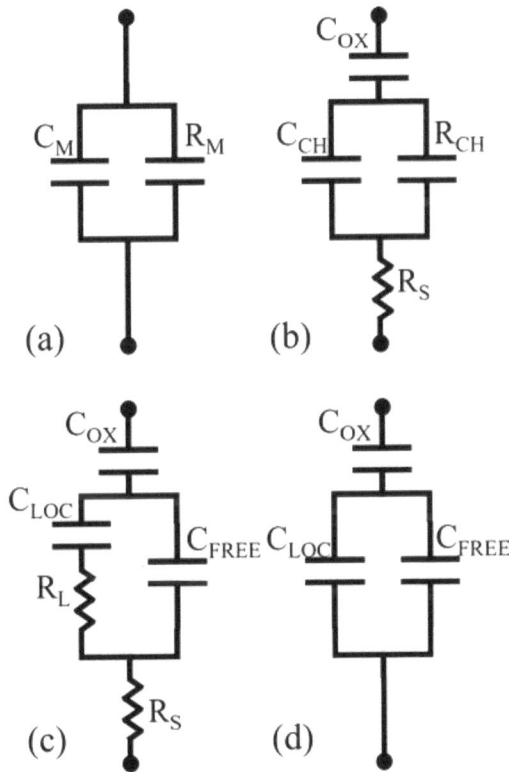

Fig. (3.7). Equivalent circuit of the MFM defect extraction method.

(2) Differential Ideality Factor Technique Method

The Differential Ideality Factor Technique (DIFT) [21] is a simpler method for extracting defect state densities compared to the MFM method. It primarily relies on the device's transfer characteristic curve for extraction. By using the formula:

$$\eta(V_{GS}) = \left(\frac{V_{GS2} - V_{GS1}}{V_{th}}\right) \Big/ \ln\left(\frac{I_{D,sub}(V_{GS2})}{I_{D,sub}(V_{GS1})}\right) \tag{3.15}$$

$$\Delta C_{LOC}(V_{GS}) = \int_{V_{GS}}^{V_{GS} + \Delta V_{GS}} C_{OX} \cdot \frac{d\eta(V_{GS})}{d(V_{GS})} dV_{GS} \tag{3.16}$$

$$g(V_{GS}) = \frac{\Delta C_{LOC}(V_{GS})}{q^2 \cdot T} \tag{3.17}$$

can determine the defect state density, where η is the ideality factor and T is the thickness of the device's active layer. Similarly, the defect state density's variation with energy can be extracted using equation 3.14.

It is important to note that the aforementioned extraction methods were mainly developed for IGZO TFTs but are also applicable to poly-Si TFTs. This reference also suggests that the tail states extracted by MFM are more accurate, while the deep levels extracted by DIFT are more consistent. Furthermore, these methods can only extract the density of defect states corresponding to their type, meaning that n-type devices can only extract acceptor-like defects, and p-type devices can only extract donor-like defects. To obtain a complete defect state distribution, it is necessary to measure both n-type and p-type devices' CV and RV curves simultaneously.

3.3.2.2. Extraction Methods for Interface Defects

Based on the extraction of bulk defects, a rough fit of the device's transfer characteristic curve can be achieved. However, as previously mentioned in Chapter 2, not only the degradation of bulk defects but also, to a certain degree, interface state defect degradation can lead to the degradation of TFTs. Therefore, even though bulk defects can be used to fit the initial performance of poly-Si TFTs fairly well, after the device has degraded, especially after a significant amount of time has passed, fitting the entire device performance through bulk defects alone will be challenging. Thus, the extraction of interface state defects is also crucial for the reliability research of poly-Si TFTs. Interface state defects can be extracted using high-low frequency capacitance methods based on CV test results.

As early as 1992, Steven C. from the National Institute of Standards and Technology in the United States proposed this method. He discovered that interface state traps could not respond to pulse signals at high frequencies, allowing for the separation of the device's bulk defects and interface state defects based on this difference [22]. Using the formula:

$$D_{it} = \frac{1}{q}\left[\left(\frac{1}{C_{LF}} - \frac{1}{C_{OX}}\right)^{-1} - \left(\frac{1}{C_{HF}} - \frac{1}{C_{OX}}\right)^{-1}\right] \qquad (3.18)$$

$$\phi_s = \int_{V_{FB}}^{V_{GS}}\left(1 - \frac{C_{LF}}{C_{OX}}\right)dV_{GS} \qquad (3.19)$$

one can extract the interface state defects of the TFT, where C_{LF} is the capacitance distribution of the device at a lower frequency, and C_{HF} is the capacitance distribution at a higher frequency. Furthermore, by using equation (3.19), the gate voltage can be converted to energy, thus obtaining the distribution of interface state defects within the bandgap.

In summary, by analyzing the CV curve's evolution of TFTs under stress, we can assess degradation mechanisms such as charge trapping, interface state generation, and gate dielectric deterioration. These insights help in understanding the TFT's long-term stability, operational reliability, and areas where material or design improvements are needed to enhance durability.

3.4. LOW-FREQUENCY NOISE ANALYSIS

Internal noise is a critical factor that significantly impacts the performance, quality, and reliability of electronic systems. In various testing systems, the level of noise directly determines the resolution and the ability to detect the smallest signal amplitudes. In communication systems, the presence of noise directly affects the fidelity of signal transmission and the sensitivity of receivers. In digital and computer systems, randomly varying noise levels can lead to false triggering and the generation of spurious signals. The noise in a system primarily originates from its constituent electronic components, with semiconductor active devices playing a crucial role [23, 24]. The application of low-noise semiconductor devices is indispensable in new technology fields such as weak signal detection, precision metrology, infrared detection, underwater acoustic communication, and high-fidelity audio.

Noise can be categorized into two types: one is external noise, such as that generated by electromagnetic interference from adjacent circuits, AC power supply circuits, and radio transmitters; the other is internal noise arising from the physical processes of electron transport within the medium. The noise discussed here is of the latter type. Due to its random nature, noise cannot be fundamentally eliminated. However, the magnitude of semiconductor device noise, particularly low-frequency noise, reflects the intrinsic quality and reliability of the device. Low-frequency noise (LFN) measurement can serve as an effective tool for assessing the quality of semiconductor thin films, enabling a deeper understanding of the physical mechanisms of semiconductor devices [25, 26].

3.4.1. Low-Frequency Noise Measurements in Thin-Film Transistors

Low-frequency noise testing typically covers a frequency range of 1 Hz to 1 MHz. Since the spectral shape provides crucial information, the power spectral density (PSD) is the preferred measure for such tests. These measurements require specialized equipment to detect minute signals and minimize internal noise and external interference that could affect the results.

Main semiconductor parameter testing systems usually connect the equipment to a probe station to measure the LFN and I-V characteristics of devices [27, 28]. The LFN analysis system directly captures time-domain signals and analyzes the spectrum using the Fast Fourier Transform (FFT) method, ensuring accurate DC testing and minimal AC interference, which guarantees the quality of noise testing.

A schematic of a noise power spectral density measurement setup is shown in Fig. (**3.8**). When the DUT is subjected to a voltage, the current is amplified by a low-noise amplifier (LNA) and then sent to a high-precision analog-to-digital converter (ADC) for data acquisition. The acquired data is transmitted to a controller for Fast Fourier Transform (FFT) analysis, which yields the PSD of the noise current [29].

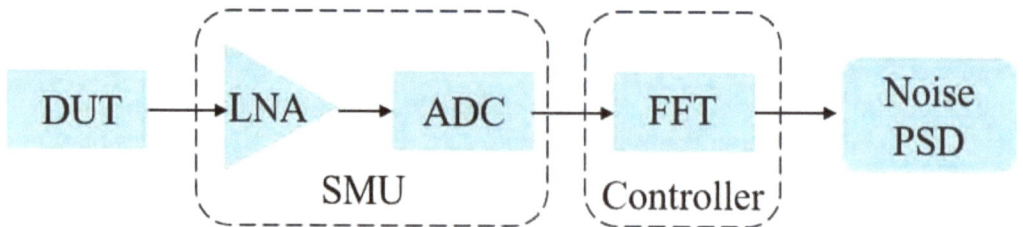

Fig. (3.8). Schematic diagram of the noise PSD measurement device.

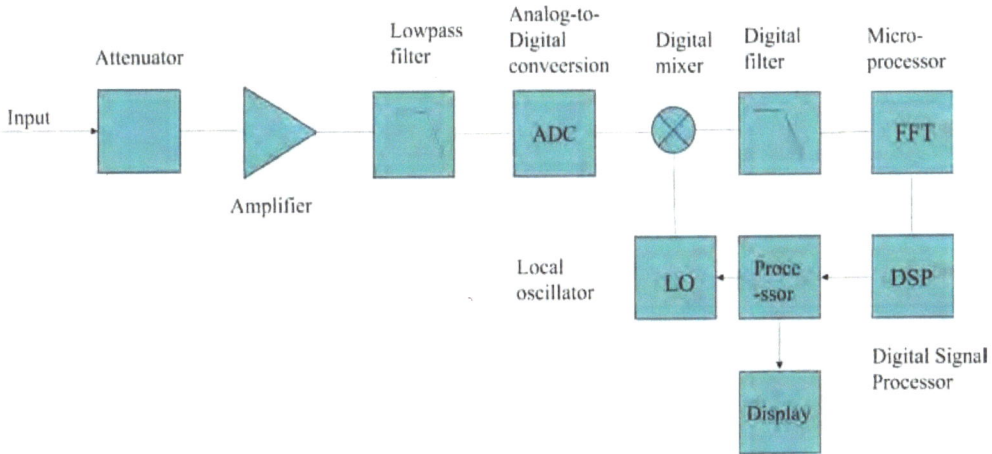

Fig. (3.9). Simplified block diagram of the FFT spectrum analysis function.

The controller module is used for measuring and analyzing frequency-domain signals. Using the FFT algorithm [30, 31], it converts the measured signal from the time domain to the frequency domain. A simplified block diagram describing the FFT spectral analysis function is shown in Fig. (**3.9**). After initial attenuation, amplification, and low-pass filtering, the signal is sampled and digitized, followed by digital signal processing.

LFN testing systems typically employ DC coupling techniques to measure the time-domain current characteristics of devices using source-measure units (SMU) [32]. The device noise is quantified directly through FFT analysis based on the Welch spectral estimation method. During FFT analysis, the density of the frequency grid is influenced by the base frequency resolution; the smaller the base frequency resolution selected, the denser the frequency grid, and the longer the data acquisition time. Additionally, the Welch method results in less fluctuation in the curve. The noise test bandwidth can achieve high precision up to 100 kHz for high frequencies and up to 40 Hz for ultra-low frequencies. The time-domain signal acquisition time is less than 1 μs, and the minimum resistance for noise testing is 500 Ω. The frequency resolution for noise testing is 0.1 Hz for high precision and 0.001 Hz for ultra-low frequencies. The ADC has a quantization precision of 24 bits and a sampling rate of 1.8 MHz. The background noise during measurement is as low as 10^{-28} A^2/Hz, meeting the experimental requirements for device noise precision. Different cable types are used for different signal types corresponding to different device pins. As shown in Fig. (**3.10**), the gate control signal requires a TXN noise-specific cable and a low-noise test module for connection to the SMU.

Test and logic signals are connected to the SMU using universal cables, and ground signals are connected to the common ground terminal with universal cables.

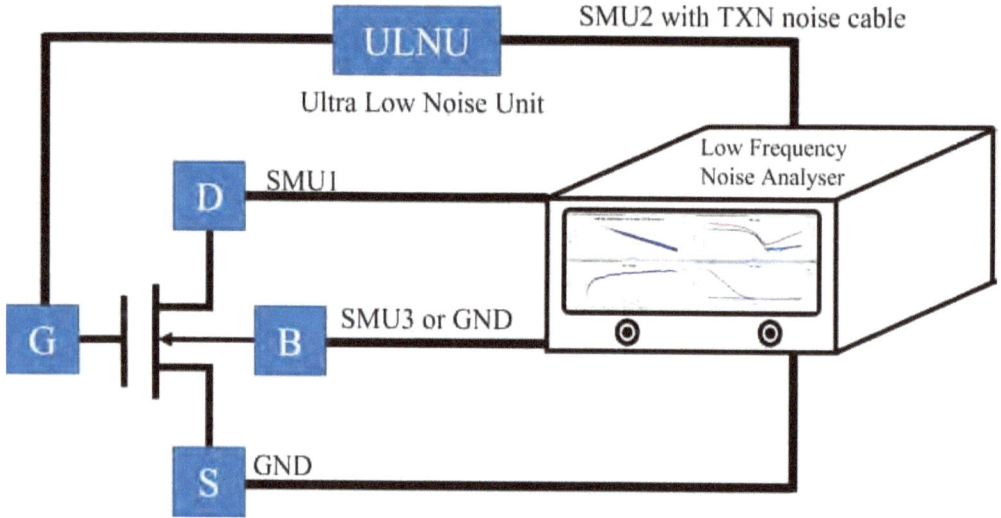

Fig. (3.10). Test instrument to SMU connection diagram.

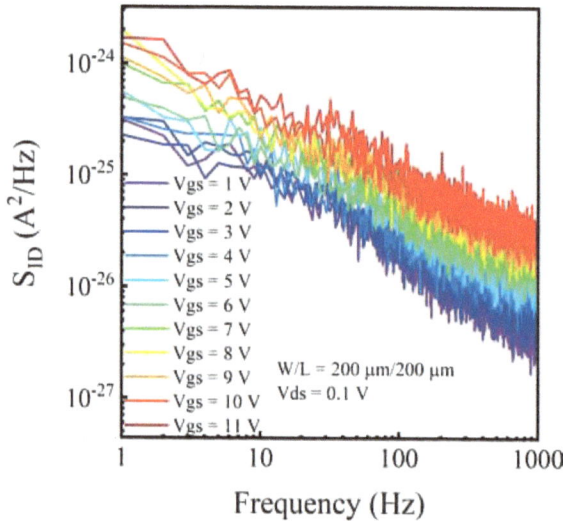

Fig. (3.11). The relationship between S_{ID} and f for a tested TFT.

Fig. (**3.11**) illustrates the relationship between the drain current noise power spectral density (S_{ID}) and frequency (f) for a tested TFT. The test conditions were at room temperature, with the device having a width-to-length ratio of 200 μm / 200 μm and the device operating in the linear region with a drain-source voltage (V_{ds}) of 0.1 V. The frequency range measured was from 1 to 100 kHz, with observations focused on the range from 1 to 100 Hz.

In low-f tests, different gate-source voltages V_{gs} ranging from 1 to 12 V were applied, each corresponding to a different drain current and, as depicted in the figure, to twelve distinct curves representing different S_{ID}. The noise levels were observed to be between 10^{-23} and 10^{-27} A^2/Hz, with S_{ID} being inversely proportional to f, indicating that the low-frequency noise of the TFT follows a $1/f$ noise law [33, 34]. It can also be observed from the figure that under different gate voltage conditions, the S_{ID} increases with the increase of V_{gs}. This is because as the gate voltage increases, the Fermi level gradually approaches the conduction band, allowing trap states near the Fermi level to more easily capture and release charge carriers, thereby increasing the noise.

3.4.2. Carrier Number Fluctuation Theory in Thin-Film Transistors

In 1957, McWorther proposed the carrier number fluctuation model (ΔN model) based on the quantum mechanical tunneling of carriers between the channel and gate oxide traps in semiconductors, which leads to random exchange of charge [35]. The free carriers in the channel are captured and released by the boundary traps of the gate oxide, causing fluctuations in the surface potential, which in turn leads to fluctuations in the inversion charge density and changes in the drain current. The power spectral density of carrier number fluctuations caused by the random capture and release of carriers from a single impurity center follows a Lorentzian spectrum. By considering the capture and release of carriers by traps at different locations and superimposing the Lorentzian spectrum formulas for different time constants, the noise power spectral density can be obtained as [35]:

$$S_N(f) = 4\overline{\Delta N^2} \int_1^\infty \frac{\tau g(\tau)\, d\tau}{1 + (2\pi f)^2 \tau^2} = 4\overline{\Delta N^2} \int_1^\infty \frac{\tau g(\tau)\, d\tau}{1 + \omega^2 \tau^2} \tag{3.20}$$

where $g(\tau)$ is the probability function of the time constant τ. $g(\tau)d\tau$ represents the probability of the time constant being in the interval $\tau - \tau + d\tau$. This function also satisfies the normalization condition. The distribution can be assumed to be [35]:

$$g(\tau)d\tau = \begin{cases} \dfrac{\dfrac{d\tau}{\tau}}{ln\left(\dfrac{\tau_2}{\tau_1}\right)}, & \tau_1 \le \tau \le \tau_2 \\ 0, & others \end{cases} \tag{3.21}$$

Substitute into equation (3.20) to obtain [35]:

$$S_N(f) = \overline{2\Delta N^2} \Big/ \pi f ln\left(\tfrac{\tau_2}{\tau_1}\right) \left[arctg\left(\omega\tau_2\right) - arctg\left(\omega\tau_1\right)\right] \tag{3.22}$$

When $1/\tau_2 \le \omega \ 1/\tau_1$, the function can be simplified as [35]:

$$S_N(f) = \frac{\overline{\Delta N^2}}{ln\left(\tfrac{\tau_2}{\tau_1}\right)}\frac{1}{f} \tag{3.23}$$

It can be observed that within this frequency range, the noise power spectral density $S_N(f)$ varies with $1/f$, being inversely proportional to the frequency f, indicating that the noise in this frequency range is $1/f$ noise. From the aforementioned derivation, it can be concluded that for traps to be the source of noise, the distribution of time constants must have the form of equation (3.21), which implies that τ is continuously distributed over a relatively large range. Currently, there are two primary physical models that can satisfy the required distribution of time constants: the tunneling model and the thermal activation model.

(1) Tunneling Model

It is assumed that capture and release occur *via* tunneling, and the time constant for carrier exchange is given by [36]:

$$\tau = \tau_0(E)e^{\frac{z}{\lambda}} \tag{3.24}$$

where z is the distance from the oxide trap to the interface, λ is the tunneling decay distance, approximated by the Wentzel-Kramers-Brillouin (WKB) theory [36].

$$\lambda = \left[\frac{4\pi}{h} \sqrt{2m^* \phi_B} \right]^{-1} \qquad (3.25)$$

where m^* is the effective mass of the carrier, ϕ_B is the interface barrier height, and h is the Planck constant. Calculations show that for the Si/SiO$_2$ system, $\lambda \approx 1$ Å and with a time constant τ_0 of 10^{-10} seconds, frequencies of 0.01 Hz and 1 MHz yield z values of 2.6 nm and 0.7 nm, respectively. Therefore, oxide traps near the channel interface that are too fast do not generate $1/f$ noise, and those more than 3 nm from the channel interface that are too slow to produce $1/f$ noise.

(2) Thermal Activation Model

Observations have revealed that the noise waveform in small-sized devices exhibits random telegraph noise (RTS), with the time constant for trap capture of carriers being exponentially related to energy [37]:

$$\tau_{th} = \tau_{0,th} e^{\frac{E}{kT}} \qquad (3.26)$$

where E is the activation energy. Test results indicate that the distribution range of time constants is essentially consistent with the range for $1/f$ noise. Studies on RTS noise have shown that carrier capture and release assisted by thermally activated phonons play a significant role. Additionally, by analyzing the relationship between the average capture and release time constants of RTS and the gate voltage, the characteristics of the defects causing RTS can be determined.

By inserting Equation 3.26, the derivation yields [37]:

$$S_{V_{fb}} = \frac{q^2 kT \lambda N_t}{f^\gamma WLC_{ox}^2} \qquad (3.27)$$

where S_{vfb} is the flatband noise power spectral density, N_t is the density of defect states in the gate dielectric near the Fermi level (in units of cm^{-3}eV^{-1}), W is the device channel width, L is the device channel length, C_{ox} is the gate oxide capacitance per unit area, and T is the device operating temperature. If the defect state density is non-uniform in-depth, the frequency exponent γ deviates from 1; γ is less than 1 when the defect state density near the gate oxide/channel interface is higher than within the gate oxide, and γ is larger than 1 when it is lower [25, 38, 39]. Since S_I/I^2 is inversely proportional to the signal-to-noise ratio, a normalization method that divides the current noise PSD by the square of the current is commonly used in noise analysis. In the linear region, where $g_m = I_D/(V_{gs} - V_{th})$, the normalized drain current noise power spectral density is approximately [40-42]:

$$\frac{S_{ID}}{I_D^2} = \frac{q^2 kT\lambda N_t}{f^\gamma C_{ox}^2 WL (V_{gs} - V_{th})^2} \tag{3.28}$$

where N_t is the density of defect states in quasi-Fermi level in gate dielectrics.

3.4.3. Mobility Fluctuation Theory in Thin-Film Transistors

Hooge proposed that $1/f$ noise originates from lattice scattering that causes fluctuations in the mobility of carriers in the channel, a concept known as the mobility fluctuation model ($\Delta\mu$ model). According to Hooge's empirical formula, the noise power spectral density of the drain current is given by [43]:

$$\frac{S_I(f)}{I^2} = \frac{S_R(f)}{R^2} = \frac{S_V(f)}{V^2} = \frac{\alpha_H}{fN_{tot}} \tag{3.29}$$

where N_{tot} represents the total number of free carriers in the channel, and α_H is an empirical parameter highly related to the technology and material used. Variations in α_H, stemming from phonon scattering noise or surface roughness scattering noise, are the primary causes of mobility fluctuation noise mechanisms. Typical values for α_H range between 10^{-3} and 10^{-6}. This model has been used to characterize devices made from many uniform metals and semiconductor materials, all of which conform to Hooge's formula, exhibiting bulk effects. Hooge's formula can have different expressions [27, 44, 45]; for junction devices, if the condition qD = kTμ is met, fluctuations in the diffusion coefficient are the cause of mobility fluctuations, and the empirical formula can be expressed as [43]:

$$\frac{S_\mu(f)}{\mu^2} = \frac{S_D(f)}{D^2} = \frac{\alpha_H}{fN_{tot}} \tag{3.30}$$

For the FET devices, if $Q_i = C_{ox}(V_{gs} - V_{th})$ is met, the empirical formula can be expressed as [46, 47]:

$$\frac{S_{ID}}{I_D^2} = \frac{\alpha_H q}{fWLQ_i} = \frac{\alpha_H q}{fC_{ox}^2 WL (V_{gs} - V_{th})} \tag{3.31}$$

where Q_i is the charge corresponding to unit area, and the overdrive voltage is $V_{gs} - V_{th}$. Therefore, the normalized noise power spectral density is inversely proportional to the overdrive voltage.

3.4.4. Carrier Number Correlated with Mobility Fluctuations Theory in Thin-Film Transistors

The widely accepted model for $1/f$ noise in transistors is the carrier number fluctuation and induced mobility fluctuation model (ΔN-$\Delta\mu$ model) [48, 49]. This model posits that carriers captured and released between the channel and the gate oxide traps not only shift the flat band voltage, causing fluctuations in carrier number but also that the defect states in the gate oxide, after capturing and releasing carriers, undergo charge fluctuations that affect the mobility through Coulomb scattering. These mobility fluctuations are related to the fluctuations in the inversion charge density, which is closely tied to the capture and release of carriers in the gate oxide traps, thereby generating $1/f$ noise.

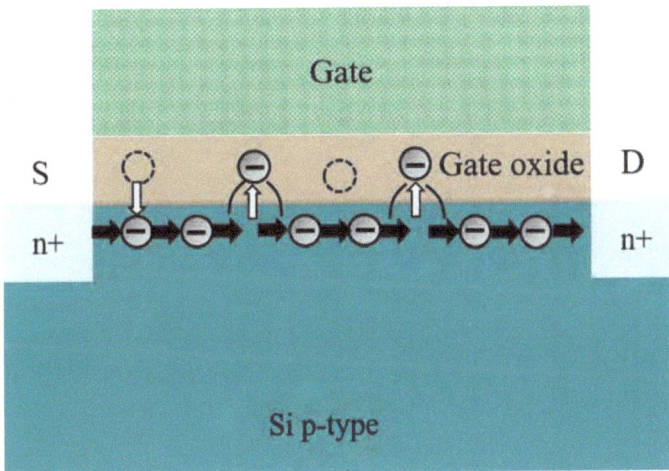

Fig. (3.12). The charge exchange process between the channel and the gate oxide.

As shown in Fig. (**3.12**), charge exchange occurs between the channel and the gate oxide, leading to fluctuations in the number of carriers. Trap states in the gate oxide near the energy level EF capture carriers that tunnel into the channel, and similarly, trap states near the energy level EF emit charges that tunnel into the inverted channel, becoming free carriers and contributing to 1/f noise through this charge exchange mechanism.

For transistors, the ΔN-$\Delta\mu$ model can be expressed as [49]:

$$\frac{S_{ID}}{I_{DS}^2} = \left(1 \pm \alpha_c \mu_{eff} C_{ox} \frac{I_{DS}}{g_m}\right)\left(\frac{g_m}{I_{DS}}\right)^2 S_{V_{fb}} \tag{3.32}$$

where g_m represents the device's transconductance, C_{ox} is the gate capacitance per unit area, μ_{eff} is the effective carrier mobility, I_{ds} is the current flowing through the device, α_c is the Coulomb scattering parameter, and S_{vfb} represents the flatband voltage noise power spectral density. This formula is applicable to devices operating in the linear region. Similarly, based on tunneling theory, S_{vfb} can be expressed as [49]:

$$S_{V_{fb}} = \frac{q^2 kT \lambda N_t}{fWLC_{ox}^2} \tag{3.33}$$

where W is the device channel width, L is the device channel length, f is the device frequency, λ is the tunneling decay coefficient in the gate oxide, with a typical value for SiO$_2$ of about 0.1 nm, C_{ox} is the gate oxide capacitance per unit area, T is the device operating temperature, and N_t is the density of defect states in the gate oxide near the Fermi level. In the analysis of device noise data, Equation (3.32) is typically used to fit the experimental data to obtain the parameter S_{vfb}, and then Equation (3.29) is used to extract N_t, allowing for the characterization and analysis of TFT device parameters [26, 50].

In summary, by analyzing the LFN curve's changes over stress time, we can assess degradation mechanisms like trap generation, mobility fluctuation, oxide quality deterioration, and interface instability. These insights help evaluate the long-term stability and reliability of TFT devices under various stress conditions, providing valuable feedback for optimizing materials, design, and manufacturing processes.

3.5 THIN-FILM QUALITY ANALYSIS

For TFTs, the material of the active layer directly affects the performance of the device. To further investigate the specific impact of the physical and chemical states of thin films on device performance, various characterization techniques are typically employed to analyze the thin films. Thin-film characterization refers to the process of analyzing the physical, chemical, and optical properties of thin films. The purpose of thin-film characterization is to understand the composition, structure, and properties of thin films. By characterizing the thin films through multiple methods, a more in-depth understanding of the carrier transport mechanisms within the thin films and the degradation mechanisms under various stresses can be achieved at the microscopic level. This understanding provides a theoretical basis for optimizing the performance of TFTs. Thin-film characterization typically includes composition analysis, structural analysis, physical property analysis, mechanical property analysis, and thermal property analysis. Composition analysis, chemical analysis methods, and electron probe microanalysis are other means to determine the composition of the thin film. Structural analysis employs atomic force microscopy (AFM), scanning electron microscopy (SEM), X-ray diffraction, electron diffraction, and other means to understand the crystal structure and surface morphology of the thin film. Physical property analysis uses electrical testing, magnetic testing, optical testing, and other means to determine the electrical, magnetic, and optical properties of the thin film. Mechanical property analysis uses mechanical testing to understand the mechanical properties of the thin film, such as hardness and elasticity. Thermal property analysis uses thermogravimetric analysis, thermal expansion testing, and other methods to determine the thermal properties of the thin film, such as thermal stability and coefficient of thermal expansion. Other property analyses include studies on photoelectrochemical properties, electronic transport properties, catalytic properties, and more. Below is a brief introduction to several commonly used characterization techniques for TFTs.

3.5.1 Atomic Force Microscope Analysis

The atomic force microscope (AFM) is a high-resolution microscopy technique capable of observing and measuring the surface of samples with high resolution. It operates based on the interactions between a probe and the sample surface, as shown in Fig. (**3.13**). The probe has an extremely sharp tip that can move to make contact with the sample surface. In an AFM instrument, the probe is typically combined with a laser beam that is directed onto the tip of the probe and then reflected back. As the probe moves across the sample surface, the reflection of the laser beam is detected by a detector. The unevenness of the sample surface affects the position and intensity of the reflected laser beam. The measurement system monitors the changes in the position and intensity of the reflected laser beam to maintain a constant distance between the probe and the sample. This feedback mechanism ensures that the probe follows the contours of the sample surface, allowing for high-resolution microscopy in three dimensions. By recording the movement of the probe and the changes in the reflected laser beam during scanning, a high-resolution image of the sample surface can be generated, revealing the topography and other properties of the sample surface.

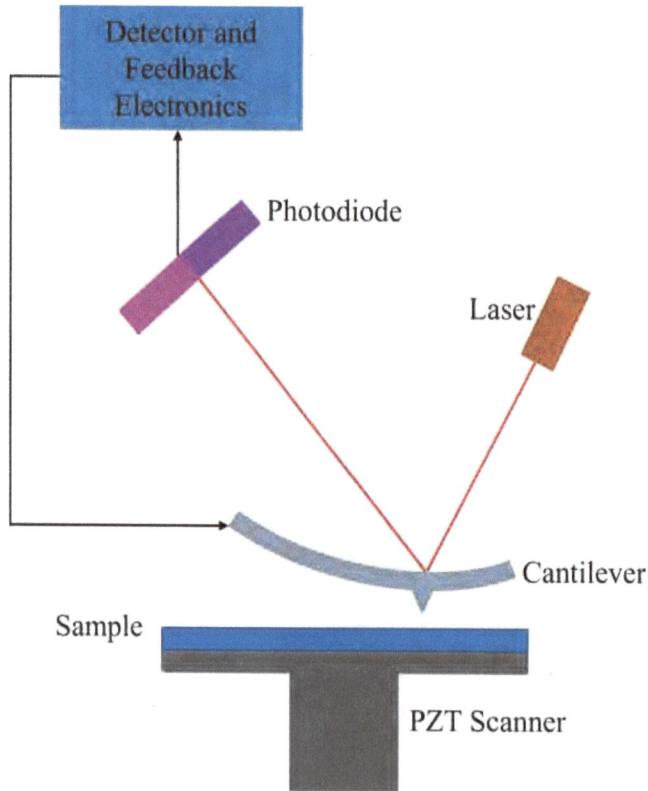

Fig. (3.13). Schematic diagram of atomic force microscope testing.

AFM utilizes the interaction of a tiny laser beam and its reflection, leveraging the interaction between the probe and the sample to perform high-resolution surface imaging. This method offers extremely high resolution and is used to study the microstructure and properties of sample surfaces.

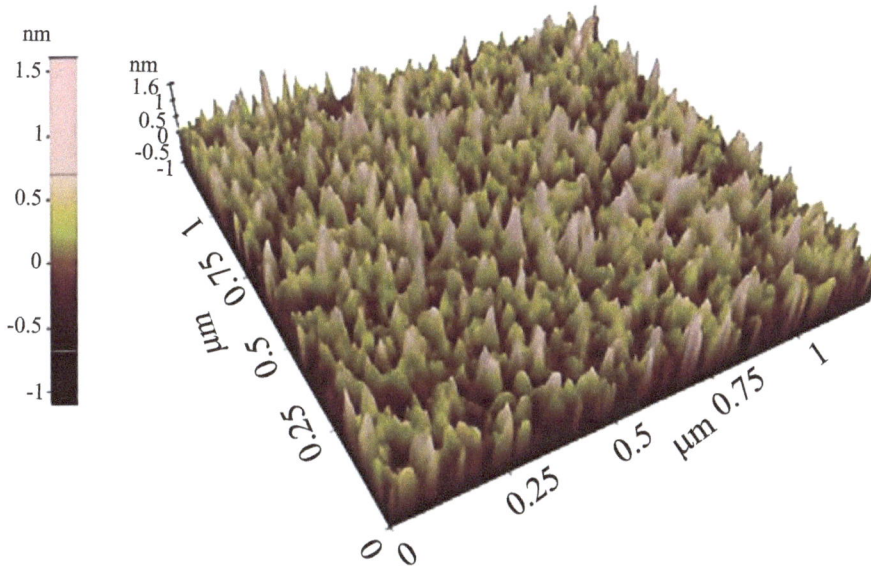

Fig. (3.14). AFM image [51].

As shown in Fig. (**3.14**), the AFM image of an Al_2O_3 thin film exhibits a root mean square (RMS) surface roughness of 0.3 nm [51]. In the field of TFTs, using an atomic force microscope to characterize the 3D topography or surface roughness of thin films helps to gain a deeper understanding of how the film structure affects the performance and reliability of the thin films.

3.5.2 Scanning Electron Microscope Analysis

The scanning electron microscope (SEM) is an advanced microscopy technique that uses an electron gun to generate a high-energy electron beam, as shown in Fig. (**3.15**), which is focused to an extremely small diameter, typically only a few nanometers. The energy of this electron beam usually ranges from several thousand volts to tens of thousands of volts, far exceeding the energy of photons, thus enabling SEM to achieve very high resolution. The electron beam is directed at the sample surface and scanned along a predetermined path. Interactions between the electron beam and the sample cause various signals to be emitted from the surface. SEM acquires information about the sample surface by detecting these signals that result from the interaction with the electron beam. SEM typically detects and records three primary signals:

1. Secondary electrons: These are electrons emitted from the sample surface that are almost perpendicular to it, providing information about the sample's surface morphology.

2. Backscattered electrons: These electrons are emitted from the sample surface at larger angles and provide information about the sample's composition.

3. X-rays: When the electron beam strikes the sample surface, the sample emits X-ray signals, which provide information about the sample's elemental composition.

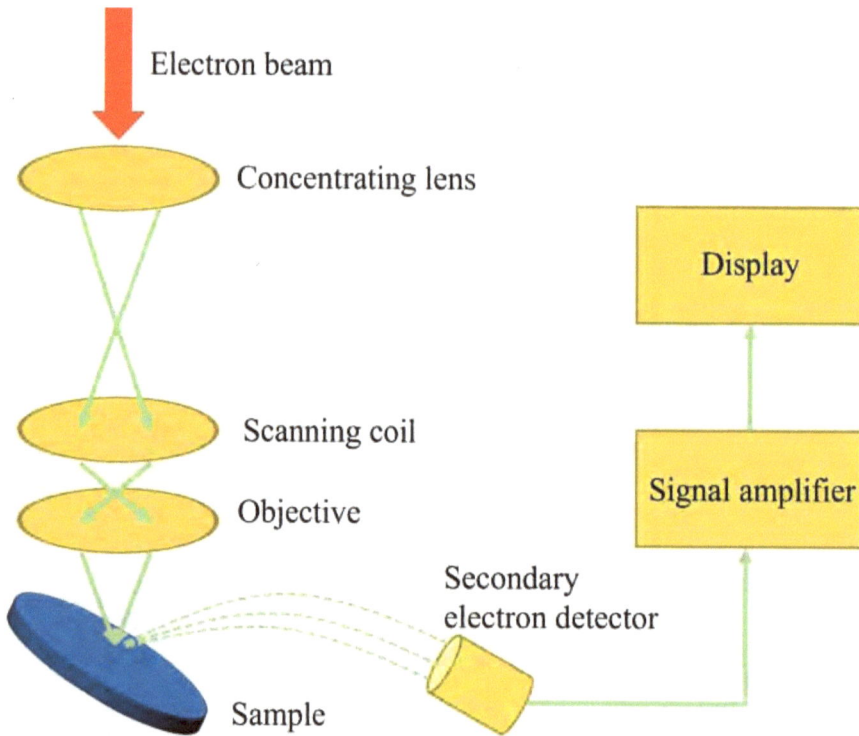

Fig. (3.15). Principle diagram of scanning electron microscope testing.

By processing and analyzing these detected signals, SEM can generate high-resolution images of the sample surface and provide detailed information about its morphology, composition, and structure.

For example, Fig. (**3.16**) shows the structure of the photoresist on a TFT device observed by SEM. After the channel crystallization, an anti-reflective coating (ARC) and photoresist are applied by spin coating. The photoresist is then patterned into gratings with a period of 800 nm using laser interference lithography. The setup of the photolithography system employs a Lloyd's mirror interferometer. The ratio of the length of the photoresist-covered area to the exposed area within one period is approximately 2:1 [52]. Under SEM characterization, microstructures can be more clearly observed, aiding in the understanding and analysis of the fabrication process of microstructures.

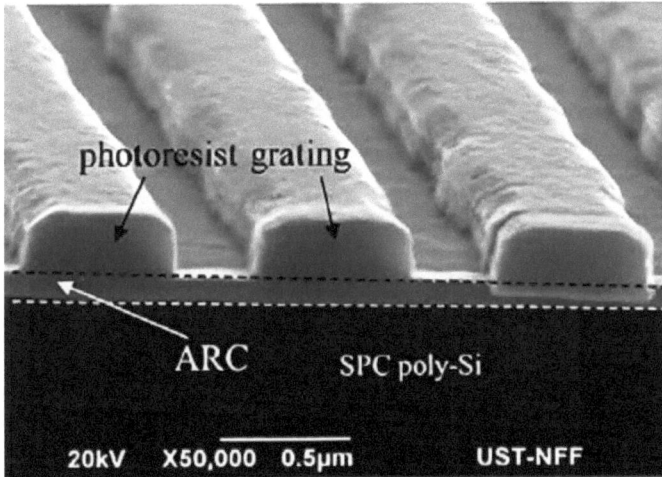

Fig. (3.16). Scanning electron microscope image [52].

3.5.3 Transmission Electron Microscopy Analysis

Transmission electron microscope (TEM) is a high-resolution imaging technique that uses a beam of electrons to probe samples, allowing for the observation of finer structures at a smaller scale than what is possible with optical microscopy and offering greater resolution, as shown in Fig. (**3.17**).

The fundamental working principle of TEM involves an electron beam generated by an electron gun that passes through the sample and is then focused and converged by a series of lenses. As the focused electron beam interacts with the atoms, lattices, or structures within the sample, various scattering, transmission, and absorption phenomena occur. Depending on the electron transmission properties of the sample, some electrons are scattered or absorbed while the rest are transmitted to the detector. Ultimately, by detecting and processing the electrons that are transmitted to the detector, a high-resolution image of the sample is obtained. By analyzing these images, researchers can gain insights into the microstructure, crystal structure, elemental composition, and distribution of the sample.

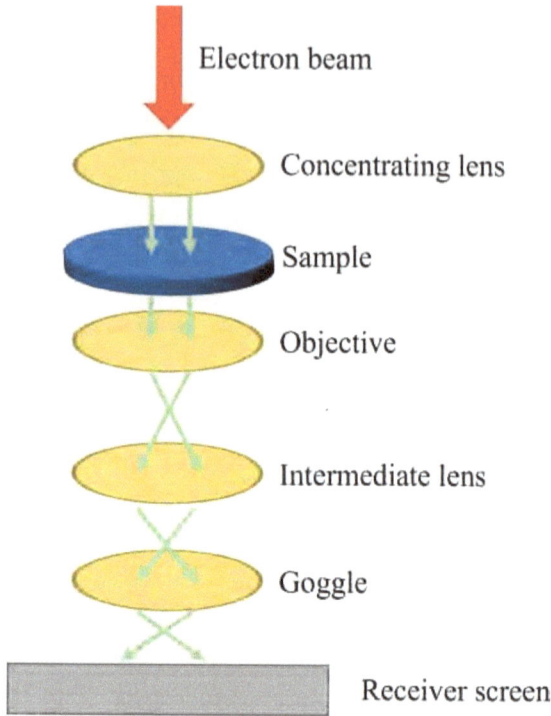

Fig. (3.17). Principle diagram of transmission electron microscopy testing.

TEM leverages the principle of interaction between the electron beam and the sample to achieve high-resolution imaging of the sample's microstructure, providing a powerful tool for research in fields such as materials science, biology, and nanotechnology.

In general, when observing the microstructure of devices, TEM is employed, as shown in Fig. (**3.18**), where the unique gate structure of the device can be observed, and precise measurements of length and width can be provided at the microscale.

3.5.4 X-ray Photoelectron Spectroscopy Analysis

X-ray Photoelectron Spectroscopy (XPS) is a surface analysis technique that provides information about the chemical state and electronic structure of the elements on a sample surface, as shown in Fig. (**3.19**). By irradiating the sample surface with X-rays and measuring the energy of the absorbed and emitted electrons, XPS yields insights into the chemical state of the elements present.

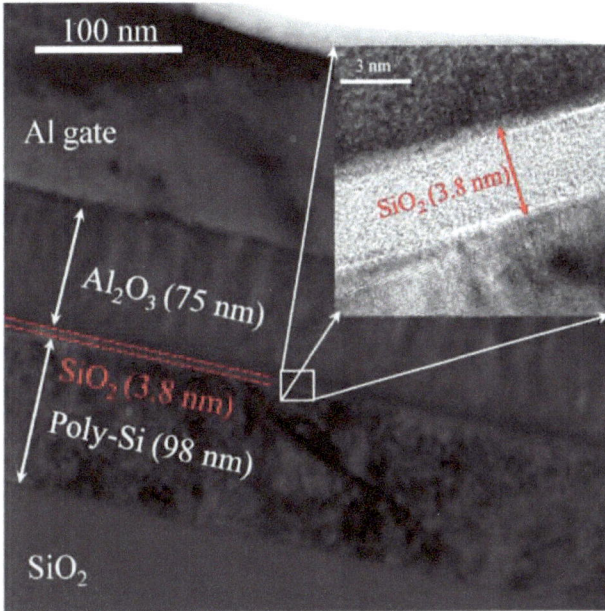

Fig. (3.18). TEM image [53].

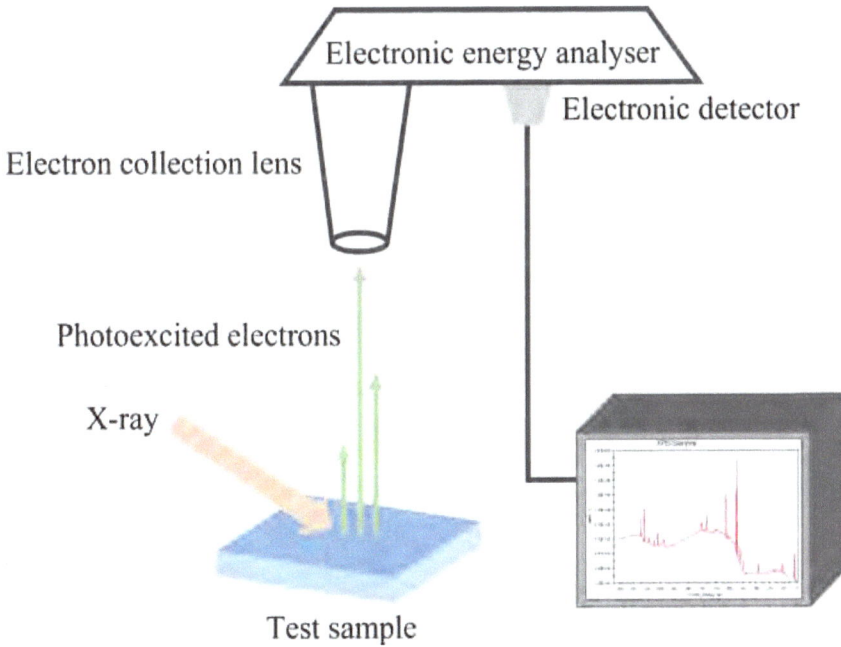

Fig. (3.19). Principle diagram of x-ray photoelectron spectroscopy testing.

An XPS instrument emits X-rays onto the sample surface, which have sufficient energy (typically between 100-2000 eV) to excite inner-shell electrons of the surface atoms. These inner-shell electrons, once excited by the X-rays, transition to higher energy levels. Some of these electrons have enough energy to escape to the surface and are known as photoelectrons. The XPS instrument measures and records the energy of these photoelectrons, from which an energy spectrum is plotted. Based on the energy of the photoelectrons, the types of elements and their chemical states on the sample surface can be determined. Analysis of the energy spectrum provides information about the relative content of surface elements, the nature of their chemical bonding, and their charge states, helping researchers understand the surface chemistry of the sample.

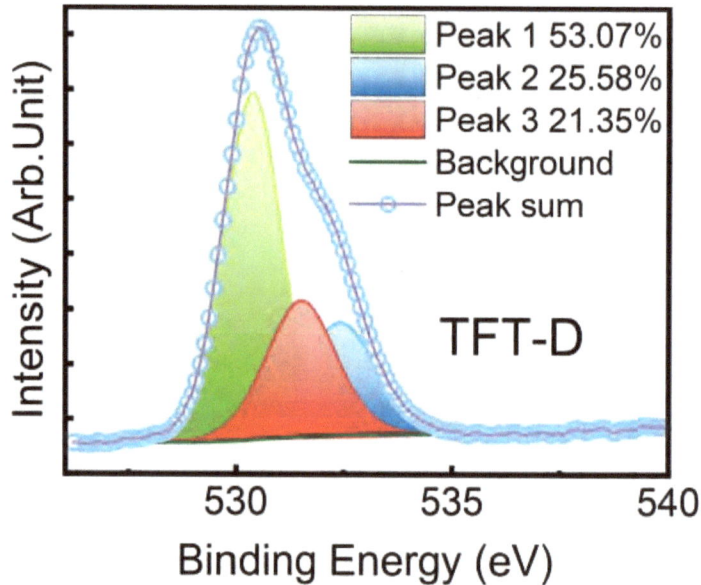

Fig. (3.20). Schematic of the XPS test spectrum [54].

In practical applications, XPS is often used to characterize the chemical states and composition ratios of elements within thin films. For example, in the case of oxygen, as shown in Fig. (**3.20**), thin films were prepared using magnetron sputtering. The electron spectrum of the oxygen atoms was deconvoluted into three sub-peaks, where the oxygen 1 peak typically represents metal-oxygen bonds, the oxygen 2 peak represents oxygen vacancies (V_os), and the oxygen 3 peak represents hydroxyl and other impurity bonds [55].

3.5.5 X-ray Powder Diffractometer Analysis

X-ray Powder Diffractometer (XRD) is a widely used technique for material characterization that determines the crystal structure and crystallographic properties of materials by measuring the diffraction patterns of incident X-rays within the material, as shown in Fig. (**3.21**). Typically, X-rays generated from a tungsten or copper target are used as the incident radiation source, which has high energy and can penetrate the sample, interacting with its atoms. Before measurement, the sample must be prepared in powder or thin-film form and placed in the path of the X-rays. The crystallinity of the sample affects the X-ray diffraction pattern. According to Bragg's Law, when X-rays strike the sample and interact with its lattice, diffraction occurs. The diffraction angle θ is determined by the following equation: $n\lambda = 2d\sin(\theta)$, where n is an integer, λ is the wavelength of the X-ray, and d is the interplanar spacing of the crystal lattice. By rotating the sample stage or the detector, a range of diffraction angles θ can be scanned. The relationship between the measured diffraction intensity and the diffraction angle is known as the XRD pattern.

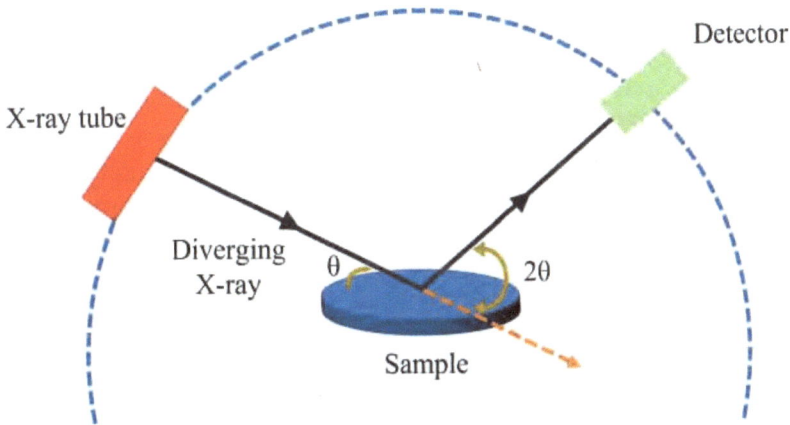

Fig. (3.21). Principal diagram of X-ray powder diffractometer analysis testing.

By comparing the experimentally measured XRD pattern with standard patterns in a database, the crystal structure, lattice parameters, and crystallographic properties of the sample can be identified.

As shown in Fig. (**3.22**) is the XRD pattern of poly-Si. For TFTs, the crystallinity of the channel directly affects the carrier transport pathways and has a significant impact on device performance. Therefore, using XRD to analyze the crystal structure of the channel is a common characterization method in the study of TFTs.

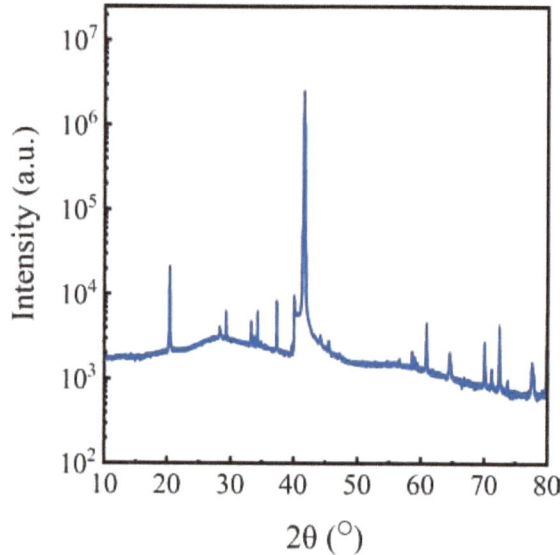

Fig. (3.22). XRD pattern of poly-Si.

3.5.6 Raman spectra Analysis

Raman spectroscopy is a non-destructive analytical method based on the Raman scattering effect. The Raman scattering effect refers to the phenomenon where, when a high-intensity incident light from a laser source is shone onto a molecular sample, as shown in Fig. (**3.23**), most of the light scatters, and a portion of the scattered light has a frequency different from that of the incident light. This scattering phenomenon provides information about molecular vibrations and rotations, which can be used for molecular structure research. When a monochromatic light with a frequency V_0 shines on a sample, the molecules cause the incident light to scatter. Most of the light only changes its direction of propagation, known as Rayleigh scattering. Another type of scattered light, which accounts for about 10^{-6} to 10^{-10} of the total scattered light intensity, not only changes its propagation direction but also its frequency; this type of scattered light is called Raman scattering. In Raman scattering, the scattered light with a frequency lower than the incident light frequency is called Stokes scattering, while the scattered light with a higher frequency is called anti-Stokes scattering. Stokes scattering is usually much stronger than anti-Stokes scattering, and Raman spectrometers typically measure Stokes scattering, which is also collectively referred to as Raman scattering.

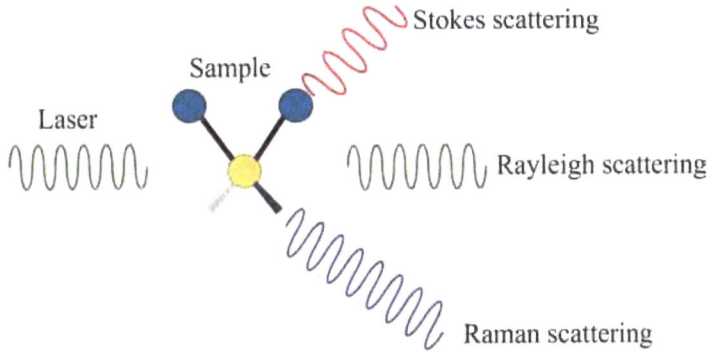

Fig. (3.23). Schematic diagram of Raman spectra analysis.

A Raman spectrometer mainly consists of a laser, a spectrometer, a detector, and a correlator, among other components. The laser generates a high-intensity laser source that causes Raman scattering when shone onto a sample; the spectrometer is used to separate the scattered light into Stokes and anti-Stokes scattering; the detector is used to measure the intensity of the scattered light; and the correlator is used to record and process the scattered signals.

Raman spectroscopy analysis can provide information about molecules in different vibrational and rotational states, as shown in Fig. (**3.24**), and it is applied in fields such as organic compound structural identification, polymer protein analysis, clinical testing, and research on crystals and materials.

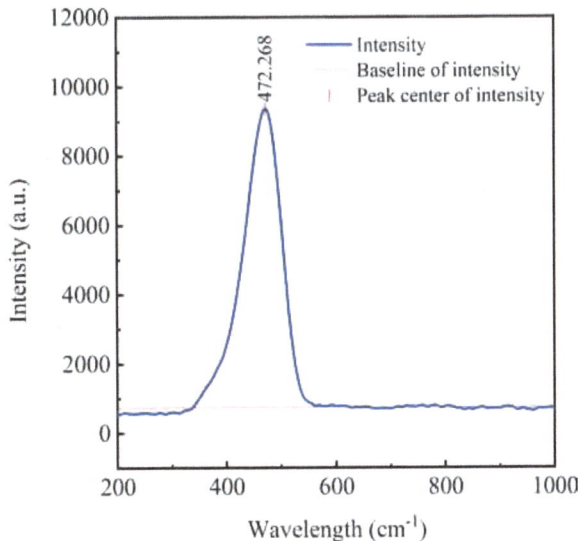

Fig. (3.24). Raman spectra.

In summary, by monitoring changes in thin-film properties through these techniques, we can assess various degradation mechanisms such as surface roughening, defect formation, grain boundary instability, phase changes, and interfacial degradation. These insights are crucial for understanding TFT device reliability, enabling the development of strategies to improve material selection, device design, and fabrication processes to enhance durability.

3.6. SIMULATION ANALYSIS

As previously briefly introduced, some basic degradation phenomena in TFTs and the defects present in the active layers of TFTs made from different materials were discussed. However, to establish physical models for device degradation under various stresses, relying solely on experimental observations is not sufficient. Degradation models inferred from experimental phenomena often need to be verified through simulation. The following will introduce two commonly used electrical and thermal simulation tools. These simulation tools can help reliability researchers reproduce the electrical and thermal parameters of devices under stress on a computer, thereby validating and refining the proposed reliability models.

3.6.1. Technology Computer-Aided Design Simulation

When performing an electrical characteristic simulation of TFTs, the technology computer-aided design (TCAD) simulation tools from SILVACO, such as ATLAS, or from SYNOPSYS, such as, are often used. This section will primarily use the ATLAS simulation engine from SILVACO as an example. It allows for custom device structure, application of stress conditions, and observation of changes in the electric field, current, and carrier behavior within the device to establish and validate the reliability models of TFTs.

3.6.1.1 Basic Operations of TCAD

To achieve simulation results that closely match real-world conditions, it is essential to construct the structure of a TFT accurately within the simulation software according to the actual situation. In SILVACO, this is primarily done through code writing to build the device structure and subsequent electrical characteristics (Fig. **3.25**).

The first step is mesh setup. The mesh defines the points at which the simulation software will calculate during the process. The density of the mesh directly affects the speed and accuracy of the computation. If the mesh is too sparse, the computation will be faster, but the accuracy of the results may be unsatisfactory.

Conversely, if the mesh is too dense, the results will be more precise, but the computation may become inefficient and even lead to insufficient memory, preventing the calculation from being completed. Therefore, setting an appropriate mesh structure is crucial for our simulations. By placing denser meshes in areas of interest, such as the active layer and contact interfaces, and sparser meshes in less critical areas like the substrate and electrodes, we can achieve more efficient simulations.

Typically, the mesh position and density are set using the 'mesh' command.

mesh width = 100

x.mesh location = 0.0 spacing = 1

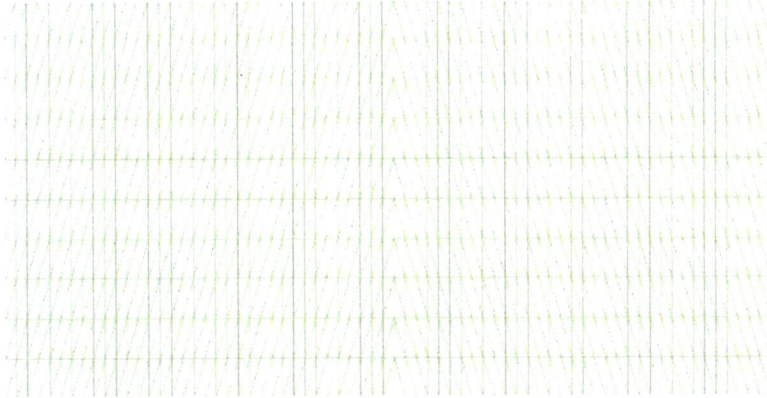

Fig. (3.25). mesh structure built in SILVACO.

Once the mesh setup is complete, you can begin to construct the device structure using code, writing statements based on the device's structure and the types of materials involved (Fig. **3.26**). It is important to note that while the ATLAS library includes initial definitions for a variety of materials, some newer materials may not be recognized by ATLAS, leading to errors. For such cases, you should refer to the ATLAS user manual. The statements for building the device structure include definitions of material types and their respective locations, which are primarily defined using the 'region' statement:

region num = 1 y.min = 0 y.max = 0.05 x.min = 0 x.max = 1 material = SiO_2

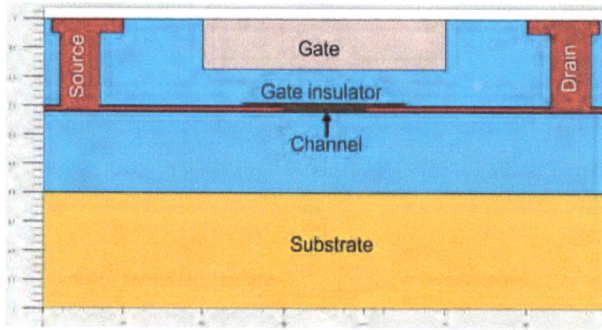

Fig. (3.26). device structure built in SILVACO.

By defining the material types and their positions, a top-gate structured IGZO TFT is constructed. After the material definitions are complete, the device's electrodes must also be defined to allow for subsequent electrical characteristic definitions. According to the device structure, the gate, source, and drain electrodes are defined, with the contact characteristics. The 'electrode' statement is used for these definitions:

electrode num = 5 name = gate y.min = 0 y.max = 0.175 x.min = 20 x.max = 50

After the device structure is constructed, to make the electrical performance characteristics more realistic, the active layer of the TFT needs to be defined separately. Taking the active layer material as metal oxide as an example, as mentioned earlier, metal oxides contain a certain amount of weak bonds and V_Os. These defects affect the orderliness of the amorphous structure of metal oxides, specifically manifesting as tail-state defects and deep-level defects in the metal oxide. Depending on whether the defects are more likely to capture electrons or holes, they can be classified as acceptor-like and donor-like defects. N_{ta} represents acceptor-like tail state defects, N_{td} represents acceptor-like deep-level defects, N_{ga} represents acceptor-like deep-level defects, and N_{gd} represents donor-like deep-level defects. Their distribution in the energy levels is described by the following formulas [56, 57]:

$$g_A(E) = N_{TA} \times \exp\left[\frac{E - E_c}{W_{TA}}\right] + N_{GA} \times \exp\left[-\left(\frac{E_{GA} - E}{W_{GA}}\right)^2\right]$$

(3.34)

$$g_A(E) = N_{TD} \times \exp\left[\frac{E_V - E}{W_{TD}}\right] + N_{GD} \times \exp\left[-\left(\frac{E - E_{GD}}{W_{GD}}\right)^2\right]$$

(3.35)

Here, $g_A(E)$ represents acceptor-like defects and $g_D(E)$ represents donor-like defects. W_{TA} and W_{TD} represent the characteristic decay amounts of the tail states, which determine the variation of the defect tail states within the bandgap. N_{GA} and N_{GD} represent the peak densities of acceptor-like and donor-like states characterized by Gaussian distribution, while W_{GA} and W_{GD} represent their characteristic decay amounts. E_{GA} and E_{GD} represent the energy levels where the peaks of these densities are located. The specific statements for defining the active layer parameters and the corresponding defect density structure are shown in Fig. (**3.27**).

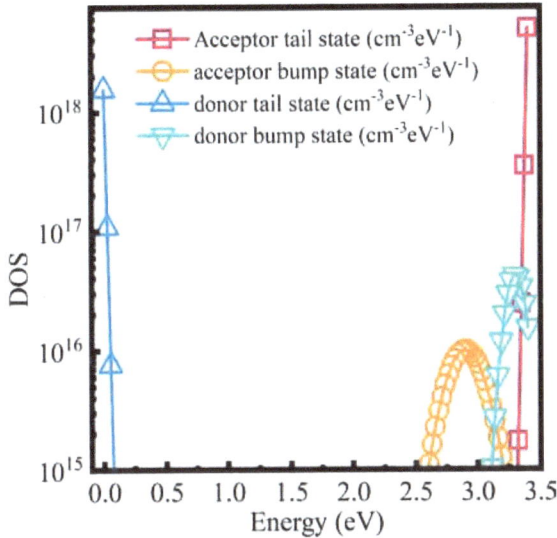

Fig. (3.27). DOS extracted in SILVACO.

After the device structure is built and the active layer parameters are defined, the transfer characteristic curve of the device can be simulated (Fig. **3.28**). The simulation statements and results are as follows: A constant voltage of 0.1V is applied to the drain of the device, and the gate is scanned from –5V to 15V in increments of 0.25V.

#transfer curve#

solve init

solve vdrain = 0.1

solve vgate = -5

log outfile = IGZO_transfer.log

solve vgate = -5 vstep = 0.25 vfinal = 15 name = gate

tonyplot IGZO_transfer.log

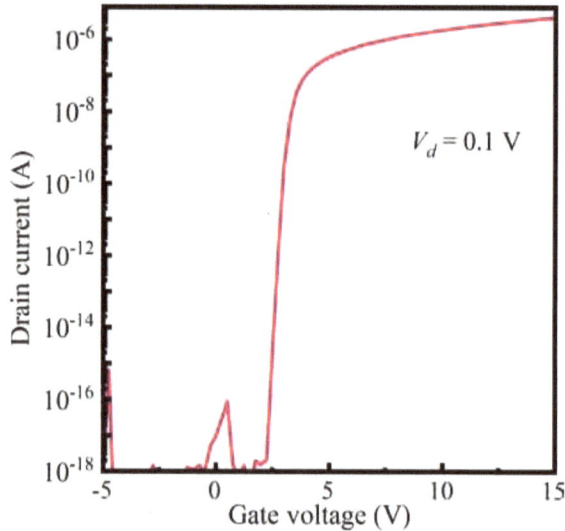

Fig. (3.28). The transfer curve simulated in SILVACO.

3.6.1.2 DC Simulation

In the section on device structure construction, the definition of defects is discussed in the active layer. Only by understanding how different defects affect the transfer characteristic curve of TFTs can we efficiently conduct reliability research. Simulation software is well-suited for this task. In the following work, DC simulation was taken as an example to observe the changes in metal oxide (MO) devices under different defect densities.

By varying different parameters in the active layer (N_{gd}, N_{ga}, N_{ta}, N_{td}) and observing the resulting transfer curves from the simulation, we can learn and understand the impact of defect changes on the transfer curve.

The effect of donor-like deep-level defects (N_{gd}) on the transfer characteristic curve is shown in Fig. (**3.29**). As the density of donor-like deep-level defects increases, the transfer characteristic curve gradually shifts in the negative direction, the V_{th} decreases, the *SS* remains essentially unchanged, and the I_{on} increases. This is because, in MO TFTs, donor-like deep-level defects are primarily composed of V_os. As mentioned in the section on V_os, an increase in the number of V_os leads to a rise in the concentration of electrons in the active layer. The device becomes easier to turn on, and the mobility also increases.

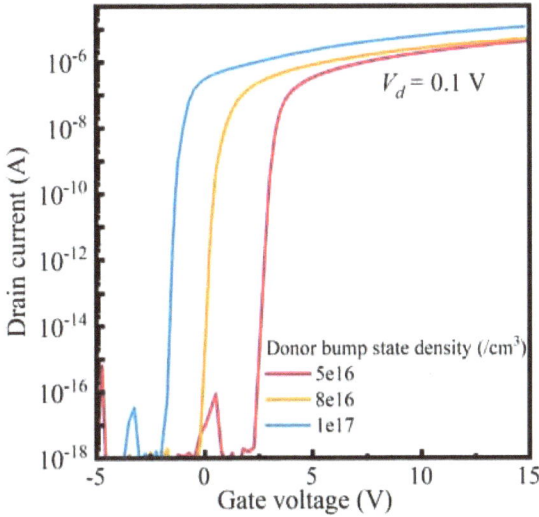

Fig. (3.29). The transfer curve under various N_{gd}.

The impact of acceptor-like shallow-level defects (N_{ta}) on the transfer characteristic curve is illustrated in Fig. (**3.30**). As the density of acceptor-like shallow-level defects increases, the I_{on} of the transfer characteristic curve gradually decreases, while the V_{th} and SS remain essentially unchanged. In metal oxides, the formation of tail states is primarily attributed to the disorder of the amorphous structure, such as the breaking of weak bonds leading to the disorder of locally ordered structures. These tail states capture charge carriers, reducing the concentration of carriers in the active layer, which results in a decrease in the device's I_{on} and mobility.

The influence of donor-like shallow-level defects (N_{td}) on the transfer characteristic curve is shown in Fig. (**3.31**). The variation in donor-like shallow-level defects has no influence on the transfer characteristic curve of MO TFTs. This is because, for most MO TFTs, the majority of carriers in the active layer are electrons, and the concentration of holes is negligible. Therefore, donor-like shallow-level defects cannot significantly affect the device's electrical characteristics by capturing holes.

The effect of acceptor-like deep-level defects (N_{ga}) on the transfer characteristic curve is depicted in Fig. (**3.32**). As the number of acceptor-like deep-level defects increases, the transfer characteristic curve of the MO TFT shifts positively, the V_{th} increases, and the SS significantly increases, while the I_{on} slightly decreases. There is currently debate regarding the origin of acceptor-like deep-level defects in metal oxides, with the prevailing view suggesting that uncoordinated oxygen atoms, also known as excess oxygen, are one of the sources of acceptor-like deep-level defects.

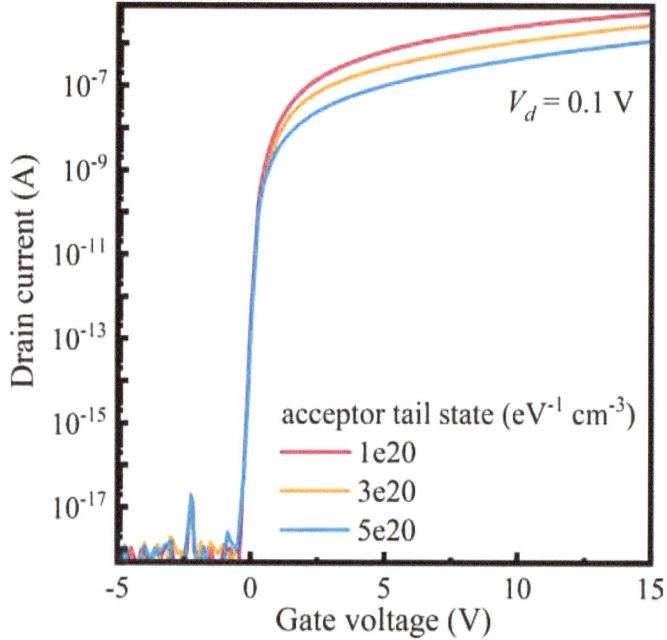

Fig. (3.30). The transfer curve under various N_{ta}.

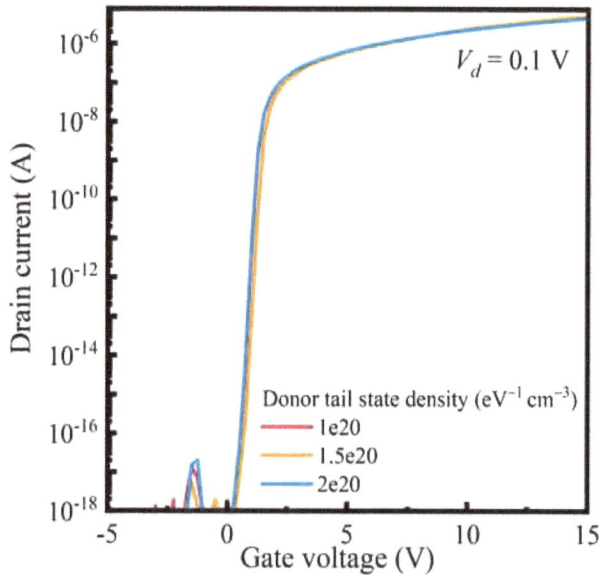

Fig. (3.31). The transfer curve under various N_{td}.

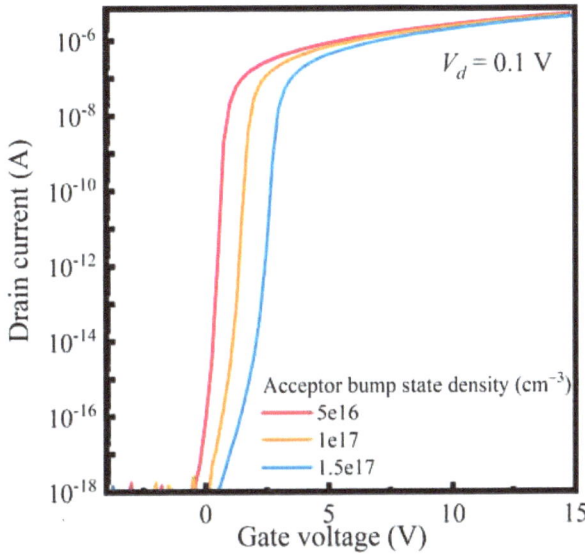

Fig. (3.32). The transfer curve under various N_{ga}.

3.6.1.3 Application of Simulation in Reliability Research

(1) Inference of Degradation Models through Observation of Defect Changes in Devices

In Section 2.5, common stress application methods in TFTs and the resulting degradation types were discussed, including gate bias stress and hot-carrier stress. These stresses are closely related to the device's degradation and the electric field distribution and carrier concentration within the active layer during the stress period.

For MO TFTs, a negative shift in the transfer curve is observed after PBS, attributed to the redistribution of V_os within the active layer. This redistribution is highly dependent on the magnitude of the vertical electric field (E_y) in the active layer (refer to Section 2.5). Therefore, by examining the E_y distribution in the active layer under different DC biases, one can investigate the location and intensity of V_o redistribution. As illustrated in Fig. (3.33), simulating the E_y distribution under various V_g and V_d conditions reveals a transition from a uniform to a non-uniform distribution with a larger field at the source and a smaller field at the drain as V_g decreases and V_d increases. The regions with an E_y are where the redistribution mechanism primarily occurs. Through electric field simulation, the range of V_o redistribution within the device structure under different stress conditions can be clearly observed, thereby refining the subsequent establishment of degradation models.

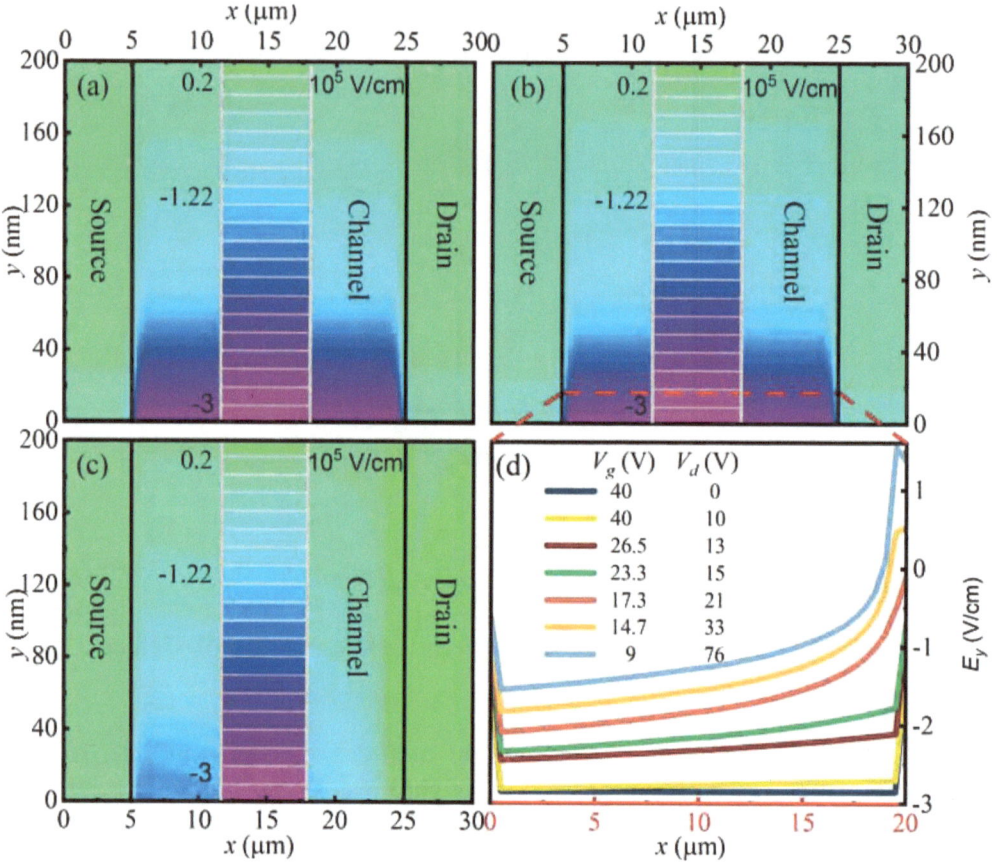

Fig. (3.33). Calculated E_y distribution in the active layer under: (a) PBS of $V_g = 40$ V; (b) under linear stress of $V_g = 40$ V and $V_d = 10$ V; and (c) under a saturation stress of $V_g = 9$ V and $V_d = 76$ V. (d) Extracted E_y in the surface of the channel from the source to the drain under various combinations of stress V_g and stress V_d.

Similarly, the location of degradation due to HC stress can be determined through electric field simulation. As shown in Fig. (**3.34**), applying a large voltage to the drain results in a significant lateral electric field (E_x) near the drain, which increases in size and range within the active layer as the voltage grows. This indicates that HC degradation predominantly occurs near the drain, with higher drain voltages leading to more severe degradation.

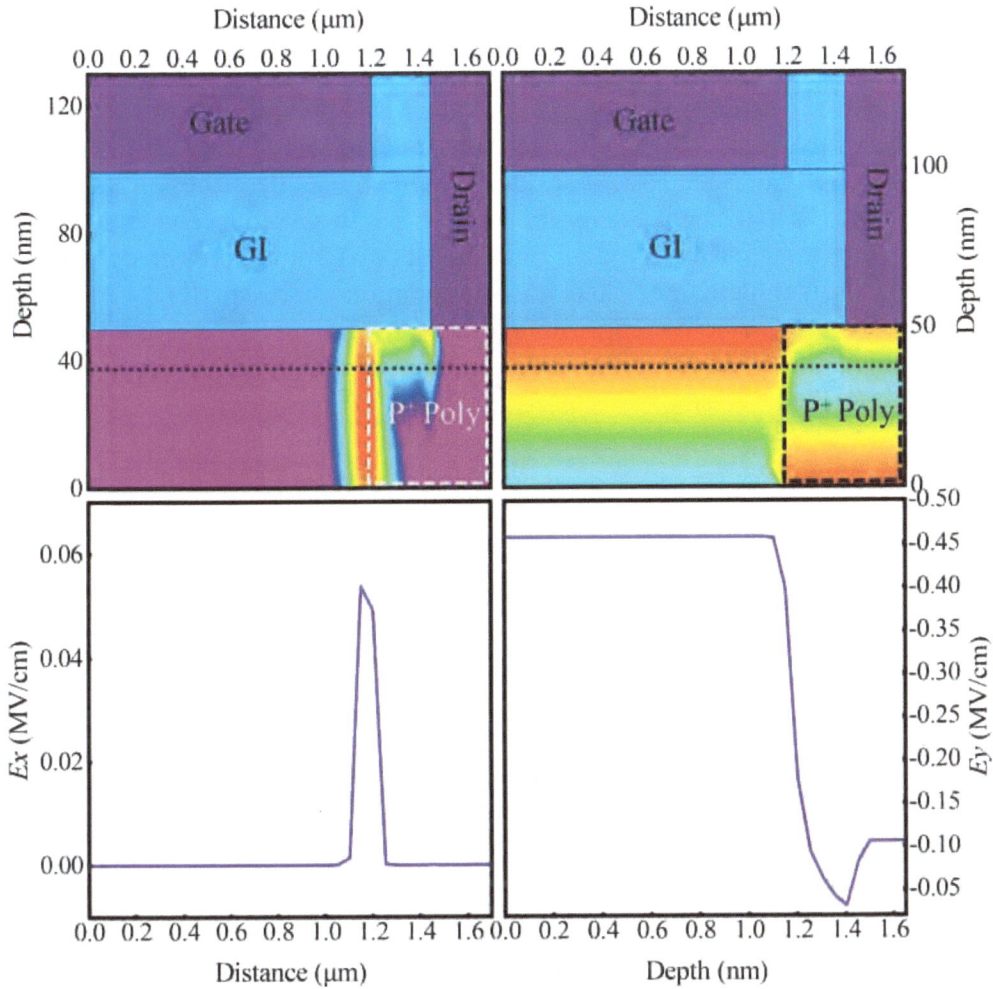

Fig. (3.34). E_x distribution in the active layer near the drain side under HC stress of $V_{gs} = -8$ V and $V_{ds} = -30$ V (b) E_y distribution near the drain side under HC stress of $V_{gs} = -8$ V and $V_{ds} = -30$ V. (c) Extracted E_x near the interface from the channel to the drain. (d) Extracted E_y near the interface from the channel to the drain.

(2) Evaluating the Accuracy of Degradation Models through Transfer Curve Fitting

In the study of TFT reliability, observing changes in internal parameters during the stress process can help pinpoint the locations of degradation and hypothesize degradation models. However, this is not sufficient to confirm the accuracy of these models. To validate the proposed degradation models, defect variations within the active layer must be controlled in simulation software to fit the experimental transfer characteristic curves.

Based on electric field simulations, an oxygen vacancy redistribution model has been proposed to explain the negative shift observed in MO TFTs after positive bias stress (PBS) (refer to Section 2.5 for details). This model describes an increase in the number of V_os in the active layer following PBS. In accordance with the model, by appropriately increasing the concentration of N_{gd} (from approximately 10^{17} to 10^{18} cm^{-3}), the simulated pre- and post-degradation curves perfectly match the experimental data, thereby confirming the accuracy of the oxygen vacancy redistribution model [58-60] in the active layer of MO TFTs under PBS (Fig. **3.35**).

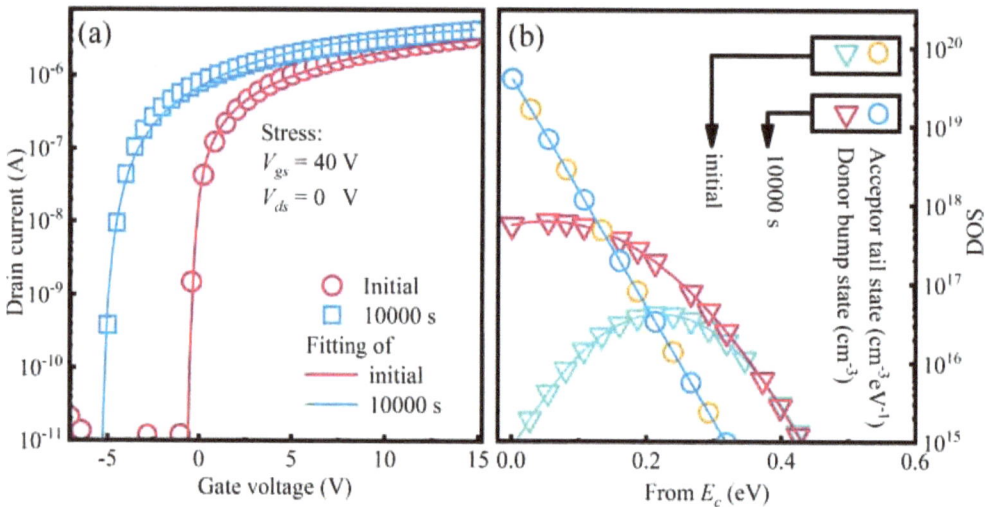

Fig. (3.35). The experimental and fitting transfer curves before and after 10000 s PBS stress. (b) Simulated DOS in a-IGZO TFT before and after 10000 s PBS stress.

Furthermore, electric field simulations indicate that during HC stress, the E_x is primarily concentrated near the drain, where HC degradation in TFTs predominantly occurs. According to the HC degradation model for TFTs (refer to Section 2.5), the defect density in the active layer near the drain increases after HC stress. For instance, in the case of MO TFTs, as illustrated in Fig. (**3.36b**), increasing the N_{ta} defect density near the drain, as shown in Fig. (**3.36b**), results in simulated pre- and post-degradation curves that closely fit the experimental data, thus validating the HC degradation model for TFTs under HC stress.

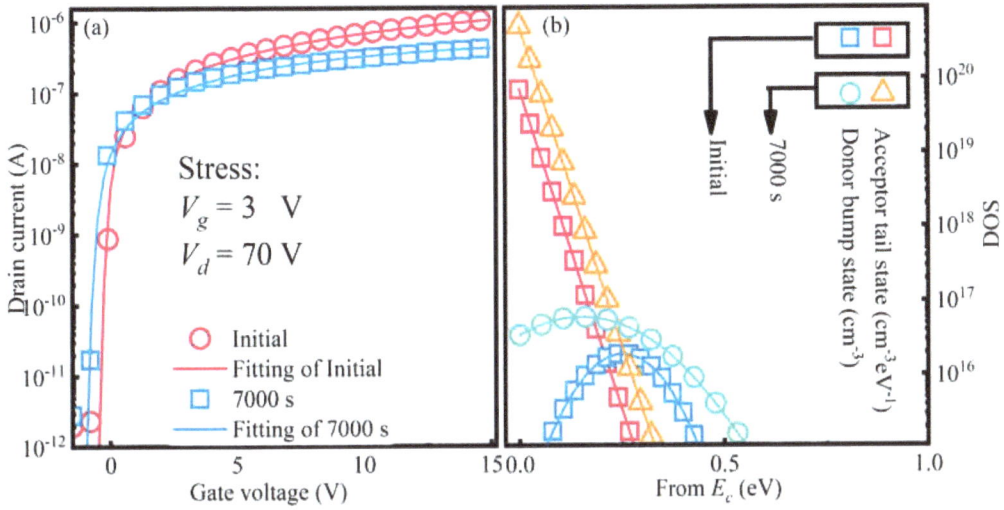

Fig. (3.36). The experimental and fitting transfer curves before and after 7000 s HC stress. (b) Simulated DOS in a-IGZO TFT before and after 7000 s HC stress.

3.6.2. Thermal Simulation

In reliability research, it is crucial not only to focus on the electrical characteristics of TFTs but also to consider the temperature distribution at different locations during device operation. Excessive temperatures can lead to the degradation of electrical properties and may impact other degradation mechanisms.

For thermal simulation of devices, COMSOL is commonly utilized. COMSOL, a Swedish company, specializes in providing multiphysics simulation solutions for engineers. Their Multiphysics platform offers a variety of modeling and simulation tools, including COMSOL Multiphysics and COMSOL Server, which are widely used in fields such as power electronics, fluid dynamics, heat conduction, and acoustics.

In COMSOL, constructing a device structure similar to that used in experiments is required. This process begins by defining boundaries to outline the device's contour, followed by specifying contact characteristics between boundaries and defining material properties. Finally, grid density is set, and the appropriate physical fields are selected for computation based on the desired results. It is important to note that since COMSOL is not specifically designed for simulating TFTs, its material library may lack many active layer materials commonly used in current TFTs. However, COMSOL offers users significant creative freedom, allowing them to define materials by consulting literature for detailed parameters.

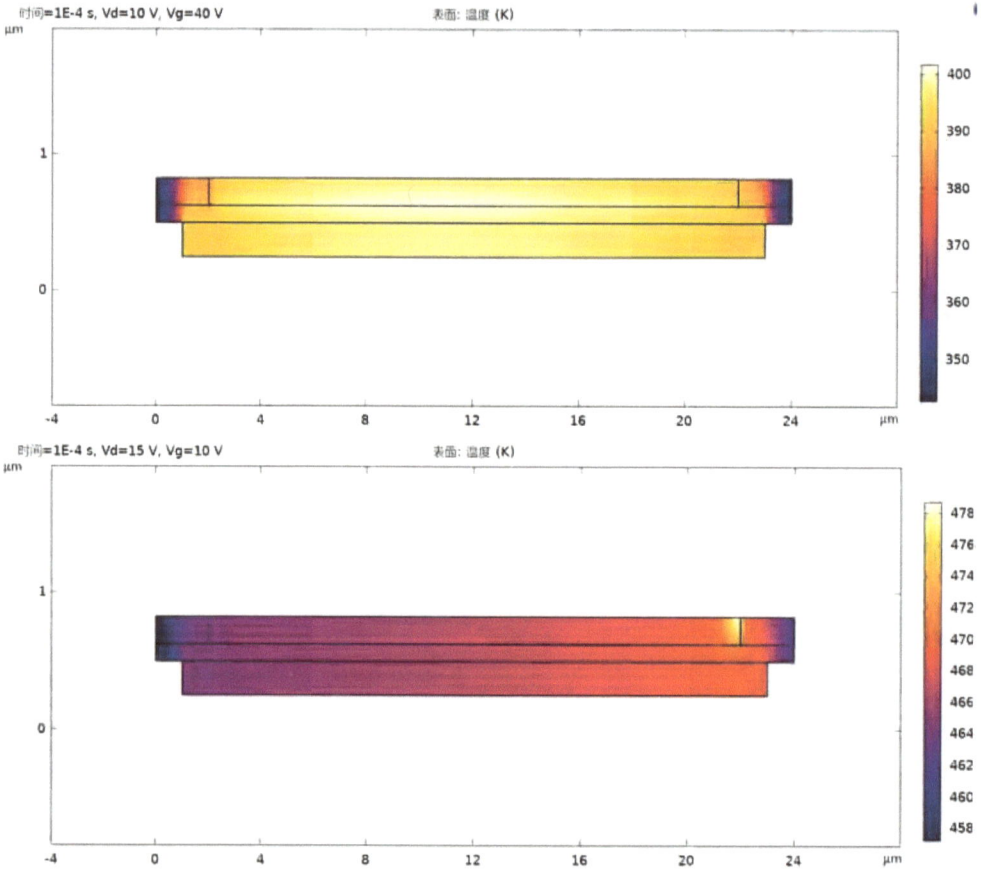

Fig. (3.37). Temperature distribution in IGZO TFTs under stress $V_g = 40$ V $V_d = 10$ V. (b) Temperature distribution in IGZO TFTs under stress $V_g = 10$ V $V_d = 15$ V.

For example, in a COMSOL simulation of an IGZO TFT (as shown in Fig. **3.37**), the device structure was constructed, including the gate, GI, source, active layer, and drain from left to right [5]. The active layer material, IGZO, was defined by the user and included parameters such as mobility, defect density and distribution, conduction and valence band density of states, and thermal conductivity. Subsequent thermal simulations of the device in both linear and saturation regions revealed different temperature distributions. When the device operates in the linear region with $V_g = 40$ V and $V_d = 10$ V, the heat generated is almost uniformly distributed across the active layer. In contrast, when operating in the saturation region with $V_d = 10$ V and $V_g = 10$ V, the heat is concentrated near the drain. [58-60]. This concentration is due to the channel pinch-off near the drain under saturation conditions, increasing the resistance near the drain and generating more heat. Since the ionization rate of HCs decreases at high temperatures, this heat can effectively mitigate the thermal degradation caused by excessive V_d (detailed discussion in Section 4.3.3).

In summary, by using TCAD simulation tools, we can gain a comprehensive understanding of how various physical, electrical, and thermal factors contribute to TFT degradation. These simulations help identify potential reliability issues early on, enabling device designers to optimize structures, materials, and operating conditions to improve the long-term performance and durability of TFTs.

CONCLUSION

This chapter presents a comprehensive examination of reliability analysis methods for TFTs, detailing various techniques to assess their performance degradation. It begins with an introduction that outlines the significance of reliability in TFT applications, particularly in display technologies. The chapter then explores degradation analysis through basic transfer curves, highlighting changes in the on-state, subthreshold, and off-state regions. It discusses CV curve analysis, which aids in understanding charge distribution and density of states. Furthermore, the chapter delves into LFN analysis, providing insights into carrier number and mobility fluctuations that impact device stability. Several advanced techniques for thin-film quality analysis, including XPS, TED, and AFM, are described to evaluate material properties and defects. Simulation methods, including technology computer-aided design and thermal simulations, are also emphasized as essential tools for predicting performance under various conditions. Overall, this chapter serves as a foundational resource for understanding the methodologies used to analyze and enhance the reliability of TFTs in modern electronic applications.

REFERENCES

[1]   Meng Zhang, Wei Zhou, Rongsheng Chen, Man Wong, and Hoi-Sing Kwok, "Characterization of DC-Stress-Induced Degradation in Bridged-Grain Polycrystalline Silicon Thin-Film Transistors", *IEEE Trans. Electron Dev.,* vol. 61, no. 9, pp. 3206-3212, 2014.
       http://dx.doi.org/10.1109/TED.2014.2341676

[2]   T.Y. Hsieh, T-C. Chang, T-C. Chen, M-Y. Tsai, Y-T. Chen, Y-C. Chung, H-C. Ting, and C-Y. Chen, "Origin of self-heating effect induced asymmetrical degradation behavior in InGaZnO thin-film transistors", *Appl. Phys. Lett.,* vol. 100, no. 23, 2012.232101
       http://dx.doi.org/10.1063/1.4723573

[3]   M. Wang, *Degradation Mechanisms of Low Temperature Poly-Si Thin-Film Transistors under Circuit Applications.*

[4]   M. Zhang, W. Zhou, R. Chen, S. Zhao, M. Wong, and H.S. Kwok, "P-6: Static Reliability of Bridged-Grain Poly-Si TFTs", *Dig. Tech. Pap.*, vol. 45, no. 1, pp. 964-967, 2014.

http://dx.doi.org/10.1002/j.2168-0159.2014.tb00250.x

[5]   M.J. Kim, and D. Choi, "Effect of enhanced-mobility current path on the mobility of AOS TFT", *Microelectron. Reliab.*, vol. 52, no. 7, pp. 1346-1349, 2012.

http://dx.doi.org/10.1016/j.microrel.2012.02.012

[6]   Y. Uraoka, T. Hatayama, T. Fuyuki, T. Kawamura, and Y. Tsuchihashi, "Reliability of low temperature poly-silicon TFTs under inverter operation", *IEEE Trans. Electron Dev.*, vol. 48, no. 10, pp. 2370-2374, 2001.

http://dx.doi.org/10.1109/16.954479

[7]   D. Zhou, M. Wang, and S. Zhang, "Degradation of Amorphous Silicon Thin Film Transistors Under Negative Gate Bias Stress", *IEEE Trans. Electron Dev.*, vol. 58, no. 10, pp. 3422-3427, 2011.

http://dx.doi.org/10.1109/TED.2011.2161635

[8]   D. Hyung Kim, H. Kwang Jung, D. Hwan Kim, and S. Yeol Lee, "Effect of interface states on the instability under temperature stress in amorphous SiInZnO thin film transistor", *Appl. Phys. Lett.*, vol. 99, no. 16, 2011.162101

http://dx.doi.org/10.1063/1.3645597

[9]   Y. Zhang, Z. Wang, M. Wang, D. Zhang, H. Wang, and M. Wong, "A Unified Degradation Model of Elevated-Metal Metal Oxide (EMMO) TFTs Under Positive Gate Bias With or Without an Illumination", *IEEE Trans. Electron Dev.*, vol. 68, no. 3, pp. 1081-1087, 2021.

http://dx.doi.org/10.1109/TED.2021.3053915

[10]   Q.J. Sun, J. Wu, M. Zhang, Y. Yuan, X. Gao, S-D. Wang, Z. Tang, C-C. Kuo, and Y. Yan, "Enhanced Electrical Performance and Bias-Stress Stability of Solution-Processed Bilayer Metal Oxide Thin-Film Transistors", *Phys. Status Solidi., A Appl. Mater. Sci.*, vol. 219, no. 22, 2022.2200311 [a].

http://dx.doi.org/10.1002/pssa.202200311

[11]   G. Zhu, M. Zhang, Z. Jiang, J. Huang, Y. Huang, S. Deng, L. Lu, M. Wong, and H-S. Kwok, "Significant Degradation Reduction in Metal Oxide Thin-Film Transistors via the Interaction of Ionized Oxygen Vacancy Redistribution, Self-Heating Effect, and Hot Carrier Effect", *IEEE Trans. Electron Dev.*, vol. 70, no. 8, pp. 4198-4205, 2023.

http://dx.doi.org/10.1109/TED.2023.3283940

[12]   Z. Jiang, M. Zhang, S. Deng, M. Wong, and H-S. Kwok, "Degradation of InSnZnO Thin-Film Transistors Under Negative Bias Stress", *IEEE Trans. Electron Dev.*, vol. 70, no. 12, pp. 6381-6386, 2023.

http://dx.doi.org/10.1109/TED.2023.3327975

[13]   M. Zhang, Y. Song, Z. Jiang, X. Xu, M. Wong, and H-S. Kwok, "Off-State Current Degradation Behavior of Polycrystalline Silicon Thin-Film Transistors under Dynamic Drain Voltage Stress", *2023 IEEE International Symposium on the Physical and Failure Analysis of Integrated Circuits (IPFA)*, pp. 1-5, 2023.

http://dx.doi.org/10.1109/IPFA58228.2023.10249125

[14] S. Han, and A.J. Flewitt, "The Origin of the High Off-State Current in p-Type Cu 2 O Thin Film Transistors", *IEEE Electron Device Lett.,* vol. 38, no. 10, pp. 1394-1397, 2017.

http://dx.doi.org/10.1109/LED.2017.2748064

[15] Ching-Fa Yeh, Shyue-Shyh Lin, Tzung-Zu Yang, Chun-Lin Chen, and Yu-Chi Yang, "Performance and off-state current mechanisms of low-temperature processed polysilicon thin-film transistors with liquid phase deposited SiO/sub 2/ gate insulator", *IEEE Trans. Electron Dev.,* vol. 41, no. 2, pp. 173-179, 1994.

http://dx.doi.org/10.1109/16.277383

[16] J. Park, K.S. Jang, D.G. Shin, M. Shin, and J.S. Yi, "Gate-induced drain leakage current characteristics of p-type polycrystalline silicon thin film transistors aged by off-state stress", *Solid-State Electron.,* vol. 148, pp. 20-26, 2018.

http://dx.doi.org/10.1016/j.sse.2018.07.009

[17] Y.H. Tai, S.C. Huang, C.W. Lin, and H.L. Chiu, "Degradation of the Capacitance-Voltage Behaviors of the Low-Temperature Polysilicon TFTs under DC Stress", *J. Electrochem. Soc.,* vol. 154, no. 7, p. H611, 2007.

http://dx.doi.org/10.1149/1.2735921

[18] K.J. Yang, and Chenming Hu, "MOS capacitance measurements for high-leakage thin dielectrics", *IEEE Trans. Electron Dev.,* vol. 46, no. 7, pp. 1500-1501, 1999.

http://dx.doi.org/10.1109/16.772500

[19] Sangwon Lee, Sungwook Park, Sungchul Kim, Yongwoo Jeon, Kichan Jeon, Jun-Hyun Park, Jaechul Park, Ihun Song, and Youngsoo Park, "Extraction of Subgap Density of States in Amorphous InGaZnO Thin-Film Transistors by Using Multifrequency Capacitance–Voltage Characteristics", *IEEE Electron Device Lett.,* vol. 31, no. 3, pp. 231-233, 2010.

http://dx.doi.org/10.1109/LED.2009.2039634

[20] Z.Y. Hsieh, M.C. Wang, S.Y. Chen, C. Chen, and H.S. Huang, "Gate-to-drain capacitance verifying the continuous-wave green laser crystallization n-TFT trapped charges distribution under dc voltage stress", *Appl. Phys. Lett.,* vol. 95, no. 25, 2009.253503

http://dx.doi.org/10.1063/1.3275728

[21] M. Bae, D. Yun, Y. Kim, D. Kong, H.K. Jeong, W. Kim, J. Kim, I. Hur, D.H. Kim, and D.M. Kim, "Differential Ideality Factor Technique for Extraction of Subgap Density of States in Amorphous InGaZnO Thin-Film Transistors", *IEEE Electron Device Lett.,* vol. 33, no. 3, pp. 399-401, 2012.

http://dx.doi.org/10.1109/LED.2011.2182602

[22] S.C. Witczak, J.S. Suehle, and M. Gaitan, "An experimental comparison of measurement techniques to extract Si-SiO2 interface trap density", *Solid-State Electron.,* vol. 35, no. 3, pp. 345-355, 1992.

http://dx.doi.org/10.1016/0038-1101(92)90238-8

[23] J. Rhayem, D. Rigaud, M. Valenza, N. Szydlo, and H. Lebrun, "1/f Noise in amorphous silicon thin film transistors: effect of scaling down", *Solid-State Electron.,* vol. 43, no. 4, pp. 713-721, 1999.

http://dx.doi.org/10.1016/S0038-1101(98)00324-4

[24]   D. Rigaud, J. Rhayem, and M. Valenza, "Low frequency noise in thin film transistors", *IEE Proc., Circ. Devices Syst.,* vol. 149, no. 1, pp. 75-82, 2002.
      http://dx.doi.org/10.1049/ip-cds:20020063

[25]   C.A. Dimitriadis, J. Brini, J.I. Lee, F.V. Farmakis, and G. Kamarinos, "1/f γ noise in polycrystalline silicon thin-film transistors", *J. Appl. Phys.,* vol. 85, no. 7, pp. 3934-3936, 1999.
      http://dx.doi.org/10.1063/1.369770

[26]   J. Song, K. Shu, J. Wang, X. Wang, N. Huo, and R. Notzel, "Origin of the Low-Frequency 1/f Noise of a Photoelectrochemical Photodetector", *IEEE Trans. Electron Dev.,* vol. 69, no. 11, pp. 6184-6187, 2022.
      http://dx.doi.org/10.1109/TED.2022.3206181

[27]   S. Martin, A. Dodabalapur, Z. Bao, B. Crone, H.E. Katz, W. Li, A. Passner, and J.A. Rogers, "Flicker noise properties of organic thin-film transistors", *J. Appl. Phys.,* vol. 87, no. 7, pp. 3381-3385, 2000.
      http://dx.doi.org/10.1063/1.372354

[28]   S. Jeon, S.I. Kim, S. Park, I. Song, J. Park, S. Kim, and C. Kim, "Low-Frequency Noise Performance of a Bilayer InZnO–InGaZnO Thin-Film Transistor for Analog Device Applications", *IEEE Electron Device Lett.,* vol. 31, no. 10, pp. 1128-1130, 2010.
      http://dx.doi.org/10.1109/LED.2010.2059694

[29]   Y. Yang, M. Zhang, L. Lu, M. Wong, and H.S. Kwok, "Low-Frequency Noise in Bridged-Grain Polycrystalline Silicon Thin-Film Transistors", *IEEE Trans. Electron Dev.,* vol. 69, no. 4, pp. 1984-1988, 2022.
      http://dx.doi.org/10.1109/TED.2022.3148697

[30]   D. Zhang, H. Wang, Y. Feng, X. Wang, G. Liu, K. Han, X. Gong, J. Liu, X. Zhan, and J. Chen, "Fast Fourier Transform (FFT) Using Flash Arrays for Noise Signal Processing", *IEEE Electron Device Lett.,* vol. 43, no. 8, pp. 1207-1210, 2022.
      http://dx.doi.org/10.1109/LED.2022.3183111

[31]   W. Shin, J. Kim, G. Jung, S. Ju, S.H. Park, Y. Jeong, S. Hong, R.H. Koo, Y. Yang, J.J. Kim, S. Han, and J.H. Lee, "In-Memory-Computed Low-Frequency Noise Spectroscopy for Selective Gas Detection Using a Reducible Metal Oxide", *Adv. Sci. (Weinh.),* vol. 10, no. 7, 2023.2205725
      http://dx.doi.org/10.1002/advs.202205725 PMID: 36646505

[32]   G. Giusi, O. Giordano, G. Scandurra, C. Ciofi, M. Rapisarda, and S. Calvi, "Automatic measurement system for the DC and low-f noise characterization of FETs at wafer level", *in 2015 IEEE International Instrumentation and Measurement Technology Conference (I2MTC) Proceedings,* pp. 2095-2100, 2015.
      http://dx.doi.org/10.1109/I2MTC.2015.7151606

[33]   T.C. Fung, G. Baek, and J. Kanicki, "Low frequency noise in long channel amorphous In–Ga–Zn–O thin film transistors", *J. Appl. Phys.,* vol. 108, no. 7, 2010, 074518.
      http://dx.doi.org/10.1063/1.3490193

[34] Z. Bai, T. Gong, X. Duan, J. Wang, K. Xiao, D. Geng, and L. Li, "Low Frequency Noise of Channel-All-Around (CAA) InGaZnO Field Effect Transistors", *IEEE Electron Device Lett.*, vol. 43, no. 12, pp. 2117-2120, 2022.
http://dx.doi.org/10.1109/LED.2022.3216609

[35] R.H. Kingston, Ed., *Semiconductor Surface Physics..* University of Pennsylvania Press, 1957.
http://dx.doi.org/10.9783/9781512803051

[36] F. Crupi, G. Giusi, G. Iannaccone, P. Magnone, C. Pace, E. Simoen, and C. Claeys, "Analytical model for the 1/f noise in the tunneling current through metal-oxide-semiconductor structures", *J. Appl. Phys.,* vol. 106, no. 7, 2009.073710
http://dx.doi.org/10.1063/1.3236637

[37] C. Surya, and T.Y. Hsiang, "A thermal activation model for $1/f\gamma$ noise in Si-MOSFETs", *Solid-State Electron.,* vol. 31, no. 5, pp. 959-964, 1988.
http://dx.doi.org/10.1016/0038-1101(88)90051-2

[38] C. Surya, and T.Y. Hsiang, "Surface mobility fluctuations in metal-oxide-semiconductor field-effect transistors", *Phys. Rev. B Condens. Matter,* vol. 35, no. 12, pp. 6343-6347, 1987.
http://dx.doi.org/10.1103/PhysRevB.35.6343 PMID: 9940867

[39] Y. Liu, S-T. Cai, C-Y. Han, Y-Y. Chen, L. Wang, X-M. Xiong, and R. Chen, "Scaling Down Effect on Low Frequency Noise in Polycrystalline Silicon Thin-Film Transistors", *IEEE J. Electron Devices Soc.,* vol. 7, pp. 203-209, 2019.
http://dx.doi.org/10.1109/JEDS.2018.2890737

[40] Y.B. Zhu, H. Xu, M. Li, M. Xu, and J-B. Peng, "Analysis of low frequency noise characteristics of praseodymium doped indium gallium oxide thin film transistor", *Wuli Xuebao,* vol. 70, no. 16, 2021.168501
http://dx.doi.org/10.7498/aps.70.20210368

[41] Y.H. Tai, C.Y. Chang, C.L. Hsieh, Y.H. Yang, W.K. Chao, and H.E. Chen, "Dependence of the Noise Behavior on the Drain Current for Thin Film Transistors", *IEEE Electron Device Lett.,* vol. 35, no. 2, pp. 229-231, 2014.
http://dx.doi.org/10.1109/LED.2013.2291565

[42] Y.H. Tai, C.Y. Chang, and C.L. Hsieh, "Effects of Illumination on the Noise Behavior of Amorphous Indium Gallium Zinc Oxide Thin Film Transistors", *J. Disp. Technol.,* vol. 12, no. 7, pp. 685-689, 2016.
http://dx.doi.org/10.1109/JDT.2016.2519045

[43] F.N. Hooge, "1/f noise", *Physica B+C,* vol. 83, no. 1, pp. 14-23, 1976.
http://dx.doi.org/10.1016/0378-4363(76)90089-9

[44] Y. Liu, H. He, R. Chen, Y.F. En, B. Li, and Y.Q. Chen, "Analysis and Simulation of Low-Frequency Noise in Indium-Zinc-Oxide Thin-Film Transistors", *IEEE J. Electron Devices Soc.,* vol. 6, no. 1, pp. 271-279, 2018.
http://dx.doi.org/10.1109/JEDS.2018.2800049

[45] Y. Yang, ""Characterization of Low-Frequency Noise in Polycrystalline Silicon Thin-Film Transistors under Different Temperature", *in 2021 9th International Symposium on Next Generation Electronics (ISNE),* pp. 1-4, 2021.

http://dx.doi.org/10.1109/ISNE48910.2021.9493595

[46]   W. Ye, Y. Liu, B. Wang, J. Huang, X. Xiong, and W. Deng, "Low-Frequency Noise Modeling of Amorphous Indium–Zinc-Oxide Thin-Film Transistors", *IEEE Trans. Electron Dev.,* vol. 69, no. 11, pp. 6154-6159, 2022.

http://dx.doi.org/10.1109/TED.2022.3206274

[47]   C.A. Dimitriadis, G. Kamarinos, and J. Brini, "Model of low frequency noise in polycrystalline silicon thin-film transistors", *IEEE Electron Device Lett.,* vol. 22, no. 8, pp. 381-383, 2001.

http://dx.doi.org/10.1109/55.936350

[48]   K.K. Hung, P.K. Ko, C. Hu, and Y.C. Cheng, "A unified model for the flicker noise in metal-oxide-semiconductor field-effect transistors", *IEEE Trans. Electron Dev.,* vol. 37, no. 3, pp. 654-665, 1990.

http://dx.doi.org/10.1109/16.47770

[49]   G. Ghibaudo, O. Roux, Ch. Nguyen-Duc, F. Balestra, and J. Brini, "Improved Analysis of Low Frequency Noise in Field-Effect MOS Transistors", In: *physica status solidi (a),* vol. 124. 1991, no. 2, pp. 571-581.

http://dx.doi.org/10.1002/pssa.2211240225

[50]   Y. Xu, T. Minari, K. Tsukagoshi, J. Chroboczek, F. Balestra, and G. Ghibaudo, "Origin of low-frequency noise in pentacene field-effect transistors", *Solid-State Electron.,* vol. 61, no. 1, pp. 106-110, 2011.

http://dx.doi.org/10.1016/j.sse.2011.01.002

[51]   R. Chen, W. Zhou, S. Zhao, M. Zhang, and H. S. Kwok, *Self-aligned Top-gate ZnO TFTs with Sputtered Al2O3 Gate Dielectric.* Kunshan, China, November, 2011.

[52]   W. Zhou, Z. Meng, S. Zhao, M. Zhang, R. Chen, M. Wong, and H-S. Kwok, "Bridged-Grain Solid-Phase-Crystallized Polycrystalline-Silicon Thin-Film Transistors", *IEEE Electron Device Lett.,* vol. 33, no. 10, pp. 1414-1416, 2012.

http://dx.doi.org/10.1109/LED.2012.2210019

[53]   M. Zhang, W. Zhou, and R. Chen, "Man Wong, and Hoi-Sing Kwok, "A Simple Method to Grow Thermal $\{\rm SiO\}_{2}$ Interlayer for High-Performance SPC Poly-Si TFTs Using $\{\rm Al\}_{2}\{\rm O\}_{3}$ Gate Dielectric,"", *IEEE Electron Device Lett.,* vol. 35, no. 5, pp. 548-550, 2014.

http://dx.doi.org/10.1109/LED.2014.2308527

[54]   X. Xu, M. Zhang, Y. Yang, L. Lu, F. Lin, M. Wong, and H-S. Kwok, "Realization of Visible-Light Detection in InGaZnO Thin-Film Transistor via Oxygen Vacancy Modulation through N 2 Treatments", *Adv. Electron. Mater.,* vol. 9, no. 10, 2023.2300351

http://dx.doi.org/10.1002/aelm.202300351

[55]   H. Xie, J. Xu, G. Liu, L. Zhang, and C. Dong, "Development and analysis of nitrogen-doped amorphous InGaZnO thin film transistors", *Mater. Sci. Semicond. Process.,* vol. 64, pp. 1-5, 2017.

http://dx.doi.org/10.1016/j.mssp.2017.03.003

[56]   S. H. Han, H. Kim, J.-H. Bae, S.-J. Choi, D. H. Kim, and D. M. Kim, "Photovoltaic Effect De-Embedded Photonic C–V Characterization of Subgap Density of States in Amorphous

Oxide Semiconductor Thin-Film Transistors", *IEEE Transactions on Electron Devices*, vol. 71, no. 11, pp. 6795-6798, 2024.

http://dx.doi.org/10.1109/ted.2024.3469161

[57] M.C. Nguyen, A.H.T. Nguyen, H. Ji, J. Cheon, J-H. Kim, K-M. Yu, S-Y. Cho, S-W. Kim, and R. Choi, "Application of Single-Pulse Charge Pumping Method on Evaluation of Indium Gallium Zinc Oxide Thin-Film Transistors", *IEEE Trans. Electron Dev.*, vol. 65, no. 9, pp. 3786-3790, 2018.

http://dx.doi.org/10.1109/TED.2018.2859224

[58] H. Kim, B. J. Kim, J. Oh, S.-Y. Choi, and H. Park, "Bi-directional threshold voltage shift of amorphous InGaZnO thin film transistors under alternating bias stress", *Semiconductor Science and Technology*, vol. 39, no. 2, 2024.

http://dx.doi.org/10.1088/1361-6641/ad1b15

[59] S. Li, M. Wang, D. Zhang, H. Wang, and Q. Shan, "A Unified Degradation Model of a-InGaZnO TFTs Under Negative Gate Bias With or Without an Illumination", *IEEE J. Electron Devices Soc.*, vol. 7, pp. 1063-1071, 2019.

http://dx.doi.org/10.1109/JEDS.2019.2946383

[60] X. Li, Z. Ma, J. Li *et al.*, "Heterojunction-engineered carrier transport in elevated-metal metal-oxide thin-film transistors", *Journal of Semiconductors*, vol. 45, no. 10, 2024.

http://dx.doi.org/10.1088/1674-4926/ 24040016

CHAPTER 4

Direct Current Voltage Stress-Induced Degradation in Thin-Film Transistors

Abstract. This chapter investigates the degradation induced by direct current（DC）voltage stress in thin-film transistors (TFTs). It provides a detailed examination of gate bias stress, hot-carrier effect, and self-heating effect, discussing their impact on silicon-based and metal oxide TFTs. The chapter aims to summarize and analyze the influence of these degradation mechanisms on the performance and reliability of TFT devices.

Keywords: DC voltage stress, Hot-carrier effect, Metal oxide TFT, Self-heating effect, Silicon-based TFT.

4.1. INTRODUCTION

In the context of thin-film transistor (TFT) reliability research, it is essential to consider the electrical reliability of TFTs, as these devices are inevitably subjected to various voltage conditions during operation. The study of TFT electrical reliability is both fundamental and widespread, focusing on the examination of device degradation under long-term application of electrical stress and the observation of the resulting degradation phenomena.

TFT electrical reliability can be categorized into two main types: direct current (DC) electrical stress reliability and alternating current (AC) electrical stress reliability. Research on DC electrical stress reliability forms the foundation of reliability studies involving the application of prolonged DC stress to the gate and drain electrodes of TFTs and monitoring the subsequent degradation. Different stress application methods can lead to various degradation scenarios, such as on-state current (I_{on}) degradation or subthreshold swing (SS) degradation. Therefore, to address TFT reliability issues, it is necessary to conduct targeted research into the physical mechanisms behind different degradation phenomena and to modify device structures or alter stress application methods to suppress device degradation.

Typically, degradation due to DC electrical stress reliability can be classified into several types: gate DC bias stress degradation, hot carrier (HC) degradation, and self-heating (SH) degradation. Gate DC bias stress degradation primarily encompasses both positive bias stress (PBS) and negative gate bias (NBS) degradation. This chapter will provide a summary and analysis of the impact of the

Meng Zhang & Mingxiang Wang

aforementioned three degradation mechanisms on silicon-based and metal oxide (MO) TFT devices.

4.2. GATE BIAS STRESS

In operation, the gate electrode of a TFT serves as the primary control element for modulating the device's on-off state and is typically subjected to various prolonged voltage stresses. Consequently, gate bias stress is often the most fundamental area of TFT reliability research. The standard method of applying stress involves imposing a long-duration gate bias while grounding the source electrode and drain electrode. Regardless of the biasing condition, either PBS or NBS, the most common degradation phenomenon is a shift in the threshold voltage (V_{th}). Typically, devices exhibit a positive shift under PBS [1-13] and a negative shift under NBS [1, 2, 4, 7, 14-22]. The mainstream explanation for gate bias stress involves carrier trapping at the interface or within the gate insulator (GI), where a high concentration of charged particles can shield the electric field of the gate, leading to premature or delayed device turn-on [4-6, 12, 14, 15].

However, the aforementioned mechanism alone cannot fully explain the entire range of degradation phenomena observed in TFTs under both PBS and NBS, especially for MO TFTs. In addition to the expected positive threshold voltage (V_{th}) shift during PBS [8, 9], some MO TFTs have exhibited a negative V_{th} shift [22]. Concurrently, the wide bandgap of MO TFTs makes it difficult to ionize holes directly under NBS. Therefore, a multitude of alternative explanations for gate bias stress degradation coexist alongside the mainstream mechanism [2, 21].

4.2.1 Positive Bias Stress

As previously mentioned in session xxx, PBS is typically characterized by the imposition of a positive voltage bias on the gate terminal of a transistor. Different types of TFT devices exhibit varying degradation phenomena and mechanisms under PBS. This section primarily discusses the PBS degradation of two types of TFT devices: silicon-based (Si-based) TFTs [1-3, 13] and MO TFTs [4-9, 12], which will be introduced separately in the following content.

4.2.1.1. Positive-Bias-Stress-Induced Degradation in Silicon-Based Thin-Film Transistor

After polycrystalline silicon (poly-Si) TFTs are subjected to PBS stress, the degradation is typically not pronounced, with the primary manifestation being a shift in the V_{th} [1]. The V_{th} shift is mainly due to electron injection into the GI layer, as previously discussed, which creates fixed charges. These fixed charges shield the electric field, leading to premature or delayed device turn-on and resulting in V_{th} shift. As mentioned in Chapter 2, degradation in SS and I_{on} is generally related to the generation of defect states. However, for poly-Si TFTs, the channel and interface do not have as many weak bonds, and PBS generally does not produce defect states or only a small number that are insufficient to cause SS and I_{on} degradation. Poly-Si TFTs typically require a larger bias voltage to exhibit significant degradation [2], which is usually analyzed in combination with additional conditions such as temperature or light exposure for reliability assessment [2].

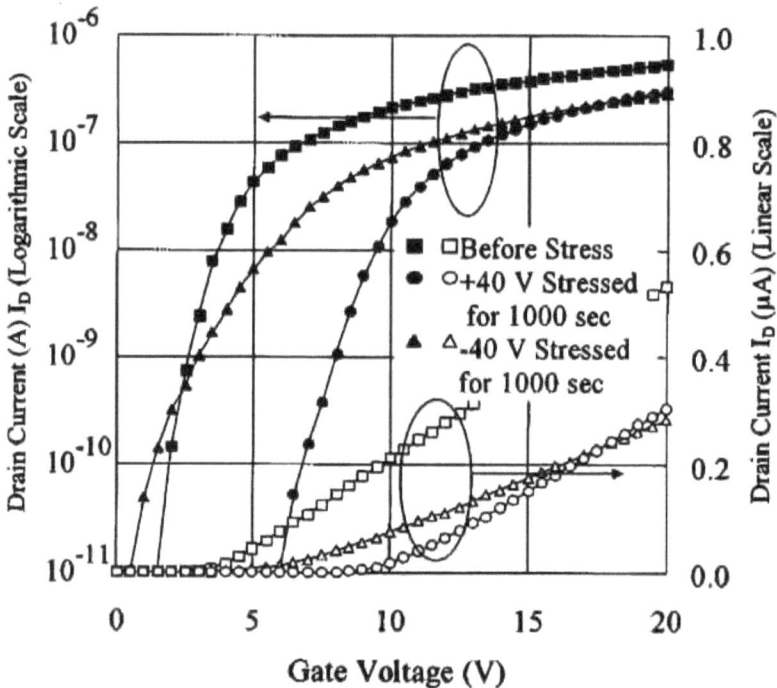

Fig. (4.1). Degradation of a typical a-Si:H TFT device under PBS and NBS stresses [1].

Thin-film transistors fabricated from hydrogenated amorphous silicon (a-Si:H) utilize silicon nitride as the GI and show different PBS reliability characteristics. Under PBS, in addition to a positive shift in V_{th}, there is also degradation in SS and I_{on}. Fig (**4.1**) displays the shift in the transfer characteristics of an a-Si:H thin-film transistor when subjected to a gate voltage stress for 1000 seconds, as observed at a constant drain voltage of 1 V, both pre- and post-stress application. It is evident that after a 40 V PBS stress, the device exhibits a significant positive shift in V_{th}, along with a noticeable change in SS, deviating notably from the typical response observed in poly-Si thin-film transistors under similar stress conditions. The primary mechanism for the positive V_{th} shift is still electron injection into the GI layer, leading to charge trapping. Charge trapping is associated with trap sites at the GI layer interface.

To further investigate the causes of *SS* degradation, an extraction analysis of *SS* was conducted, as shown in Fig. (**4.2**). According to the theory presented in Chapter 2, changes in *SS* indicate an increase in defect states in the vicinity of the conduction band within the TFT. For a-Si:H TFTs, there are many weak Si-H bonds within the material, which are prone to breaking under PBS stress, resulting in dangling bonds and other defect states. This ultimately leads to an increased density of defect states in a-Si:H. Moreover, the generation of defect states exhibits exponential characteristics with respect to time and a dependency on activation temperature, while charge-trapping behavior is characterized by a logarithmic relationship with time and exhibits minimal sensitivity to temperature variations.

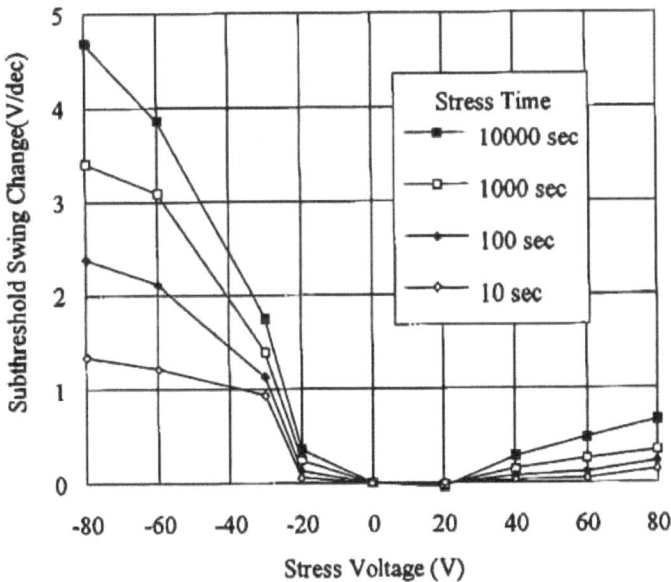

Fig. (4.2). The dependence of the *SS* change on the stress voltage for the a-Si:H TFTs stressed with different times [1].

In summary, the degradation phenomena and mechanisms related to V_{th} in Si-based TFTs are fundamentally consistent. However, a-Si:H TFTs, due to the presence of weak bonds such as Si-H, produce additional defect states under PBS stress compared to poly-Si TFTs [13]. This leads to more pronounced degradation phenomena, most notably SS and I_{on} degradation, which represent the most significant difference between the two types of Si-based TFTs.

4.2.1.2. Positive Bias Stress-Induced Degradation in Metal Oxide Thin-Film Transistor

Typically, two distinct degradation scenarios are observed in MO TFTs following PBS application. The first scenario involves a noticeable positive shift in the MO TFT transfer curve. The principle is akin to that of Si-based TFTs, primarily due to electron injection and accumulation [5, 6, 12], leading to field screening and consequently causing the device to turn on prematurely or be delayed.

However, unlike Si-based TFTs, MO TFTs also exhibit a significant negative shift in V_{th} after PBS, indicating that the degradation mechanism is primarily influenced by factors other than electrons [8, 9]. Fig. (**4.3**) presents a schematic diagram of the InSnZnO TFTs utilized in this research, highlighting a top-gate, bottom-contact configuration in Fig. (**4.3a** and **4.3b**) depicts the transfer curve degradation under a 25 V PBS, demonstrating a gradual negative shift with prolonged stress duration. This observation diverges from the common PBS-induced positive curve shift in InSnZnO TFTs, which previous literature has predominantly ascribed to electron interface trapping [6]. In contrast, the negative shift is linked to the ionization of water molecules [7, 11]. Additionally, after 100 seconds of stress, a minor hump in the current profile emerges in the subthreshold region, as indicated in the figure, suggesting the presence of two parallel transistors in the InSnZnO TFT with differing V_{th} values. To further analyze the degradation phenomena, low frequency noise (LFN) experiments were also conducted. It is important to note that a slight recovery of the device is observed during LFN measurements. For instance, subjected to a 25 V PBS for 100 seconds, the V_{on} exhibits a recovery of approximately 4.88%. Fig. (**4.3c** and **4.3d**) display the extracted I_{on} degradation and V_{on} shift as functions of stress time under various PBS conditions varying from 10 V to 30 V. Interestingly, an increase in the gate-source voltage (V_{gs}) intensifies both the I_{on} degradation and the negative V_{on} shift. The negative trend observed in the transfer curve is hypothesized to be influenced by the electric field (E_y) and the ionization of water molecules.

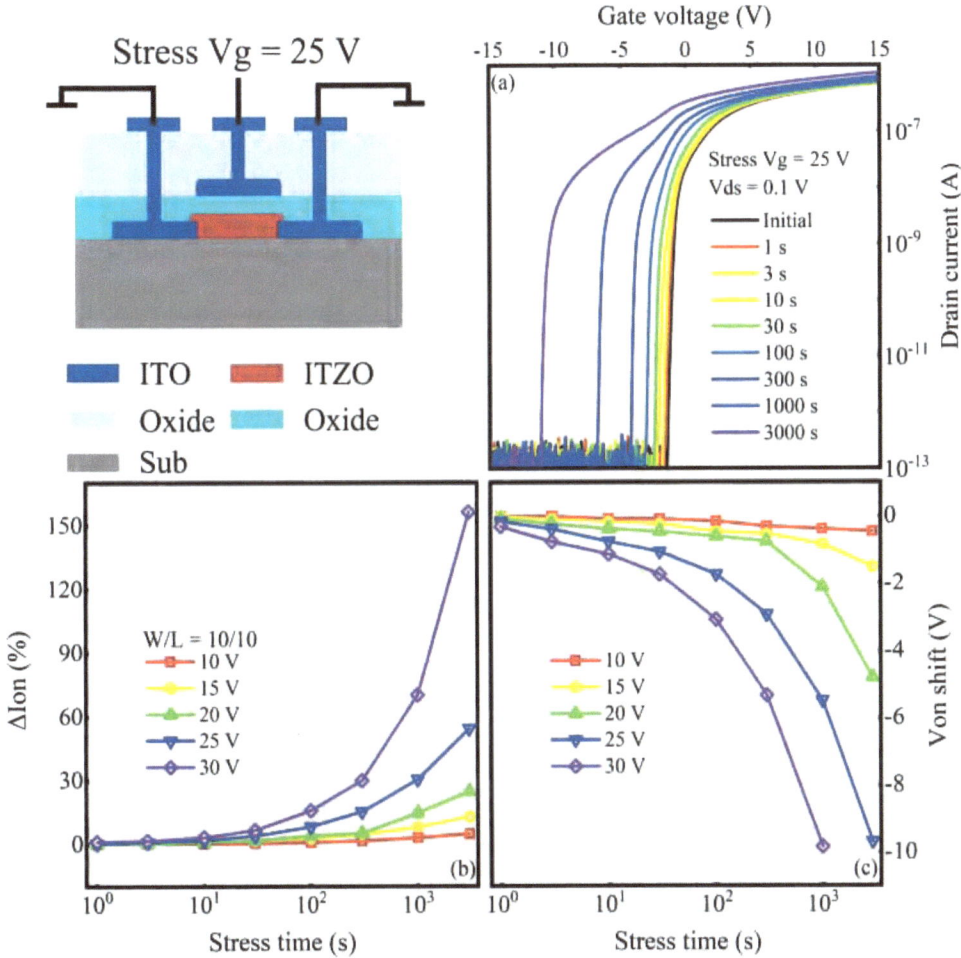

Fig. (4.3). (a) Schematic diagram illustrating the InSnZnO TFT and the applied PBS conditions. (b) Time-dependent evolution of the transfer curves of InSnZnO TFTs subjected to PBS. (c) ΔI_{on} and (d) V_{on} shifts as functions of stress time under different PBS conditions [9].

Following an assessment of transfer curve degradation, the LFN characteristics of InSnZnO TFTs under PBS were scrutinized, as depicted in Fig. (**4.4, 4.4a** and **4.2.7b**) present the S_{ID} dependent on f under a 25 V PBS. The S_{ID} measurements adhere to a $1/f^{\sim 1}$ relationship, signifying that the LFN in InSnZnO TFTs is predominantly influenced by the classical $1/f$ noise mechanism. However, a closer examination of Fig. (**4.4a** and **4.4b**) reveals an unexpected trend: the LFN level decreases within the first 300 seconds of PBS application, then increases, suggesting the involvement of different mechanisms. Fig. (**4.4c**) shows the normalized S_{ID} as a function of frequency for various PBS conditions after 30 seconds of stress. Interestingly, the LFN level decreases as V_{gs} increases. After a stress duration of 1000 seconds, the LFN level exhibits an initial decline followed by an ascent as the V_{gs} increases, as illustrated in Fig. (**4.4d**).

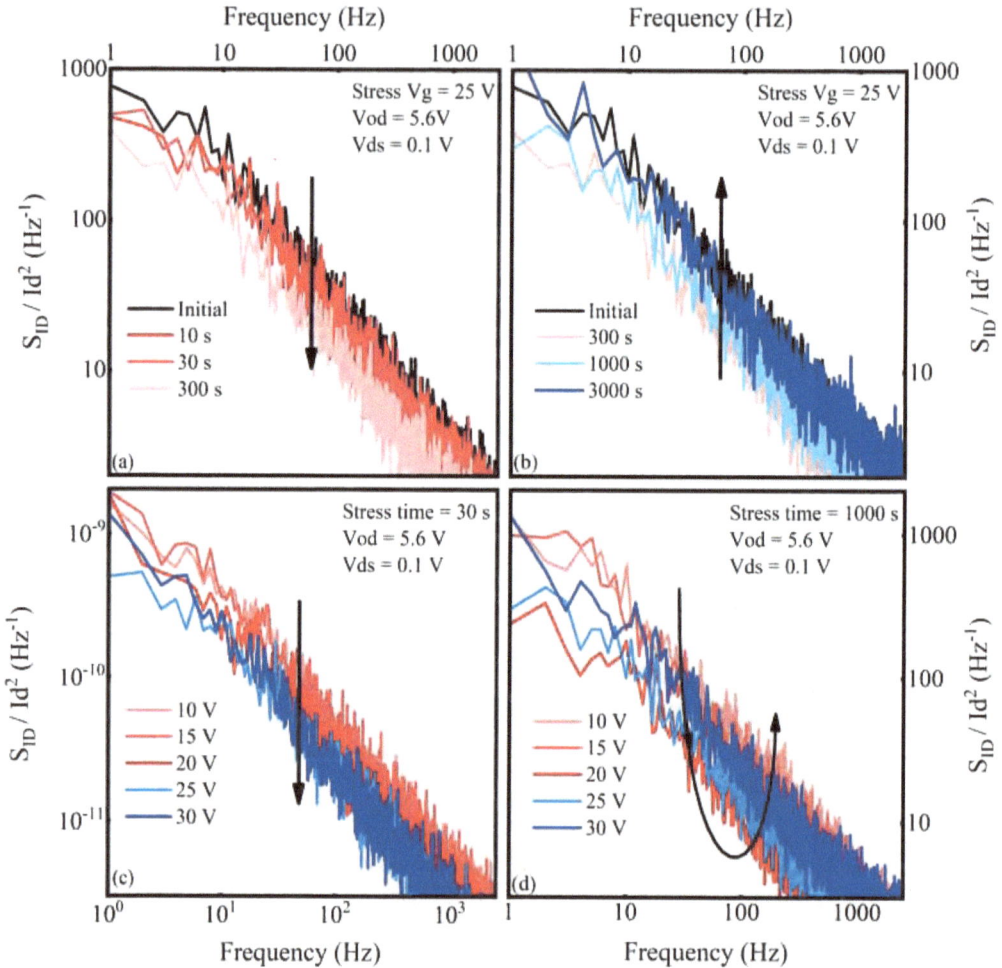

Fig. (4.4). Normalized S_{ID} as a function of f in InSnZnO TFTs under (a) PBS for up to 300 s and (b) PBS between 300 s and 3000 s. Normalized S_{ID} as a function of dependent f under various PBSs with stress time of (c) 30 s and (d) 1000 s [9].

As discussed in the previous noise analysis section, LFN testing can help differentiate between interface state defects and bulk defects in TFTs to some extent [23-27]. According to the formulas from Chapter 3, different trends can be observed for S_{id}/I_d^2 as a function of V_{od}. Under different models, there are distinct differences; S_{id}/I_d^2 exhibits a −2 power relationship in ΔN (interface state density) [26] and a −1 power relationship in $\Delta\mu$ (interface potential) [27]. It is observed that for stress V_g less than 20 V, the interface state density (N_t) counterintuitively diminishes as the stress duration extends. In contrast, when stress V_g exceeds 20 V, N_t undergoes an initial decline followed by an increase over the stress period, mirroring the pattern observed in the temporal evolution of the normalized S_{ID} over time (Fig. **4.5a** and **4.5b**). At shorter stress times, a higher stress voltage results in a lower N_t (Fig. **4.5c**). However, at longer stress times, N_t initially decreases and then rises as the V_g increases (Fig. **4.4d**), indicating a transition from interface state defects to bulk defects as the dominant defect states within the device.

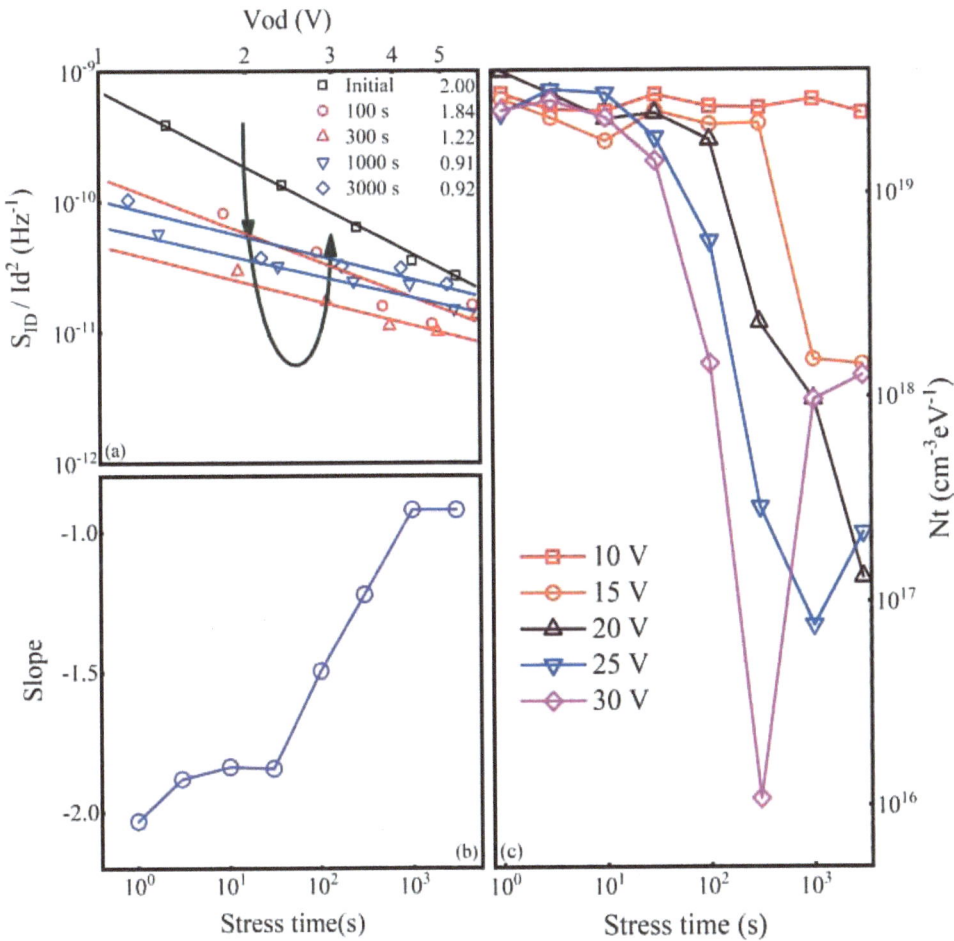

Fig. (4.5). (a) Normalized S_{ID} as a function of V_{od} in InSnZnO TFTs under PBS of 25 V at various stress times (symbols: experimental data, lines: fitted data). The f is fixed at 100 Hz. The inset shows the extracted slope as a function of stress time. (b) N_t as a function of stress time under different PBS conditions [9].

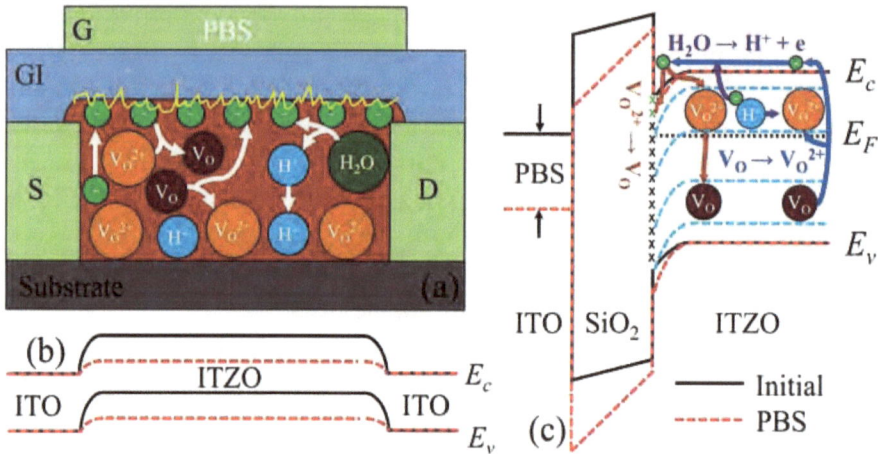

Fig. (4.6). (a) Schematic diagram illustrating the degradation model in InSnZnO TFTs under PBS. Energy band diagrams are shown (b) from the gate to the channel and (c) from the source to the drain [9].

To elucidate the degradation phenomena induced by PBS, a conceptual model has been formulated and is depicted in Fig. (**4.6**). InSnZnO thin film contains numerous oxygen vacancies (V_os) and ionized oxygen vacancies (V_o^{2+}s). These vacancies maintain a dynamic balance through [8]

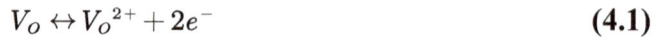

$$V_O \leftrightarrow V_O^{2+} + 2e^- \tag{4.1}$$

Upon the application of PBS, a significant E_y is induced, causing a downward bending of the energy bands as shown in Fig. (**4.6b**). This bending facilitates the accumulation of electrons at the interface between the channel and the oxide layer, which in turn leads to the neutralization of V_o^{2+} at the interface. Simultaneously, the migration of electrons from the backchannel to the interface results in the creation of additional V_o^{2+}s in bulk, thereby reducing the source/drain potential barriers, as illustrated in Fig. (**4.6c**). This mechanism contributes to the negative shift observed in the transfer curve, as depicted in Fig. (**4.3b**). Additionally, the elevated electric field also drives the ionization of V_o into V_o^{2+}, a process further discussed in [10]

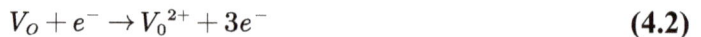

$$V_O + e^- \rightarrow V_O^{2+} + 3e^- \tag{4.2}$$

Consequently, the ionization process augments the concentration of donor-like states within the bandgap, which exacerbates the negative shift in the transfer curve, as observed in Fig. (**4.3b**).

An additional factor contributing to the negative shift in the transfer curve of InSnZnO TFTs under PBS is water ionization. The passivation layer, commonly PECVD SiO_2, may possess porosity and pinholes, allowing ambient moisture to permeate the device structure. When subjected to a high E_y, this moisture can undergo ionization, influenced by several studies [7, 11].

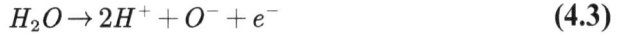

$$H_2O \rightarrow 2H^+ + O^- + e^- \qquad (4.3)$$

The ionization of water introduces additional electrons, which elevate the free carrier concentration within the channel and consequently result in a negative shift of the transfer curve, as depicted in Fig. (**4.3b**). Simultaneously, the migration of H^+ under E_y exacerbates this negative shift, as illustrated in the same figure. An increased stress V_g amplifies E_y, leading to a more pronounced negative curve shift, evident in Fig. (**4.3c** and **4.3d**). To counteract the ingress of water into the device structure, enhancing the gate dielectric and passivation layer quality presents a viable solution.

Additionally, with prolonged PBS exposure, an increasing number of H^+s and V_o^{2+} ions accumulate in the back channel of InSnZnO TFTs, leading to the formation of a parasitic transistor. The parasitic transistor, when activated prior to the primary transistor, can induce a hump current within the subthreshold region, as shown in Fig. (**4.3b**). After PBS is removed, the spontaneous rebalancing of the $\Delta\mu$ is primarily driven by scattering effects, which are significantly enhanced by defects in the bulk. "In InSnZnO TFTs without stress, interface states predominantly govern the LFN characteristics (Fig. **4.5a**). Initially, during PBS application, the additional electrons resulting from band bending and water ionization get trapped, leading to a reduction in the interface trap density, leading to a decrease in the LFN level and diminishing the influence of the ΔN model. Concurrently, the generated V_o^{2+} and H+ diffuse into the bulk [21], where they serve as donor-like states [12], enhancing electron scattering and consequently elevating the LFN level. As a result, the $\Delta\mu$ model starts to contribute. Throughout the PBS process, the ΔN and $\Delta\mu$ models compete. As stress time increases, the generation of more donor-like states in the bulk results in a continuous increase in the slope of S_{ID}/I_d^2 versus V_{od} (Fig **4.5a**). Simultaneously, the decrease in interface trap states further increases this slope (Fig. **4.5a**). At higher PBS levels, the initial decrease in LFN is primarily attributed to the mitigation of interface state density (Fig. **4.4a** and **4.5b**), with the ΔN model initially dominating (Fig. **4.5a**). As stress time continues, the accumulation of bulk defects from V_o^{2+} and H^+ increases (Fig. **4.4b** and **4.5b**), leading to the $\Delta\mu$ model becoming more significant and causing changes in the slope (Fig. **4.5a**). For lower PBS levels, an eventual reversal point emerges as the stress duration is prolonged, indicative of the stabilization of the device's initial interface states.

4.2.2. Negative Bias Stress

Device performance deterioration under NBS often occurs as a result of the imposition of a negative voltage on the gate with zero bias on the drain and source, leading to reliability concerns. This condition generally results in a negative shift in the device's V_{th}. Previous research suggested that the cause of NBS degradation was similar to that of PBS, primarily due to hole injection into the GI layer, leading to field screening and causing the device to turn on prematurely or be delayed [4, 18, 20]. However, later studies indicated that NBS degradation is caused by the injection of positively charged ions or charges, with a detailed analysis provided in the MO TFT section of this chapter. This section will be divided into two parts to discuss the common degradation and mechanisms of NBS. It will first introduce the NBS degradation in Si-based TFTs [1, 2, 14-17], followed by the typical [4, 7, 18, 20, 21] and special [19, 22] degradation cases of MO TFTs under NBS. Through the NBS degradation mechanisms of MO TFTs, we can analyze whether holes or other positive ions are responsible for the V_{th} shift.

4.2.2.1. Negative Bias Stress-Induced Degradation in Silicon-Based Thin-Film Transistor

Fig. (**4.7a**) displays the characteristic degradation pattern of the transfer curve in an a-Si TFT subjected to a V_g stress of -30 V for 3000 s, measured at $V_d = 0.5$ V and 10 V. The curve measured at $V_d = 0.5$ V is presented in both linear and semilogarithmic scales, revealing a positive shift in V_{th} and a degradation in SS. This suggests the creation of deep energy states in the upper region of the bandgap, potentially resulting from the disruption of vulnerable Si-Si bonds [14]. Additionally, the extracted μ_{FE} from the transconductance measurements shows a slight decrease (approximately 10%) after stress, indicating the creation of a tail state. Both the degradation in V_{th} and μ_{FE} contribute to the observed decrease in I_{on}. Furthermore, as depicted in the inset figure, there is a noticeable reduction in the off-state current (I_{off}) following the stress application, which is likely related to the creation of channel defect states. With the increased presence of these defect states in the channel, the Fermi level is positioned deeper within the bandgap at a given V_g in the OFF region, moving further away from the valence band edge. Consequently, fewer holes are available to contribute to I_{off} in the stressed TFTs, leading to the observed reduction in I_{off}. Thus, the degradation of both Ion and Ioff under DC stress is ascribed to the mechanism of defect state formation.

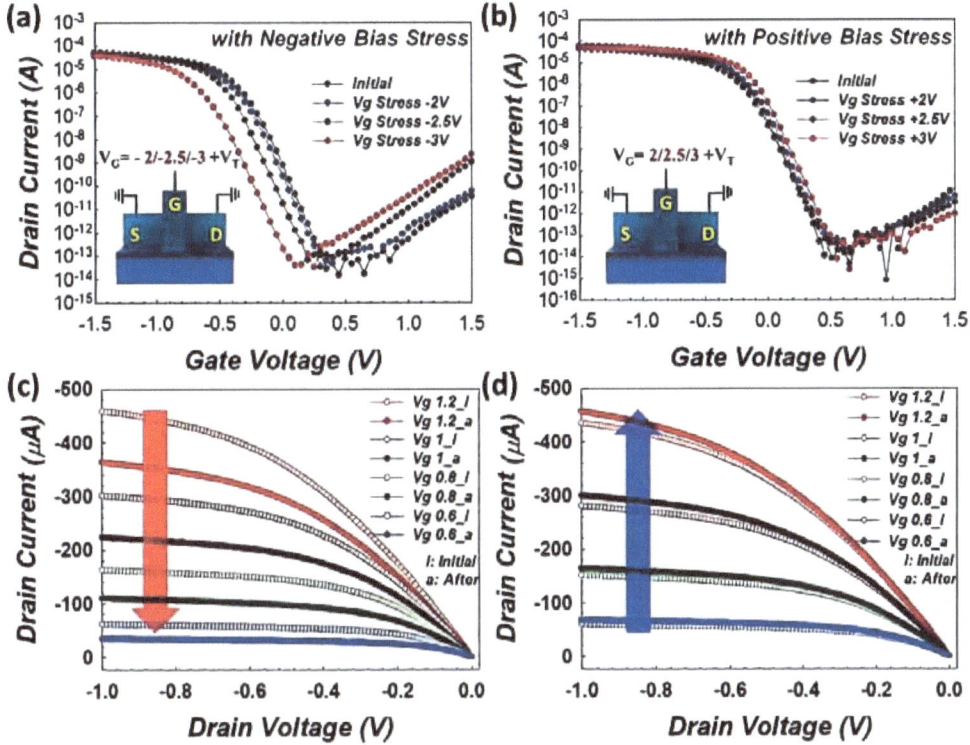

Fig. (4.7). Device I_d–V_d results after various (a) NBS and (b) PBS operations. Note that the V_{th} shift direction and leakage current exhibit different trends along with various NBS and PBS operations. The output currents of I_d–V_d with different gate biases are also shown in (c) NBS and (d) PBS operations to the initial and after-stress operations [2].

As previously mentioned, early research believed that the positive ions injected during NBS stress were holes. For poly-Si TFTs, the degradation caused by NBS generally manifests as a V_{th} shift with minimal I_{on} degradation and a slight increase in I_{off}, as shown in Fig. (**4.8**) under NBS conditions with $V_g = -37$ V. The primary mechanism is the trapping of carriers in the GI layer, resulting in a field screening effect. LFN analysis of poly-Si TFTs also reveals that a small number of interface states can cause NBS degradation. Under negative bias stress, Si-OH bonds in poly-Si TFTs can be broken, forming Si and negative charges OH⁻ [17]. Si forms interface states, and some OH⁻ are captured by oxide traps, creating fixed negative charges in the GI layer.

In summary, the degradation phenomena in Si-based TFTs under NBS are relatively similar, with the main mechanisms being carrier injection and trapping in the GI layer, leading to field screening.

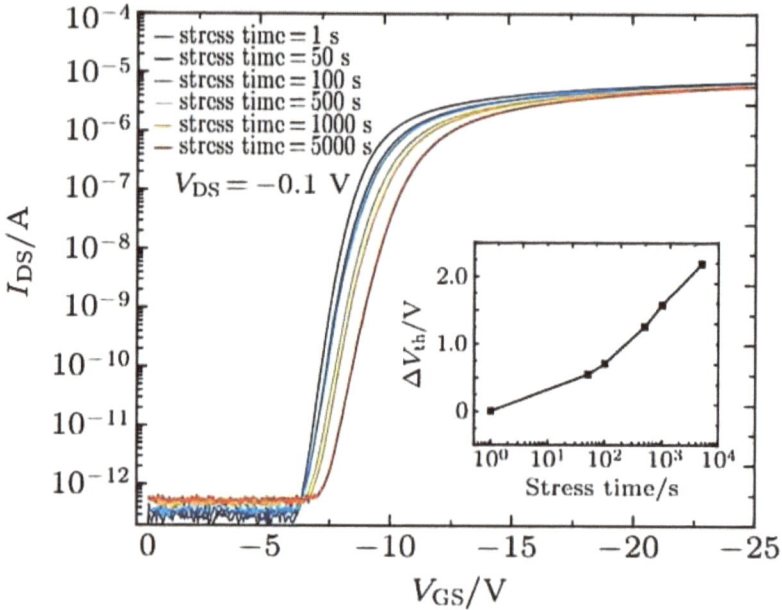

Fig. (4.8). Transfer characteristics of poly-Si TFT measured at different stress durations [17].

4.2.2.2 Negative Bias Stress-Induced Degradation in Metal Oxide Thin-Film Transistor

Unlike Si-based TFTs, the degradation mechanisms at play in MO TFTs under NBS are considerably more complex. Previous research has posited that charge trapping in the GI serves as an explanatory model for the observed V_{th} shift in transistors [4, 20, 21]. As depicted in Fig (**4.9a**), the presence of positive charges trapped at the GI interface results in the accumulation of electrons in proximity to the GI when the V_g is 0 V. Potential sources of these positive charges include holes, H+, and V_o^{2+}. However, for MO TFTs, which have a wider bandgap, holes are typically difficult to ionize. There has been doubt expressed about the extent to which charge trapping at the GI interface can explain the observed behavioral discrepancies between high- and low-mobility amorphous oxide semiconductors (AOSs). To validate the charge-trapping model, it would be necessary to satisfy specific prerequisites: AOSs with higher mobility should demonstrate elevated hole concentrations or enhanced mobility; they should exhibit augmented hydrogen or proton conductivity; and they should possess a considerably reduced hole-injection barrier at the GI interface. However, upon investigation, these criteria proved to be implausible [21]. For instance, IGO and ITZO have similar ionization potentials of approximately 7.2 eV. The analysis suggests that both materials exhibit a similar hole-injection barrier height at the GI interface. Furthermore, it was determined that the quantities of hydrogen incorporated in IGO and TZO are closely comparable. As a result, the hypothesis of charge trapping at the GI interface was discounted. Instead, attention was redirected towards investigating the likelihood of Fermi-level perturbations or alterations in the carrier concentration within the AOS material itself, as illustrated in Fig. (**4.9b**). Further research has indicated that NBS degradation in MO TFTs is primarily driven by V_Os. A unified model has been proposed to elucidate this correlation, addressing the degradation phenomena linked to both NBS and negative bias illumination stress (NBIS). This model posits that the formation of negative charges and V_o^{2+} at the channel/GI interface drives the respective positive and negative shifts in V_{th} [19].

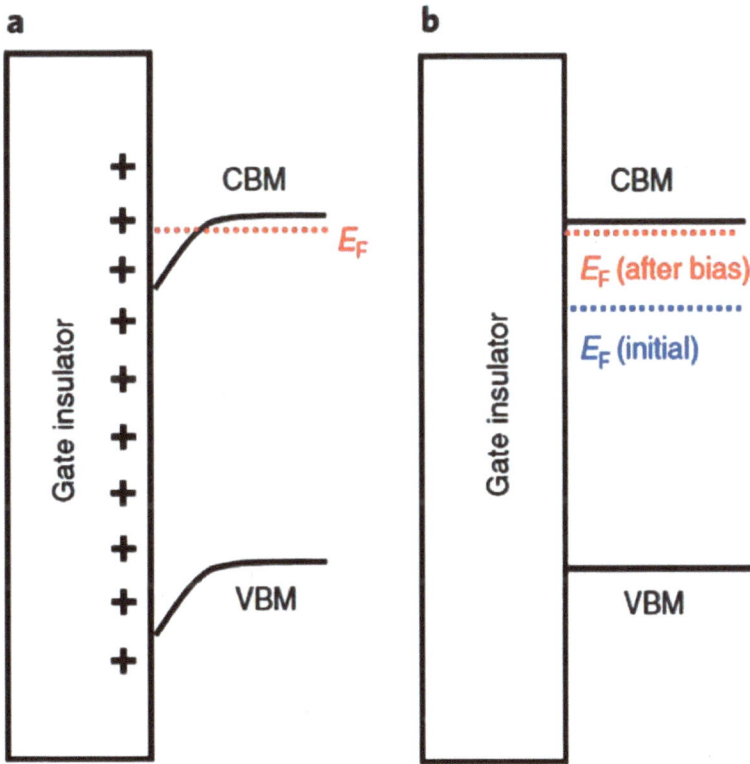

Fig. (4.9). Schematic of two possible NBS mechanisms. (a), Gate-insulator charge-trapping model. (b), AOS Fermi-level shift model [21].

Recent studies have revealed that the NBS degradation in MO TFTs also involves the HC mechanism [22]. As illustrated in Fig (**4.10**), taking the InSnZnO TFT as an example, the MO TFT demonstrates the various degradation phenomena and mechanisms under NBS. During the NBS experiments in InSnZnO TFTs, a two-stage degradation pattern is observed: I_{on} first shows a slight increase in the initial stage, followed by a marked decrease in the subsequent stage.

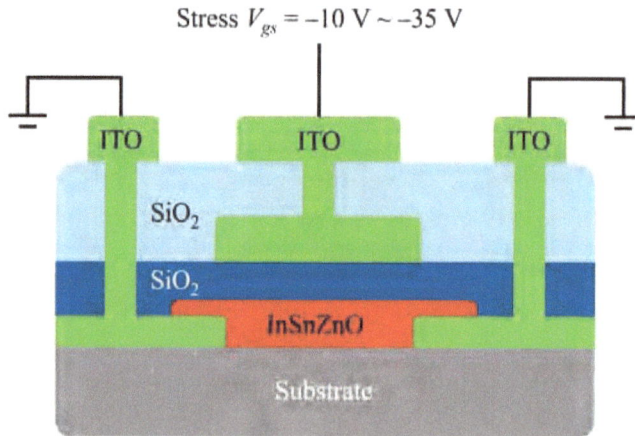

Fig. (4.10). Schematic of the InSnZnO TFT and the NBS conditions [22].

As shown in Fig. (**4.11a**), the temporal progression of the transfer curve for an InSnZnO TFT under NBS at $V_{gs} = -35$ V reveals a distinct trend. I_{on} experiences a substantial decrease as the stress duration increases, whereas the SS exhibits a minor change. This degradation behavior resembles the HC degradation observed in poly-Si TFTs, suggesting that an HC effect may also be at play in InSnZnO TFTs under NBS.

A detailed analysis of the transfer curve degradation indicates that the NBS in InSnZnO TFTs presents a two-stage degradation pattern. Initially, I_{on} shows a slight increase before experiencing a dramatic decrease, as illustrated in Fig. (**4.11b**). Notably, the recovery of I_{on} after NBS removal is minimal. For instance, after 10 seconds of NBS at $V_{gs} = -30$ V, I_{on} recovers only by 6.67% after 24 hours. InGaZnO TFTs also display an initial surge in I_{on} under NBS, a phenomenon believed to stem from the bending of the energy bands. The degradation behavior under various NBS conditions is also analyzed (Fig. **4.11b**). A reduction in $|V_{gs}|$ results in both a smaller initial increase in I_{on} and a less pronounced decrease in I_{on} during the second stage. Additionally, the degradation response under NBS is significantly distinct from that observed under PBS, where the transfer curve experiences a persistent negative shift.

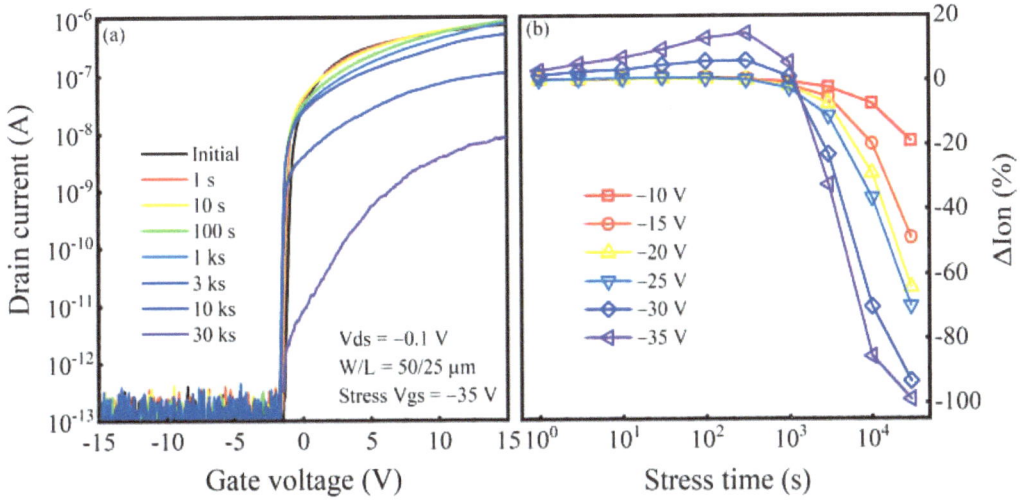

Fig. (4.11). (a) Temporal evolution of the transfer curve for an InSnZnO TFT under stress $V_{gs} = -35$ V. (b) I_{on} degradation as a function of stress time under various NBS conditions [22].

It is thus cautiously deduced that the latter phase of degradation might be correlated with an HC effect, potentially instigating the creation of defects, thereby reducing I_{on}. In contrast, the initial stage of degradation might be due to energy band distortion, which alters carrier concentration and consequently increases I_{on}.

To validate the aforementioned inference, a device simulation utilizing Silvaco ATLAS was performed, with a particular emphasis on examining the lateral electric field (E_x) and the vertical electric field (E_y). Fig. (4.12a) depicts the E_x distribution under a $V_{gs} = -35$ V. It is evident that the negative V_{gs} results in a significant E_x within the channel, in close proximity to both the drain and source regions. The distribution of E_x on either side is roughly 0.5 μm in width (w). For further clarity, Fig. (4.12b) shows the extracted E_x profile along the channel, extending from the drain to the source. The $|E_x|$ increases as one moves closer to either the drain or the source, and a higher $|V_{gs}|$ results in a larger $|E_x|$. The high E_x can accelerate carriers into HCs, leading to defect creation and degradation of device performance [28, 29, 30]. One might question the source of these carriers, considering the channel remains off during NBS.

To distinguish the defect types produced during the second stage of HC degradation, the density of states (DOS) is performed. Fig. (4.13a) displays the experimental transfer curves with those fitted post-NBS, showing a close match between the fitted and experimental curves, indicating a good fit. Fig. (4.13b) illustrates the simulated DOS for InSnZnO TFTs before and after NBS, indicating that NBS predominantly leads to a proliferation of acceptor tail states [31].

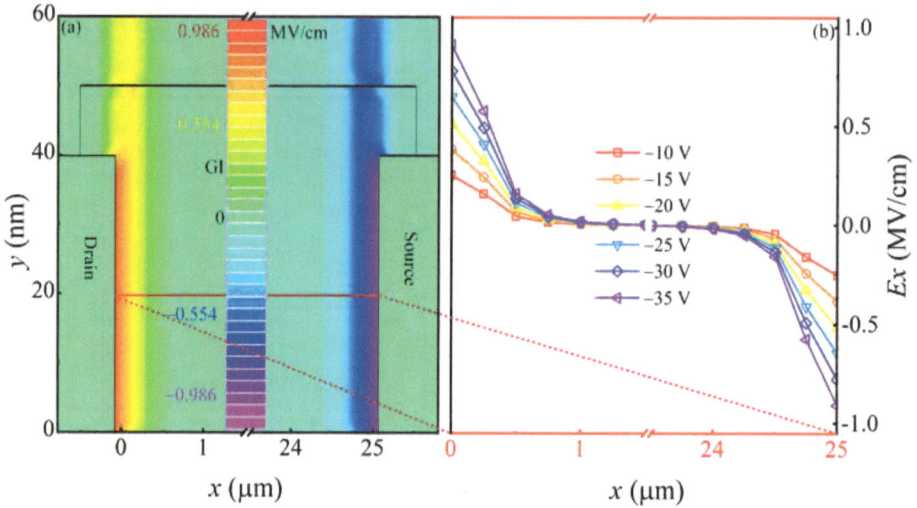

Fig. (4.12). Calculated E_x distribution under a stress voltage $V_{gs} = -35$ V. (b) Extracted E_x values at various stress V_{gs}s along the middle of the channel from the drain to the source [22].

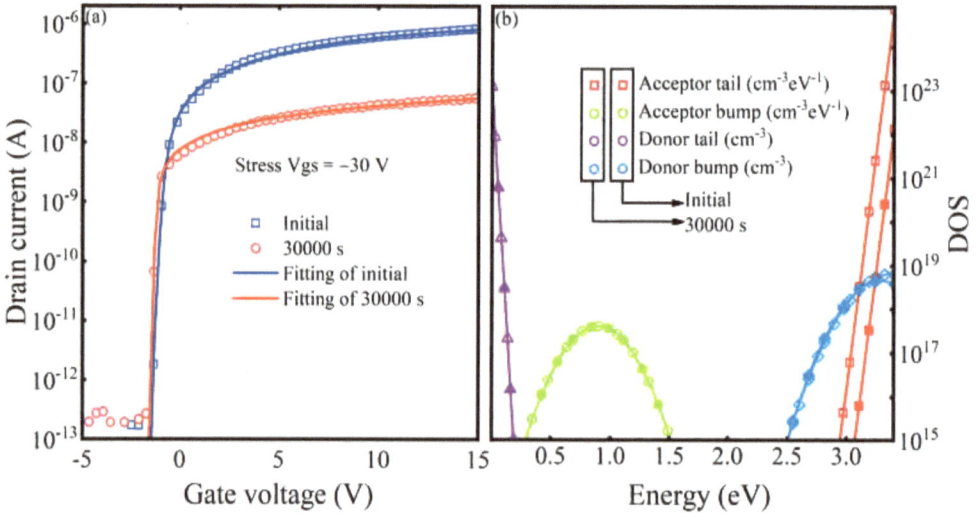

Fig. (4.13). (a) Experimental and fitted transfer curves before and after 30,000 seconds of stress at $V_{gs} = -30$ V. (b) Simulated DOS in InSnZnO TFTs before and after 30000 s of stress at $V_{gs} = -30$ V [22].

To elucidate the dual-stage degradation mechanism, a model has been formulated, drawing insights from experimental data and simulation outcomes. Upon the application of NBS to the gate electrode of nSnZnO TFTs, an E_y is induced, which causes the energy bands in the semiconductor to bend upwards, as depicted in Fig. (**4.14a**). This upward bending prompts neutral donor states to move beyond the Fermi level, releasing electrons and transitioning into positively charged donor states. The consequent surge of free electrons in the channel enhances the I_{on} and partially fills interface traps between the channel and the GI, resulting in a modest improvement in the *SS*. An increased magnitude of $|V_{gs}|$ amplifies the $|E_y|$, leading to more pronounced band bending and, consequently, a more significant increase in I_{on}, as illustrated in Fig. (**4.14b**).

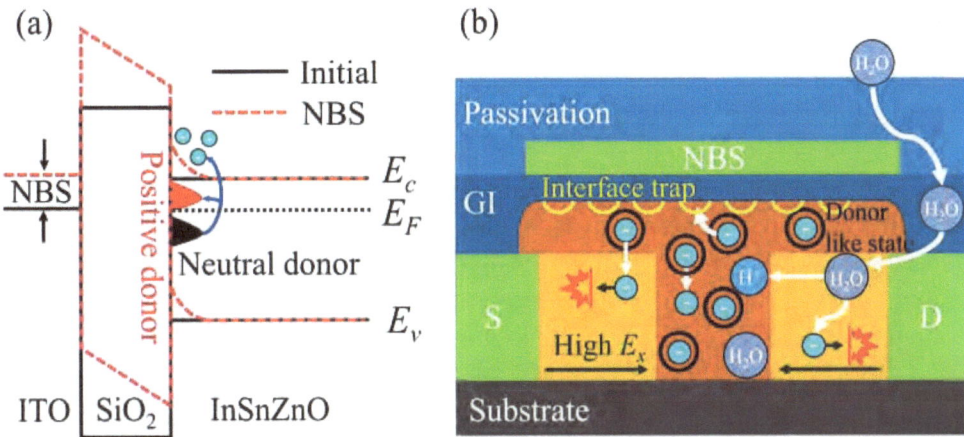

Fig. (4.14). (a) Energy band diagram from the gate to the channel before and after NBS in InSnZnO TFTs. (b) Schematic diagram of the degradation model in InSnZnO TFTs under NBS [22].

Beyond the increase in on-state current I_{on} due to energy band bending, moisture is also implicated as a significant contributing factor. The InSnZnO TFTs examined in this study are encapsulated and passivated with PECVD SiO_2. It has been established that moisture can permeate the channel through the porous nature of the PECVD SiO_2 layer, as illustrated in Fig. (**4.14**). Our preceding research, corroborated by XPS analysis, confirmed the infiltration of moisture within both the GI and the active layer. When subjected to the intensified electric field generated by NBS, H_2O is believed to participate in the chemical reaction outlined in equation (4.3).

The application of NBS induces a substantial upward energy band shift in InSnZnO, which in turn elevates the barrier height at the interface between the ITO and the InSnZnO layer. This increase in barrier height results in a pronounced E_x within the active layer, particularly in the vicinity of the source and drain regions. The free carriers, produced by both energy band bending and the ionization of H_2O, serve as precursors for HC degradation. Under the influence of a strong E_x, these carriers gain sufficient energy to become HCs. They then initiate the breakdown of weaker bonds within the InSnZnO thin film, leading to the formation of a significant concentration of acceptor-tail defects near the source and drain areas, as depicted in Fig. (**4.13a** and **4.13b**). This defect generation contributes to the observed degradation in the I_{on}, as shown in Fig. (**4.11**). A larger $|V_{gs}|$ results in a greater $|E_x|$ (Fig. **4.14**), causing a more significant decrease in I_{on} (Fig. **4.11b**). It is crucial to distinguish that the DC HC degradation behavior in InSnZnO TFTs under NBS is distinct from that observed in poly-Si TFTs, where HC degradation is typically triggered by an inverted channel mechanism. In the case of InSnZnO TFTs, the initial surge in I_{on} is primarily attributed to the increase in free carrier concentration resulting from energy band bending and the ionization of H_2O. Conversely, the subsequent decline in I_{on} during the second stage is predominantly a consequence of HC degradation instigated by free carriers and a high E_x. Moreover, the accumulation of H^+ ions, a byproduct of H_2O ionization, within the channel diminishes the potential barrier between the channel and the electrodes, further enhancing I_{on}. Nonetheless, the reformation of H_2O from its ionized state is challenging, which prolongs the recovery time for I_{on}. Given this degradation mechanism, it is postulated that employing a more impermeable oxide film, such as one fabricated through ALD, could potentially enhance the NBS reliability of InSnZnO TFTs.

Fig. (**4.15a**) shows the normalized S_{ID} as a function of various stress times under a $V_{gs} = -35$ V. The fitting data align well with the experimental results, indicating that the $\Delta N{-}\Delta\mu$ model [26, 27] is appropriate for InSnZnO TFTs. Fig. (**4.15b**) presents the extracted α_C as a function of stress time, which increases slightly for stress times less than 3000 seconds and rises more rapidly thereafter, mirroring the degradation behavior of I_{on} (Figs. **4.11** - **4.15c**) displays the estimated N_t from LFN measurements, showing an increase in N_t with stress time. An accentuated increase in I_{on} is observed at higher magnitudes of $|V_{gs}|$. Furthermore, the subthreshold noise, which is indicative of the tail states within the channel, is extracted and depicted in Fig. (**4.16**). This noise component escalates over the stress period, signifying an augmentation of tail states subsequent to NBS application.

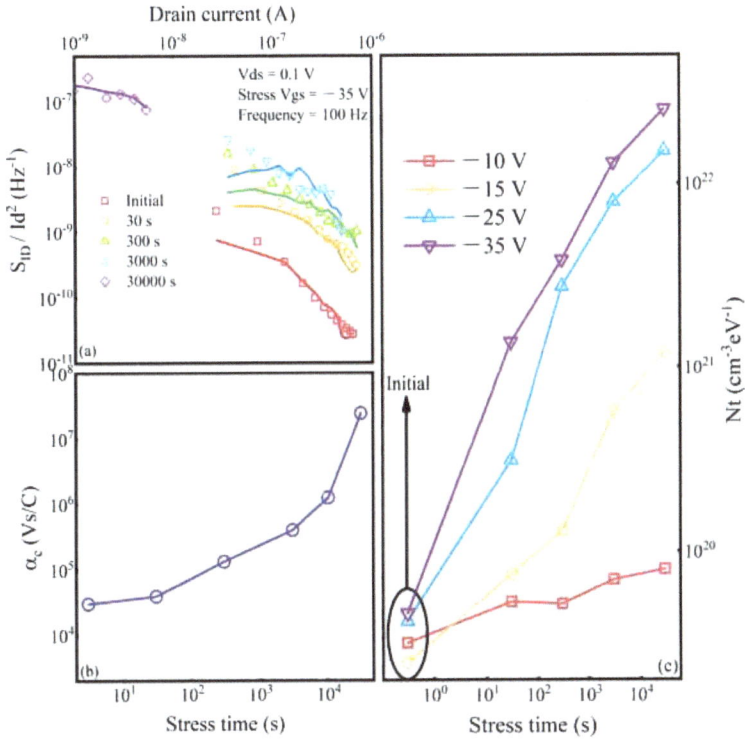

Fig. (4.15). (a) Normalized S_{ID} as a function of I_d after various stress times (symbols: experimental data, lines: fitted data). (b) Extracted α_c a function of stress time under a stress voltage $V_{gs} = -35$ V. (c) Calculated N_t as a function of stress time under different NBS conditions [22].

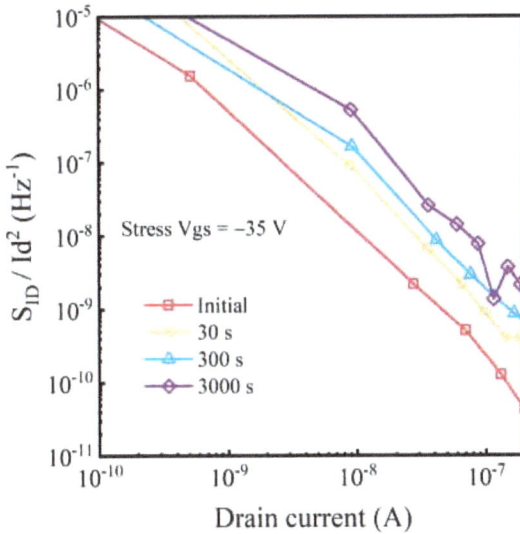

Fig. (4.16). Extracted excess noise after different stress durations [22].

At the initiation of NBS, there is a notable liberation of carriers due to the ionization of moisture and the transition of neutral donor states to ionized states [12], which increase I_{on} and fill the interface trap states. This improvement at the interface results in a slight increase in α_C. The E_x distribution is primarily concentrated in the channel region near the source and drain, causing HC-induced defects to be mainly located in these areas. This localized defect generation further enhances α_C. Additionally, defects that originate in the bulk can be projected to the surface, resulting in a progressive escalation of the interface N_t throughout the duration of the NBS process.

In summary, the mechanisms driving NBS degradation in MO TFTs have progressed from initial hole-dominant field-screening effects to more complex processes involving oxygen vacancy redistribution and, ultimately, to mechanisms incorporating the impact of hydrogen complexes. The understanding of NBS-induced degradation phenomena and mechanisms in Si-based TFTs and MO TFTs has become increasingly sophisticated and comprehensive.

4.3 HOT-CARRIER EFFECTS

The HC effect is one of the most prevalent phenomena in the reliability studies of TFTs, as it is virtually omnipresent wherever significant electric fields exist. Section 2.5 has provided a preliminary overview of the degradation caused by HC stress in TFTs with different materials. However, a detailed discussion on the specific origins of HCs and their impact on TFTs with various active layer materials and device structures has yet to be addressed. Additionally, methods to mitigate the effects of HC degradation on device performance have been a focal point of reliability research and are thoroughly examined in subsequent sections.

4.3.1. Generation of Hot Carrier

As mentioned previously in session xx, the generation of HCs is intrinsically linked to the presence of electric fields. [32-36] When an electric field exists within a TFT, carriers in the active layer absorb energy from the field and subsequently transfer this energy to the lattice of the active layer in the form of phonons. On average, the number of phonons emitted by carriers exceeds the number absorbed. When the system reaches a steady state, the energy acquired by carriers from the electric field over time is equivalent to the energy imparted to the lattice, and the system is in thermal equilibrium, with no generation of HCs.

However, when the electric field within the TFT is strong, the energy gained by carriers from the field greatly exceeds the energy transferred to the lattice. At this point, the carriers possess much higher energy than they would in thermal equilibrium, and these high-energy carriers are referred to as HCs [37-46].

From the definition of HCs, it is evident that their generation requires two conditions:

(1) Carriers are exposed to a strong electric field.

(2) There is a sufficient acceleration distance for carriers to gain energy.

Since HCs carry a high amount of energy, they can easily damage the lattice structure within the active layer of a TFT during their motion, leading to the creation of defects in the active layer [6-12]. Consequently, HCs can cause significant degradation in the electrical performance of TFT devices.

The HC effect manifests differently in TFTs with various active layer materials. The most commonly used active layer materials in commercial TFTs are poly-Si [32-36] and metal oxides [6-15]. Therefore, the following discussion will focus on these two types of TFTs to elaborate on the impact of HC effects and how to suppress them.

4.3.2. Impact on Silicon-Based Thin-Film Transistors

The typical degradation phenomena in poly-Si TFTs under DC HC stress are depicted in Fig. (**4.17**), where the transfer curve primarily shows a decrease in I_{on} and I_{off}, a slight increase in V_{th}, and essentially no change in SS. The main mechanism behind this degradation is the damage caused by HCs to the grain boundaries and the Si-SiO$_2$ interface near the drain, leading to an increase in defects within the active layer and the formation of additional trap energy barriers.

HC effects typically occur when the drain-source voltage (V_{ds}) is relatively high and the gate voltage is slightly above the V_{th}. A high V_{ds} significantly shields the vertical electric field (E_y) near the drain, reducing the carriers in that area and causing the device to "pinch off." At this point, the electric field generated by the high V_{ds} is primarily concentrated in the pinch-off region, creating an electric field greater than 10^5 V/cm. Carriers near the drain gain considerable energy and are accelerated into HCs. A large number of HCs can then collide with the poly-Si lattice and grain boundaries or inject into interface states, resulting in numerous defects. The accumulation of defects near the drain further raises the potential barrier, leading to a significant reduction in the TFT's I_{on}, as shown in Fig. (**4.17**).

There is a substantial discussion on the types of defects produced by the HC (HC) effect, with two mainstream perspectives [32-36, 47]:

(1) The HC effect, under the influence of a high electric field, causes degradation by colliding with grain boundaries near the drain.

(2) The HC effect, under the influence of a high electric field, causes degradation by injecting into interface states and within the GI layer.

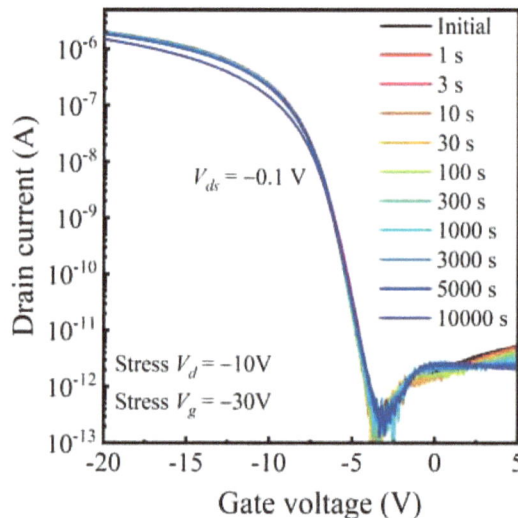

Fig. (4.17). Time evolution of transfer curve of poly-Si TFT under HC stress.

The first viewpoint suggests that poly-Si TFTs exhibit HC degradation primarily through the deterioration of I_{on}, while damage to interface states often leads to noticeable degradation in the subthreshold region. The second viewpoint observes that after optimizing the interface states in poly-Si, the HC effect is significantly suppressed, and a noticeable shift in V_{th} and degradation of SS during substantial degradation confirms that interface state damage contributes to device degradation [48].

Suppressing HC degradation has always been a key focus in reliability research, typically approached from two directions. The first method involves reducing the drain voltage, which often requires changes to the signal application method and may limit the practical application scenarios of the TFT. The second approach involves improving the device structure to weaken the E_x within the active layer of the device, which can often be achieved through the following two structural modifications to enhance the HC reliability of the device.

(1) LDD poly-Si TFT

The LDD poly-Si TFT is a classic example of a device structure designed to mitigate HC degradation by altering its structure to reduce the E_x in the active layer [49-51]. Compared to standard poly-Si TFTs, the key distinction of the LDD poly-Si TFT lies in its manufacturing process. A lightly doped layer is formed between the active layer and the drain through ion implantation. This lightly doped layer, acting as a conductor with resistance between a conductor and a semiconductor, serves to share the electric field provided by the drain, effectively acting as a small resistor. Although this design sacrifices a fraction of the mobility, it significantly enhances the reliability of the device under HC stress (for details, see section 8.2).

(2) BG poly-Si TFT

The bridge-gate (BG) poly-Si TFT also aims to reduce the E_x near the drain end of the active layer to decrease the HC effect, [52-54] similar to the LDD polycrystalline transistor. However, unlike the LDD poly-Si TFT, the BG poly-Si TFT not only improves the device's reliability under HC stress but also enhances its mobility. This dual benefit makes it a more desirable option for certain applications (for details, see Section 8.3).

4.3.3. Impact on Metal Oxide Thin-Film Transistors

In contrast to the HC degradation observed in poly-Si TFTs, the transfer characteristics of MO TFTs exhibit not only a decrease in I_{on} but also a positive shift and an increase in SS after HC stress. It is noteworthy that the degradation phenomena vary among different types of MO TFTs. The following discussion will delve into the HC effects in three distinct types of MO TFTs and explore methods to suppress these effects within the context of MO TFTs.

4.3.3.1. InGaZnO Thin-Film Transistors

Fig. (**4.18**) illustrates a bottom-gate amorphous IGZO TFT with a standard structure, and its manufacturing process is as follows: Initially, a 300 nm thick Al/Mo/Al film is sputtered onto a glass substrate to serve as the gate electrode. Subsequently, a 300 nm thick SiN_x and a 100 nm thick SiO_x are deposited via PECVD to form the GI. Then, a 50 nm thick IGZO film is deposited using an IGZO target with a molar ratio of In:Zn:Ga = 1:1:1. The IGZO film, defined as the channel layer through photolithography and etching processes, is followed by the deposition of a 200 nm thick SiOx etch stop layer through PECVD. Contact holes are formed through photolithography and etching processes, establishing a channel width-to-length ratio of 10 μm:10 μm. A 350 nm thick Ti/Al/Mo film is then sputtered and patterned through photolithography and etching. A 500 nm thick SiO_x/SiN_x is deposited through PECVD and used as a passivation layer. Finally, the device is annealed in ambient air.

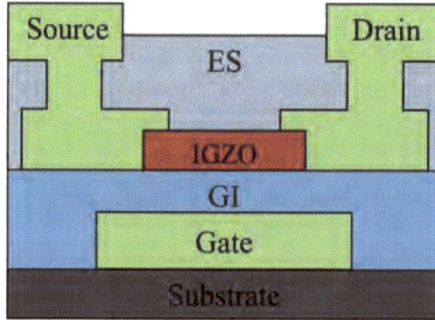

Fig. (4.18). Device structure of ES IGZO.

The degradation phenomena of the bottom-gate amorphous IGZO TFT under HC stress are examined. The transfer characteristics exhibit a noticeable positive shift after HC stress (Fig. **4.19**), with an increase in V_{th} and *SS* [38-40, 43, 55].

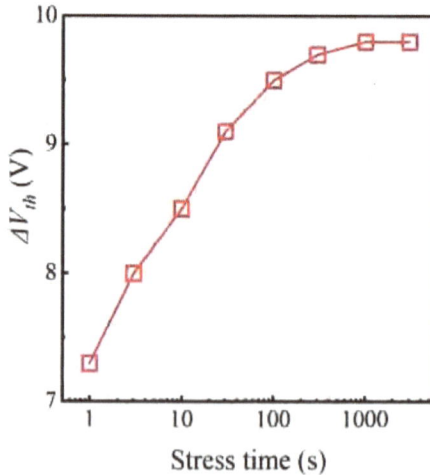

Fig. (4.19). V_{th} shift of ES IGZO TFT under HC stress.

When the device operates under strong saturation conditions at $V_g = 3$ V and $V_d = 70$ V, the active layer near the drain end becomes pinched off. The channel resistance near the drain increases, indicating that the applied voltage V_d is primarily distributed within the active layer near the drain. With the device in the on-state and containing a certain number of carriers, carriers near the drain are exposed to a strong electric field, accelerated into HCs, and cause damage to the lattice near the drain, generating defects. Some HCs are also injected into the etch stop layer, forming negatively charged fixed charges. [38-40, 43, 55]. These defects and fixed charges elevate the potential barrier near the drain in the active layer, making it more difficult for the device to turn on, manifesting as a positive shift in the transfer characteristics and an increase in *SS*.

4.3.3.2. Elevated-Metal Metal Oxide Thin-Film Transistors

Fig. (**4.20**) illustrates the basic structure of the elevated-metal metal oxide (EMMO) IGZO TFT. Unlike conventional IGZO TFTs, where the conductive electrode is directly connected to the active layer, the bridge-gate EMMO IGZO TFT utilizes an IGZO film rich in V_os as a bridge to connect the conductive electrode with the channel. (For detailed manufacturing processes, refer to section 1.2.4).

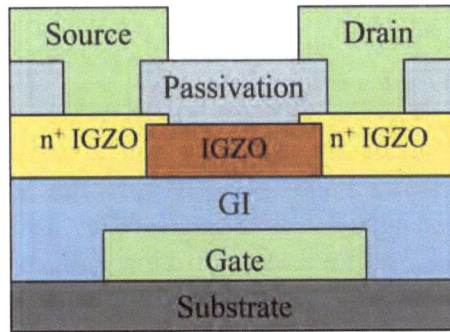

Fig. (4.20). Device structure of EMMO TFT.

Similarly, when an EMMO TFT is subjected to HC stress with $V_d = 70$ V and $V_g = 3$ V, the degradation phenomena are depicted in Fig. (**4.21**). After 10000 s of HC stress, the transfer characteristics of the bridge-gate IGZO TFT device show a significant decrease in I_{on}, while the *SS* and V_{th} remain largely unchanged. Clearly, HCs also increase the number of acceptor defects in the active layer of the bridge-gate IGZO TFT. However, compared to the ES IGZO TFT, the degree of HC degradation in the bridge-gate IGZO TFT is significantly reduced. The specific reasons for this are not yet clear, but it is hypothesized that the unique structure of the EMMO TFT may reduce the lateral and vertical electric fields within the active layer, thereby decreasing the generation and injection of HCs.

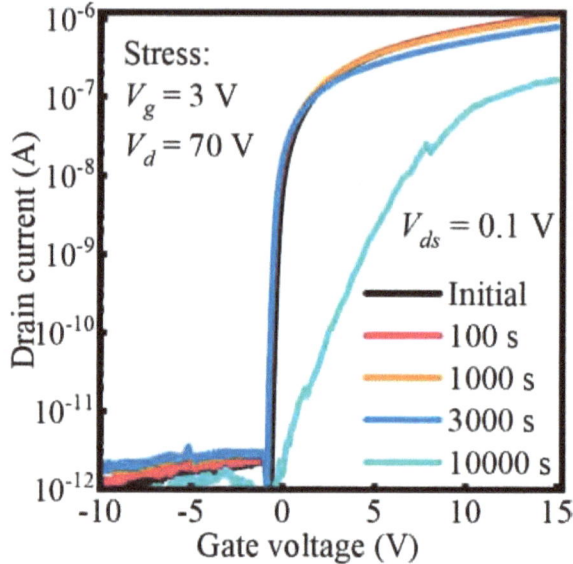

Fig. (4.21). Time evolution of transfer curve of EMMO TFT under HC stress.

4.3.3.3. InSnZnO Thin-Film Transistors

Fig. (**4.22**) presents the device structure of a top-gate ITZO TFT. The manufacturing process is as follows: Initially, a 50 nm layer of ITO is deposited onto the substrate using sputtering and patterned to form the source and drain. Subsequently, a 50 nm ITZO film is deposited through co-sputtering with ITO and zinc oxide targets, followed by the patterning of the active layer. Then, a 200 nm layer of SiO2 is deposited as the GI using PECVD, and a layer of ITO is deposited via sputtering. The gate is patterned, and a 200 nm layer of SiO2, deposited through PECVD, is used as a passivation layer. Contact holes are etched, and a layer of ITO is deposited and patterned to form the gate and source/drain electrodes. Finally, the wafer is annealed in air at 300°C for 2 hours.

Due to the higher mobility of ITZO devices compared to IGZO, carriers are more easily accelerated into HCs under the same electric field. When a DC stress of $V_g = 2$ V and V_d ranging from 7 to 22 V is applied to the device, significant HC degradation occurs, as shown in Fig. (**4.23**). At a low V_d of 7 V, the transfer characteristics of the ITZO TFT show little change after stress. However, when V_d is increased to 12 V, the transfer characteristics shift positively after 3000 seconds of stress, with an increase in V_{th} and SS, similar to the typical HC degradation observed in conventional IGZO TFTs. However, when V_d is increased to 22 V, the transfer characteristics exhibit severe positive shift and increased SS within a second, but after 1 second of continued stress at V_d = 22 V, the device shows recovery from HC degradation, with the transfer characteristics returning to near pre-stress conditions after 3000 seconds. This phenomenon is not observed in other MO TFTs [56].

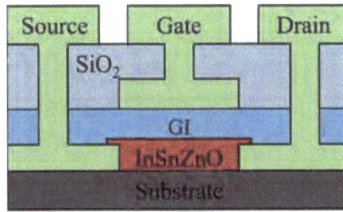

Fig. (4.22). Device structure of ITZO TFT.

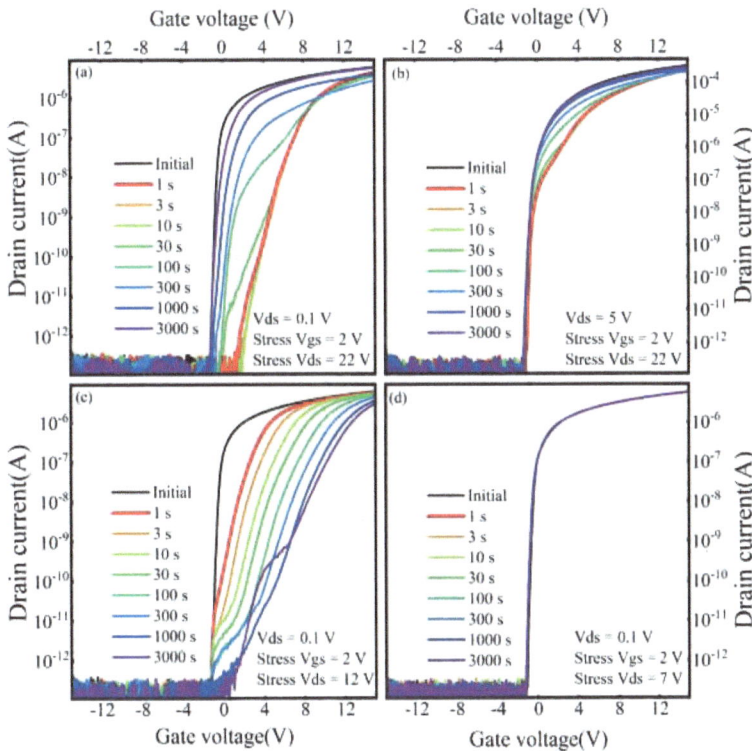

Fig. (4.23). Time evolution of transfer curve of an InSnZnO TFT under an HC stress with V_{ds} = 22 V and V_{gs} = 2 V, measured at (a) V_{ds} = 0.1 V and (b) V_{ds} = 5 V. Time evolution of transfer curve of InSnZnO TFTs (c) under stress with V_{ds} = 12 V and V_{gs} = 2 V, and (d) under stress with V_{ds} = 7 V and V_{gs} = 2 V.

When an HC stress of $V_g = 2$ V and $V_d = 22$ V is applied to the ITZO TFT, the device operates in strong saturation, with pinch-off occurring near the drain. The resistance in the active layer near the drain increases and the applied voltage is concentrated at the pinch-off point. As shown in Fig. (**4.24**), a strong E_x is present at the pinch-off point near the drain, and its magnitude and area of distribution increase with the applied voltage. Since the device is in the on-state with a certain number of free carriers in the active layer, these carriers, when exposed to the strong E_x near the drain, are accelerated into HCs. These HCs can damage the lattice structure near the drain, increasing the number of acceptor defects in the active layer and leading to an increase in SS and a decrease in I_{on} [56]. Additionally, some HCs may inject into the GI, forming negative fixed charges that make the device harder to turn on, manifesting as a positive shift in the transfer characteristics [56]. These observations align with those conclusions drawn from HC stress in ES IGZO TFTs.

However, considering only the E_x cannot explain the recovery phenomenon observed in ITZO TFTs after HC degradation. Simulations of the device under stress were conducted to examine the longitudinal electric field, as shown in Fig. (**4.24**), with its direction and magnitude extracted [56]. The longitudinal electric field, caused by a high voltage at the drain, is primarily distributed within the GI and directed from the active layer toward the GI. The magnitude of this field increases with the drain voltage. The presence of the longitudinal electric field facilitates the emission of HCs from the GI back into the active layer, leading to device recovery. The larger the longitudinal electric field, the greater the HC recovery effect. As the drain voltage increases, the HC degradation becomes more severe, and as the degradation reaches saturation, the recovery due to the longitudinal electric field begins to dominate. Therefore, the higher the drain voltage, the earlier the device exhibits HC recovery and the greater the degree of recovery after degradation.

4.4. SELF-HEATING EFFECT

The SH effect refers to the phenomenon where a significant channel current generates Joule heat within the channel when the device is in the on-state, leading to an increase in the device's internal temperature. Generally, two conditions are required for the SH effect to occur. Firstly, there must be a sufficiently large current capable of generating substantial Joule heat. Secondly, the device should have poor thermal conductivity, preventing the generated heat from dissipating quickly to the surroundings, resulting in a high-temperature buildup within the device. SH degradation, on the other hand, refers to the performance degradation caused by the increase in device temperature due to Joule heat generated by the I_{on}. This is primarily manifested by a decrease in I_{on}, an increase in I_{off}, and an increase in V_{th} and SS.

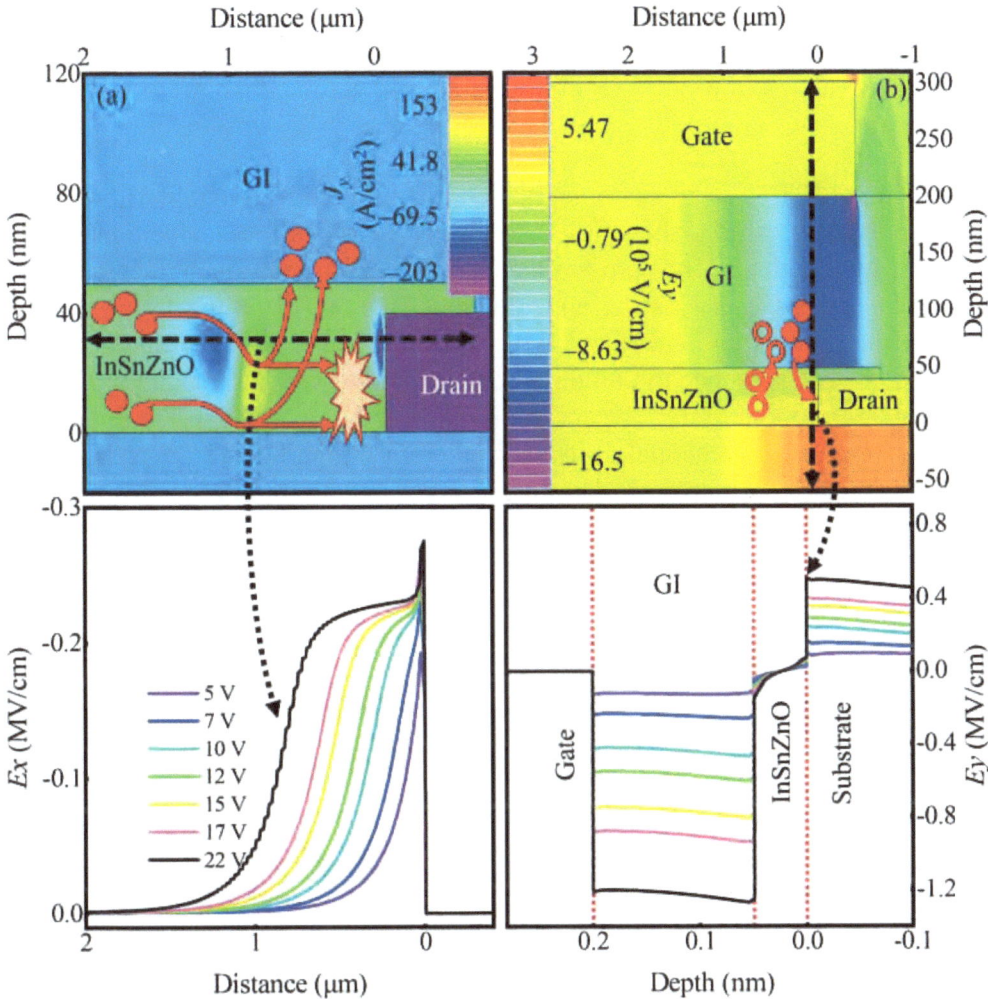

Fig. (4.24). E_x distribution in the active layer near the drain side under HC stress of $V_{gs} = 2$ V and $V_{ds} = 22$ V, and schematic diagram of HC degradation mechanism. (b) E_y distribution near the drain side under HC stress of $V_{gs} = 2$ V and $V_{ds} = 22$ V, and schematic diagram of the recovery process. (c) Extracted E_x near the interface from the channel to the drain. (d) Extracted E_y close to the drain from the gate electrode to the substrate.

The SH degradation in TFTs is a complex process involving the interaction of multiple physical mechanisms. This process predominantly occurs when a TFT device is subjected to a direct current bias. Due to the poor thermal conductivity of the glass substrate used in TFTs, when a significant bias is applied, the heat power generated by the current is also large, but the Joule heat cannot dissipate quickly, causing the temperature in the channel region to rise rapidly. As the channel temperature increases, the physical and chemical processes within the device change. At high temperatures, atoms and molecules in the channel become more active and prone to migration and diffusion. Additionally, high temperatures accelerate the interaction between electrons and the lattice, leading to an increase in electron energy.

Taking Si-based TFTs as an example, it is commonly believed that Si-H bonds can be broken at temperatures above 350 °C. Therefore, when the device temperature is high, many dangling bonds are produced at the Si/SiO$_2$ interface, increasing the interface state density across the entire channel region. Dangling bonds refer to atomic or molecular fragments that have not formed stable bonds with other atoms or molecules. The presence of dangling bonds can alter key parameters of the TFT device, such as conductivity, V_{th}, and mobility. For instance, dangling bonds may lead to an increase in channel resistance, thereby reducing the device's current driving capability. Furthermore, dangling bonds can also trigger electron trapping effects, making the capture and release processes of electrons in the channel more complex, further affecting device performance. The larger the current in the channel, the more pronounced the SH effect. SH degradation can accelerate device aging and shorten device lifespan, severely impacting device reliability.

4.4.1. Self-Heating Effect in Polycrystalline Silicon Thin-Film Transistors

TFTs operating in a DC state are prone to DC SH degradation, a common phenomenon that can negatively impact device performance and reliability. Research has shown that the SH effect is particularly prominent in poly-Si TFTs, primarily due to their substrates often being insulating materials like glass, with insulating layers such as silicon dioxide, which have low thermal conductivity. Consequently, when a large current flows through the device, SH is more likely to occur [57].

The SH degradation in poly-Si TFTs is predominantly controlled by the Joule heat generated by high power [58, 59]. As current passes through the TFT, Joule heat induced by the current is generated within the device, leading to a localized increase in temperature. Over time, the elevated internal temperature can increase the migration speed of electrons and holes while the effective mobility length of the carriers is reduced, leading to changes in the current density distribution. Since an increase in temperature can alter the electrical conductivity of materials, parameters such as the V_{th} can drift, causing a gradual decline in the performance of the TFT.

Fig. (4.25) [60] illustrates the degradation of a typical poly-Si TFT's transfer characteristics under DC SH stress (power density P = 75.8 mW/μm²) at V_{ds} = −0.1V and −5V. It can be observed that the device performance degrades in both I_{on} and V_{th}. Weak silicon-silicon and silicon-hydrogen bonds at the grain boundaries break down under SH stress, generating dangling bonds throughout the channel and fixed charges in the gate oxide, leading to device performance degradation [61]. As shown in Fig. (4.23), after 1000 seconds of SH stress, the I_{on} of the standard TFT degrades by −28.4%.

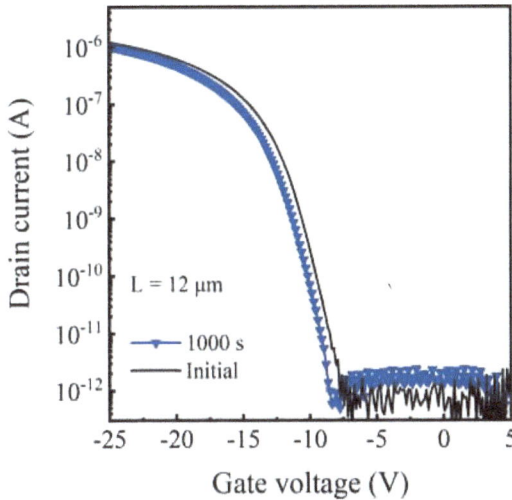

Fig. (4.25). Degradation diagram of the transfer curve of ordinary polysilicon TFT under SH electrical stress at V_{ds} = −0.1 V and −5 V.

TFTs also encounter the issue of AC SH degradation when operating in an AC state, which arises from the thermal effects produced by frequent changes in voltage and current. During AC operation, TFTs experience cyclical variations in voltage and current, leading to the repeated generation of Joule heat within the device. The frequent amplification and cutoff of current result in periodic changes in the device's internal temperature, making it difficult to maintain thermal equilibrium. Additionally, AC SH can affect the electrical conductivity of the device materials, and the constant temperature fluctuations can accelerate material aging.

For poly-Si TFTs, when a large voltage is applied between the gate and the drain, a significant current flows from the drain to the source. The Joule heat generated by this large current can trigger SH effects, causing device degradation. Therefore, to isolate the effects of AC SH during analysis, large voltages are typically applied to both the gate and the drain.

Gate-drain synchronous pulse stress is a specific stress mode where pulse voltages are applied simultaneously to the gate and drain to simulate the high-stress conditions that a TFT may experience in actual operation. This stress mode can lead to significant changes in TFT performance parameters, such as V_{th}, I_{on}, g_m, and SS.

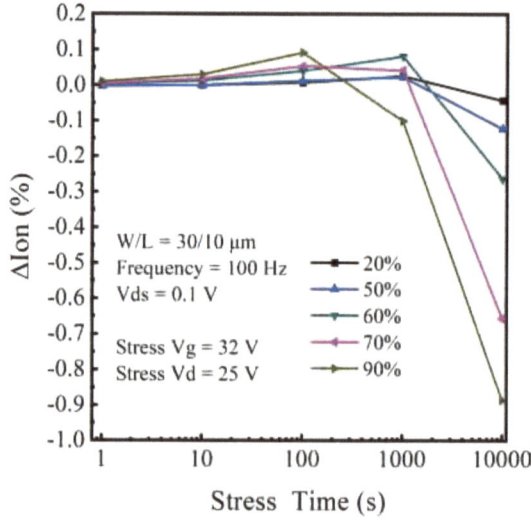

Fig. (4.26). Dependence of I_{on} degradation on stress time under different pulse duty cycles.

A study was conducted on N-channel poly-Si TFTs under different synchronous dynamic stresses with a fixed peak stress power density of 0.24 mW/μm². It was found that under the selected peak power density in DC SH conditions, severe SH degradation occurs. With the peak power density held constant, the SH effect is comparable for all devices studied under different AC stresses. The degradation mechanisms under AC stress were investigated by analyzing the evolution of device parameters, including I_{on}, V_{th}, and GIDL current defined at $V_d = 5\text{V}$ and $V_g = -12\text{V}$.

As illustrated in Fig. (**4.26**) [62], it depicts the dependence of ion degradation ΔI_{on} on stress time under various pulse duty cycles. Devices subjected to higher duty cycles exhibit a greater ΔI_{on}, which is not unexpected. This is because, for the same stress time, a higher duty cycle implies a longer cumulative time when the device is under SH stress. However, as shown in Fig. (**4.27**) [62], when ΔI_{on} is plotted against duty cycle rather than total stress time, the curves for all different pulse duty cycles almost coincide. As indicated in the inset, degradation is independent of the duty cycle only when the duty cycles are the same. Although not presented here, similar behavior was observed for V_{th} degradation. Thus, within the low-f range (< 50 kHz), device degradation is related to the operating time rather than the stress time.

Research findings suggest that, for a channel width-to-length ratio (W/L) of 30/10 μm, the time constant characterizing the channel temperature rise under SH stress is approximately 1 μs [63]. Consequently, under low-f stress, the device channel temperature has sufficient time to follow the voltage pulse changes, subjecting the device to SH effects during intervals of high V_g and V_d levels [59]. This correlation between degradation and the duration of action strongly implies that SH is the dominant degradation mechanism under low-f stress.

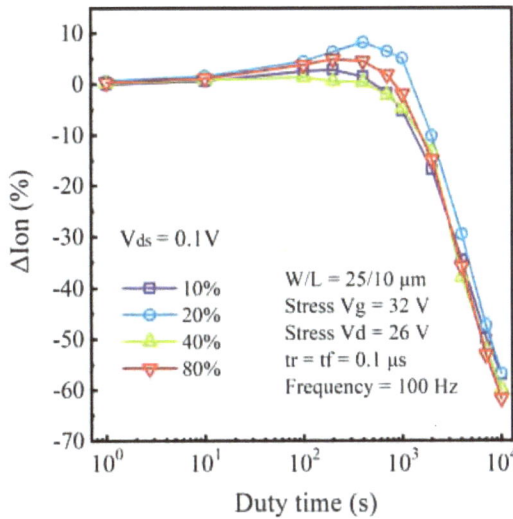

Fig. (4.27). Duty cycle dependence of I_{on} degradation at different pulse duty cycles. Illustration ΔI_{on} vs. duty cycle curve, the same stress action time is 7000s and 1000s.

4.4.2. Self-Heating Effect in Metal Oxide Thin-Film Transistors

The SH principle in MO TFTs is similar to that in poly-Si TFTs, where both result in performance degradation due to Joule heat generated by large currents causing an increase in the device's internal temperature. However, MO TFTs differ from poly-Si in that their oxide channels typically contain defects such as V_os. As carriers pass through the channel, they interact with these defects, leading to energy loss and heat release, which further promotes the occurrence of SH effects. In addition to heat accumulation within the channel, MO TFTs are also influenced by the thermal environment surrounding the device. Typically, there are heat dissipation structures or other materials around the device designed to dissipate the heat generated by the MO TFT. However, if the heat dissipation is insufficient or the thermal conductivity of the channel material is low, the degree of SH will increase. Furthermore, the SH effect in MO TFTs is closely related to the physical properties of the material, such as thermal conductivity and electron mobility. Some materials may have low thermal conductivity, making the SH effect more pronounced. For instance, a-IGZO, as a MO semiconductor with extremely low thermal conductivity, is comparable to SiO_2. Therefore, studying the degradation of a-IGZO under SH stress can provide a better analysis of MO TFT behavior.

The degradation pattern of conventional bottom-gate ESL structured a-IGZO TFTs under DC SH stress exhibits a two-stage degradation phenomenon. Initially, the transfer characteristics curve drifts towards the positive gate voltage direction upon the application of stress. As the duration of stress accumulates, the transfer characteristics curve shifts towards lower gate voltages, with a pronounced negative deviation observed in the subthreshold region exceeding that in the saturated on-state region. This shift engenders a notable peak in the hump current along the curve. Fig. (**4.28**) shows the transfer characteristics curve of a-IGZO under DC stress conditions ($V_g = 40$ V, $V_d = 25$ V). Moreover, the degradation under constant high current DC SH stress is dominated by the gate voltage, while the drain voltage's role is to increase the channel current, thereby raising the temperature in the channel and increasing the negative drift degradation in the second stage [57, 64].

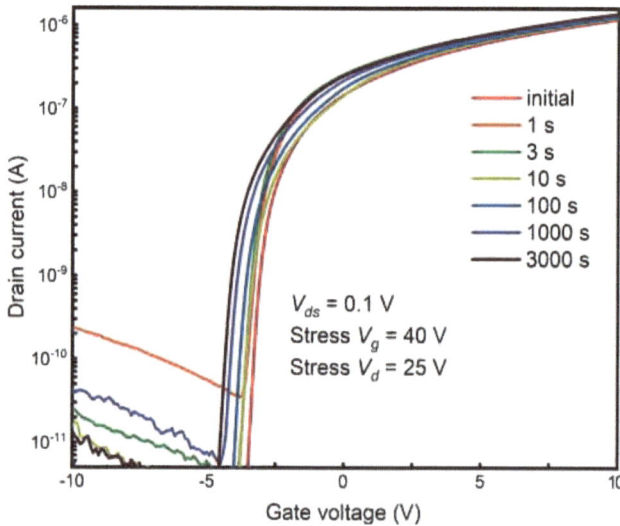

Fig. (4.28). Transfer curve of a-IGZO TFT as a function of time with applied DC SH stress (V_G=40V, V_D=25V) under conditions of $V_{DS} = 0.1$V.

The degradation pattern of traditional a-IGZO TFTs under DC SH stress is characterized by a two-stage process. The first stage involves a positive drift in the transfer characteristics curve, and the second stage involves a negative drift. During the second stage, the negative drift in the subthreshold region is greater and occurs earlier than in the on-state region. The first stage's positive drift is due to the electron capture effect, while the second stage's negative drift is attributed to the diffusion of H2O adsorbed in the passivation layer into the a-IGZO, where it releases electrons and becomes positively charged H_2O^+ or H+. Under the vertical electric field generated by a positive V_g, these positive charges are pushed to the interface between the a-IGZO and the ESL, where they accumulate and create a parasitic channel. As the stress time increases, further accumulation of positive charges at the parasitic channel site occurs while the main channel still experiences electron capture. This causes the parasitic channel to turn on earlier than the main channel, resulting in a hump current. Comparing the degradation of devices under DC SH stress and quasi-DC PBS, it is observed that the second stage's negative drift is more significant under SH stress, and the greater the V_d, the more pronounced the negative drift degradation. This is because SH stress spontaneously generates Joule heat, which increases the internal temperature of the device. The higher the temperature increase, the more intensified the $H_2O \rightarrow H_2O^+ + e^-$ process, thereby increasing the negative drift degradation in the second stage [57, 64].

Research into the AC SH effects in MO TFTs has also been conducted. For the dynamic SH analysis of IGZO TFTs, a static drain bias and pulsed gate voltage dynamic stress are typically applied in the linear region to avoid potential HC effects and systematically study the pure dynamic SH effects.

Fig. (**4.29a**) illustrates the dynamic stress setup for an a-IGZO TFT, with the VG pulse frequency, duty cycle, peak time (t_p), and base time (t_b) set to 10 kHz, 80%, 80 μs, and 20 μs, respectively. (b) Shows the progression of the effective stress time (t_{eff}) on its transfer characteristics.

Fig (**4.29a**) illustrates the application of dynamic stress, utilizing a constant V_d of 21 V alongside V_g pulses that vary from 0 V to 24 V. This configuration ensures a high current suitable for GOA and Micro-LED applications [66, 67]. The transistor's operation is characterized by a cyclic transition between the off-state and the linear operating region, thereby mitigating the risk of HC effects. At the peak of the V_g pulse, the power density reaches approximately 19 μW/μm², surpassing the previously reported threshold for SH initiation [68]. The terms 'peak time' and 'base time' correspond to the periods of heating and cooling, respectively. The aggregate duration, denoted as t_p, is designated as the effective stress time (t_{eff}).

Fig. (**4.29b**) presents the measured I_d characteristics at V_d of 0.1 V, which demonstrate commendable performance metrics in relation to the V_g. Specifically, the μ_{FE} and SS are recorded at 12.5 cm²/Vs and 170 mV/dec, respectively. With the escalation of the stress period, the characteristic transfer curve initially exhibits a minor upward adjustment within the initial 50s of the teff (Stage I). This is succeeded by a slow downward trend in response to the degradation of the SS, as the device's V_g shifts towards more negative values (Stage II). Ultimately, the device's behavior transitions to that of a 'conductor,' signifying the final stage of degradation (Stage III) and reflecting a change in the degradation dynamics.

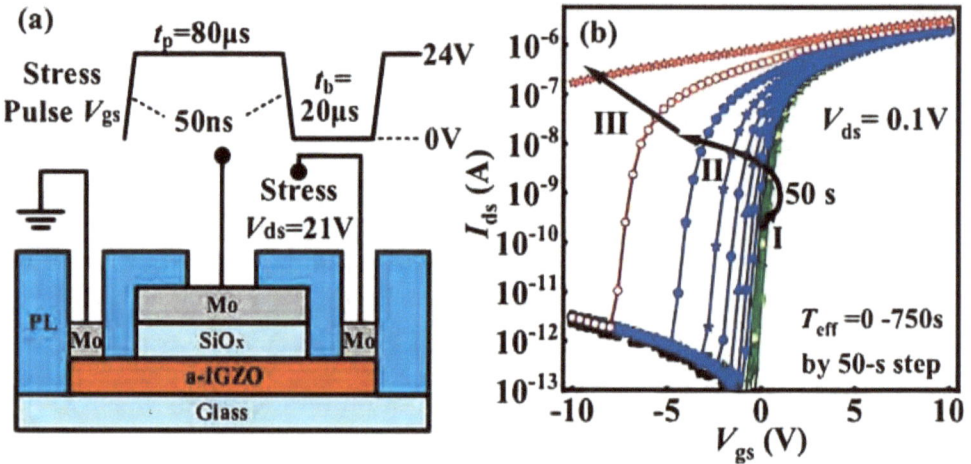

Fig. (4.29). (a) Schematic diagram of a-IGZO TFT under dynamic stress. The frequency, duty cycle, tp, and tb of the VG pulse are set to 10kHz, 80%, 80μs, and 20μs, respectively. (b) The progressive transformation of the transfer characteristics in teff [65].

The degradation pattern of traditional a-IGZO TFTs under AC SH stress is characterized by three distinct behaviors due to pure dynamic SH stress. The observed ΔV_{th} shifts, ranging from a minor increase to a progressive decrease and culminating in a permanently conductive state, are indicative of distinct degradation mechanisms. Initially, the slight increase in ΔV_{th} is attributed to the SH effect, which enhances electron capture in the GI. Subsequently, the gradual decrease in ΔV_{th} is linked to the SH-induced creation of donor defects within the channel. Finally, the transition to a normally-on state signifies the occupation of the center channel by these donor ions. The progression through these stages is governed by the complex interplay of heating and cooling cycles during dynamic stress, with the degradation process being predominantly influenced by the heating phase. This nonlinear competition underscores the greater susceptibility of the device to thermal stress-induced damage, as opposed to the cooling phase, as elucidated in [65].

4.4.3. Distinguish Between the Self-Heating Effect and Hot Carrier Effect

In the degradation of TFTs, both the HC effect and SH effect occur when voltages are applied to both the gate and drain terminals, and both can lead to a degradation in I_{on}. Additionally, during SH stress, due to the high drain voltage, both HC and SH degradation mechanisms can be present simultaneously. As a result, it can be challenging to discern the primary degradation mechanism at play.

As mentioned in Section 4.3, the HC effect typically requires a strong electric field and sufficient acceleration distance. A key factor in the occurrence of the HC effect in TFTs is the formation of a depletion region within the channel. Most of the voltage drop occurs across this depletion region, creating a strong enough electric field to generate HC. Therefore, one can initially distinguish the mechanisms based on the different ways of applying voltage. Typically, the HC effect predominantly occurs under stress conditions where $V_{ds} > V_{gs} - V_{th}$. Thus, the stress voltage V_{gs} is usually only slightly higher than V_{th}. To avoid the generation of HC while minimizing SH, one can significantly increase V_{gs}, ensuring that $V_{ds} < V_{gs} - V_{th}$, preventing the formation of a depletion region in the channel.

Furthermore, the degradation phenomena of a TFT can also be used to determine the degradation mechanism. For example, in Si-based devices, SH degradation primarily results from the breaking of numerous weak bonds throughout the channel under SH stress, leading to the formation of a large number of dangling bonds. These dangling bonds act as deep-level states, affecting carrier transport. The creation of these defect states causes a shift in the device's V_{th} and degradation of the SS. In contrast, the HC effect mainly generates a significant number of tail state defects, with the primary degradation phenomena concentrated in the on-state region of the device, while the subthreshold region remains largely unchanged.

In summary, SH degradation and HC degradation are both significant mechanisms in TFTs. The main difference lies in the degradation processes and impacts; SH degradation is primarily due to defect state generation in the channel and degradation of the SS caused by SH stress, whereas HC degradation results from damage and degradation in the active layer or gate oxide layer induced by HC injection.

CONCLUSION

Table 1. Changes of key parameters of TFT under different stress modes.

Type of TFTs	Stress Mode	V_{th}	μ	*SS*
Poly-Si TFTs	PBS	Increase	No degradation	No degradation
	NBS	Decrease	No degradation	No degradation
	HC	Increase	Decrease	No degradation
	SH	Increase	Decrease	Decrease
MO TFTs	PBS	Increase/ Decrease	No degradation	No degradation
	NBS	No degradation/ Decrease	No degradation / Decrease	No degradation
	HC	Increase	Decrease	Decrease
	SH	Increase/ Decrease	Decrease	Decrease

This chapter offers a systematic exploration of DC voltage stress-induced degradation in poly-Si and metal oxide MO TFTs. As shown in the Table **1**, above, it delves into various stress scenarios, specifically gate bias stress, HC effects, and SH phenomena, analyzing their significant impact on device performance and stability. By thoroughly examining the mechanisms and implications of these stresses, the chapter enhances our understanding of the degradation processes that can compromise the reliability of TFTs. The insights gained provide a robust theoretical framework for the design and optimization of TFTs in advanced electronic applications, ensuring their durability and effectiveness under a range of operating conditions. Ultimately, this chapter underscores the importance of addressing these degradation mechanisms to achieve long-lasting performance in modern electronic devices.

RFERENCES

[1]   Y.H. Tai, J.W. Tsai, H.C. Cheng, and F.C. Su, "Instability mechanisms for the hydrogenated amorphous silicon thin-film transistors with negative and positive bias stresses on the gate electrodes", *Appl. Phys. Lett.,* vol. 67, no. 1, pp. 76-78, 1995.

http://dx.doi.org/10.1063/1.115512

[2]   P.H. Chen, Y.Z. Zheng, T.H. Yeh, and T.Y. Nieh, "Investigating DC and AC degradation behaviors to P-type low temperature polycrystalline silicon thin film transistor with fin-like structure", *J. Phys. D Appl. Phys.,* vol. 56, no. 43, p. 435101, 2023.

http://dx.doi.org/10.1088/1361-6463/ace835

[3]   T.Y. Hsieh, T.C. Chang, T.C. Chen, and M.Y. Tsai, "Review of Present Reliability Challenges in Amorphous In-Ga-Zn-O Thin Film Transistors", *ECS J. Solid State Sci. Technol.,* vol. 3, no. 9, pp. Q3058-Q3070, 2014.

http://dx.doi.org/10.1149/2.013409jss

[4]   M. Zhang, "P-15: Gate-Bias-Stress-Induced Instability in Hybrid-Phase Microstructural ITO-Stabilized ZnO TFTs", *SID Symposium Digest of Technical Papers,* vol. 50, 2019pp. 1267-1270

http://dx.doi.org/10.1002/sdtp.13164

[5]   Seonghyun Jin, Tae-Woong Kim, Young-Gug Seol, M. Mativenga, and Jin Jang, "Reduction of Positive-Bias-Stress Effects in Bulk-Accumulation Amorphous-InGaZnO TFTs", *IEEE Electron Device Lett.,* vol. 35, no. 5, pp. 560-562, 2014.

http://dx.doi.org/10.1109/LED.2014.2311172

[6]   Y.S. Kang, Y-H. Lee, W-S. Kim, Y-J. Cho, J.K. Park, G.T. Kim, and O. Kim, "Relationship between Detrapping of Electrons and Negative Gate Bias during Recovery Process in a-InGaZnO Thin Film Transistors", *Phys. Status Solidi., A Appl. Mater. Sci.,* vol. 216, no. 4, p. 1800621, 2019. [a].

http://dx.doi.org/10.1002/pssa.201800621

[7]   F.H. Chen, T.M. Pan, C.H. Chen, J.H. Liu, W.H. Lin, and P.H. Chen, "Two-step Electrical Degradation Behavior in α-InGaZnO Thin-film Transistor Under Gate-bias Stress", *IEEE Electron Device Lett.,* vol. 34, no. 5, pp. 635-637, 2013.

http://dx.doi.org/10.1109/LED.2013.2248115

[8]   Y. Zhang, Z. Wang, M. Wang, D. Zhang, H. Wang, and M. Wong, "A Unified Degradation Model of Elevated-Metal Metal Oxide (EMMO) TFTs Under Positive Gate Bias With or Without an Illumination", *IEEE Trans. Electron Dev.,* vol. 68, no. 3, pp. 1081-1087, 2021.

http://dx.doi.org/10.1109/TED.2021.3053915

[9]   Z. Jiang, M. Zhang, S. Deng, Y. Yang, M. Wong, and H.S. Kwok, "Evaluation of Positive-Bias-Stress-Induced Degradation in InSnZnO Thin-Film Transistors by Low Frequency Noise Measurement", *IEEE Electron Device Lett.,* vol. 43, no. 6, pp. 886-889, 2022.

http://dx.doi.org/10.1109/LED.2022.3165558

[10] S. Choi, H. Kim, C. Jo, H-S. Kim, S-J. Choi, D.M. Kim, J. Park, and D.H. Kim, "The Effect of Gate and Drain Fields on the Competition Between Donor-Like State Creation and Local Electron Trapping in In–Ga–Zn–O Thin Film Transistors Under Current Stress", *IEEE Electron Device Lett.,* vol. 36, no. 12, pp. 1336-1339, 2015.

http://dx.doi.org/10.1109/LED.2015.2487370

[11] C.L. Fan, F-P. Tseng, B-J. Li, Y-Z. Lin, S-J. Wang, W-D. Lee, and B-R. Huang, "Improvement in reliability of amorphous indium–gallium–zinc oxide thin-film transistors with Teflon/SiO 2 bilayer passivation under gate bias stress", *Jpn. J. Appl. Phys.,* vol. 55, no. 2S, p. 02BC17, 2016.

http://dx.doi.org/10.7567/JJAP.55.02BC17

[12] H.U. Huzaibi, N. Lu, M.M. Billah, D. Geng, and L. Li, "Investigation of Hump Behavior of Amorphous Indium-Gallium-Zinc-Oxide Thin-Film Transistor Under Positive Bias Stress", *IEEE Trans. Electron Dev.,* vol. 69, no. 2, pp. 549-554, 2022.

http://dx.doi.org/10.1109/TED.2021.3135249

[13] Y.X. Wang, M-C. Tai, T-C. Chang, S-P. Huang, Y-Z. Zheng, C-C. Wu, Y-S. Shih, Y-A. Chen, P-J. Sun, I-N. Lu, H-C. Huang, and S.M. Sze, "Suppression of Edge Effect Induced by Positive Gate Bias Stress in Low-Temperature Polycrystalline Silicon TFTs With Channel Width Extension Over Source/Drain Regions", *IEEE Trans. Electron Dev.,* vol. 67, no. 12, pp. 5552-5556, 2020.

http://dx.doi.org/10.1109/TED.2020.3033516

[14] D. Zhou, M. Wang, and S. Zhang, "Degradation of Amorphous Silicon Thin Film Transistors Under Negative Gate Bias Stress", *IEEE Trans. Electron Dev.,* vol. 58, no. 10, pp. 3422-3427, 2011.

http://dx.doi.org/10.1109/TED.2011.2161635

[15] B. Tala-Ighil, H. Toutah, A. Rahal, K. Mourgues, L. Pichon, F. Raoult, O. Bonnaud, and T. Mohammed-Brahim, "Gate bias stress in hydrogenated and unhydrogenated polysilicon thin film transistors", *Microelectron. Reliab.,* vol. 38, no. 6-8, pp. 1149-1153, 1998.

http://dx.doi.org/10.1016/S0026-2714(98)00098-5

[16] N. Liang, D. Zhang, M. Wang, H. Wang, Y. Yu, and D. Qi, "Investigations on the Negative Shift of the Threshold Voltage of Polycrystalline Silicon Thin-Film Transistors Under Positive Gate Bias Stress", *IEEE Trans. Electron Dev.,* vol. 68, no. 2, pp. 550-555, 2021.

http://dx.doi.org/10.1109/TED.2020.3041568

[17] C.Y. Han, Y. Liu, Y.R. Liu, Y.Y. Chen, L. Wang, and R.S. Chen, "Negative gate bias stress effects on conduction and low frequency noise characteristics in p-type poly-Si thin-film transistors*", *Chin. Phys. B,* vol. 28, no. 8, p. 088502, 2019.

http://dx.doi.org/10.1088/1674-1056/28/8/088502

[18] H.C. Chen, G-F. Chen, P-H. Chen, S-P. Huang, J-J. Chen, K-J. Zhou, C-W. Kuo, Y-C. Tsao, A-K. Chu, H-C. Huang, W-C. Lai, and T-C. Chang, "A Novel Heat Dissipation Structure for Inhibiting Hydrogen Diffusion in Top-Gate a-InGaZnO TFTs", *IEEE Electron Device Lett.,* vol. 40, no. 9, pp. 1447-1450, 2019.

http://dx.doi.org/10.1109/LED.2019.2927422

[19] S. Li, M. Wang, D. Zhang, H. Wang, and Q. Shan, "A Unified Degradation Model of a-InGaZnO TFTs Under Negative Gate Bias With or Without an Illumination", *IEEE J. Electron Devices Soc.,* vol. 7, pp. 1063-1071, 2019.

http://dx.doi.org/10.1109/JEDS.2019.2946383

[20] T.K. Ha, Y. Kim, S.H. Yu, G.T. Kim, H. Jeong, J.K. Park, and O. Kim, "Abnormal threshold voltage shift by the effect of H2O during negative bias stress in amorphous InGaZnO thin film transistors", *Solid-State Electron.,* vol. 174, p. 107916, 2020.

http://dx.doi.org/10.1016/j.sse.2020.107916

[21] Y.S. Shiah, K. Sim, Y. Shi, K. Abe, S. Ueda, M. Sasase, J. Kim, and H. Hosono, "Mobility–stability trade-off in oxide thin-film transistors", *Nat. Electron.,* vol. 4, no. 11, pp. 800-807, 2021.

http://dx.doi.org/10.1038/s41928-021-00671-0

[22] Z. Jiang, M. Zhang, S. Deng, M. Wong, and H.S. Kwok, "Degradation of InSnZnO Thin-Film Transistors Under Negative Bias Stress", *IEEE Trans. Electron Dev.,* vol. 70, no. 12, pp. 6381-6386, 2023.
http://dx.doi.org/10.1109/TED.2023.3327975

[23] T.C. Fung, G. Baek, and J. Kanicki, "Low frequency noise in long channel amorphous In–Ga–Zn–O thin film transistors", *J. Appl. Phys.,* vol. 108, no. 7, p. 074518, 2010.
http://dx.doi.org/10.1063/1.3490193

[24] Y. Yang, M. Zhang, L. Lu, M. Wong, and H.S. Kwok, "Low-Frequency Noise in Bridged-Grain Polycrystalline Silicon Thin-Film Transistors", *IEEE Trans. Electron Dev.,* vol. 69, no. 4, pp. 1984-1988, 2022.
http://dx.doi.org/10.1109/TED.2022.3148697

[25] Y. Xu, T. Minari, K. Tsukagoshi, J. Chroboczek, F. Balestra, and G. Ghibaudo, "Origin of low-frequency noise in pentacene field-effect transistors", *Solid-State Electron.,* vol. 61, no. 1, pp. 106-110, 2011.
http://dx.doi.org/10.1016/j.sse.2011.01.002

[26] S. Christensson, I. Li-Jndstrm, and C. Svensson, "Low Frequency Noise In Mos Transistors" I. Theory. *Solid-State Electronics,* vol. 11, pp. 797-812, 1968.

[27] F.N. Hooge, "1/f noise sources", *IEEE Trans. Electron Dev.,* vol. 41, no. 11, pp. 1926-1935, 1994.
http://dx.doi.org/10.1109/16.333808

[28] Meng Zhang, Wei Zhou, Rongsheng Chen, Man Wong, and Hoi-Sing Kwok, "Characterization of DC-Stress-Induced Degradation in Bridged-Grain Polycrystalline Silicon Thin-Film Transistors", *IEEE Trans. Electron Dev.,* vol. 61, no. 9, pp. 3206-3212, 2014.
http://dx.doi.org/10.1109/TED.2014.2341676

[29] Y. Uraoka, T. Hatayama, T. Fuyuki, T. Kawamura, and Y. Tsuchihashi, "Reliability of low temperature poly-silicon TFTs under inverter operation", *IEEE Trans. Electron Dev.,* vol. 48, no. 10, pp. 2370-2374, 2001.
http://dx.doi.org/10.1109/16.954479

[30] M.J. Kim, and D. Choi, "Effect of enhanced-mobility current path on the mobility of AOS TFT", *Microelectron. Reliab.,* vol. 52, no. 7, pp. 1346-1349, 2012.
http://dx.doi.org/10.1016/j.microrel.2012.02.012

[31] K. Mashooq, J. Jo, and R.L. Peterson, "Extraction of SnO Subbandgap Defect Density by Numerical Modeling of p-Type TFTs", *IEEE Trans. Electron Dev.,* vol. 69, no. 5, pp. 2436-2442, 2022.
http://dx.doi.org/10.1109/TED.2022.3162803

[32] T. Kawamura, M. Matsumura, T. Kaitoh, T. Noda, M. Hatano, T. Miyazawa, and M. Ohkura, "A Model for Predicting On-Current Degradation Caused by Drain-Avalanche Hot Carriers in Low-Temperature Polysilicon Thin-Film Transistors", *IEEE Trans. Electron Dev.,* vol. 56, no. 1, pp. 109-115, 2009.
http://dx.doi.org/10.1109/TED.2008.2008376

[33]   W. Yan, Z. Yu, J. Guo, D. Shi, J. Xue, and W. Xue, "Recovery behaviors in n-channel LTPS-TFTs under DC stress", *Microelectron. Reliab.,* vol. 81, pp. 117-120, 2018.
http://dx.doi.org/10.1016/j.microrel.2017.12.026

[34]   L. Mariucci, G. Fortunato, R. Carluccio, A. Pecora, S. Giovannini, F. Massussi, L. Colalongo, and M. Valdinoci, "Determination of hot-carrier induced interface state density in polycrystalline silicon thin-film transistors", *J. Appl. Phys.,* vol. 84, no. 4, pp. 2341-2348, 1998.
http://dx.doi.org/10.1063/1.368302

[35]   M. Xue, M. Wang, Z. Zhu, D. Zhang, and M. Wong, "Degradation Behaviors of Metal-Induced Laterally Crystallized n-Type Polycrystalline Silicon Thin-Film Transistors Under DC Bias Stresses", *IEEE Trans. Electron Dev.,* vol. 54, no. 2, pp. 225-232, 2007.
http://dx.doi.org/10.1109/TED.2006.888723

[36]   A. Hatzopoulos, N. Archontas, N.A. Hastas, C.A. Dimitriadis, G. Kamarinos, N. Georgoulas, and A. Thanailakis, "Change in Transfer and Low-Frequency Noise Characteristics of n-Channel Polysilicon TFTs Due to Hot-Carrier Degradation", *IEEE Electron Device Lett.,* vol. 25, no. 6, pp. 390-392, 2004.
http://dx.doi.org/10.1109/LED.2004.828555

[37]   W.F. Chung, T-C. Chang, C-S. Lin, K-J. Tu, H-W. Li, T-Y. Tseng, Y-C. Chen, and Y-H. Tai, "Oxygen-Adsorption-Induced Anomalous Capacitance Degradation in Amorphous Indium-Gallium-Zinc-Oxide Thin-Film-Transistors under Hot-Carrier Stress", *J. Electrochem. Soc.,* vol. 159, no. 3, pp. H286-H289, 2012.
http://dx.doi.org/10.1149/2.075203jes

[38]   T.Y. Hsieh, T-C. Chang, Y-T. Chen, P-Y. Liao, T-C. Chen, M-Y. Tsai, Y-C. Chen, B-W. Chen, A-K. Chu, C-H. Chou, W-C. Chung, and J-F. Chang, "Hot-Carrier Effect on Amorphous In-Ga-Zn-O Thin-Film Transistors With a Via-Contact Structure", *IEEE Electron Device Lett.,* vol. 34, no. 5, pp. 638-640, 2013.
http://dx.doi.org/10.1109/LED.2013.2248341

[39]   T. Takahashi, R. Miyanaga, M.N. Fujii, J. Tanaka, K. Takechi, H. Tanabe, J.P. Bermundo, Y. Ishikawa, and Y. Uraoka, "Hot carrier effects in InGaZnO thin-film transistor", *Appl. Phys. Express,* vol. 12, no. 9, p. 094007, 2019.
http://dx.doi.org/10.7567/1882-0786/ab3c43

[40]   N. On, B.K. Kim, S. Lee, E.H. Kim, J.H. Lim, and J.K. Jeong, "Hot Carrier Effect in Self-Aligned In–Ga–Zn–O Thin-Film Transistors With Short Channel Length", *IEEE Trans. Electron Dev.,* vol. 67, no. 12, pp. 5544-5551, 2020.
http://dx.doi.org/10.1109/TED.2020.3032383

[41]   D. Lin, W-C. Su, T-C. Chang, H-C. Chen, Y-F. Tu, J. Yang, K-J. Zhou, Y-H. Hung, I-N. Lu, T-M. Tsai, and Q. Zhang, "Effects of Redundant Electrode Width on Stability of a-InGaZnO Thin-Film Transistors Under Hot-Carrier Stress", *IEEE Trans. Electron Dev.,* vol. 67, no. 6, pp. 2372-2375, 2020.
http://dx.doi.org/10.1109/TED.2020.2990135

[42]   D. Lin, W-C. Su, T-C. Chang, H-C. Chen, Y-F. Tu, K-J. Zhou, Y-H. Hung, J. Yang, I-N. Lu, T-M. Tsai, and Q. Zhang, "Degradation Behavior of Etch-Stopper-Layer Structured a-

InGaZnO Thin-Film Transistors Under Hot-Carrier Stress and Illumination", *IEEE Trans. Electron Dev.,* vol. 68, no. 2, pp. 556-559, 2021.

http://dx.doi.org/10.1109/TED.2020.3047015

[43]   T.Y. Hsieh, T-C. Chang, Y-T. Chen, P-Y. Liao, T-C. Chen, M-Y. Tsai, Y-C. Chen, B-W. Chen, A-K. Chu, C-H. Chou, W-C. Chung, and J-F. Chang, "Characterization and Investigation of a Hot-Carrier Effect in Via-Contact Type a-InGaZnO Thin-Film Transistors", *IEEE Trans. Electron Dev.,* vol. 60, no. 5, pp. 1681-1688, 2013.

http://dx.doi.org/10.1109/TED.2013.2253611

[44]   S.W. Lee, P.J. Jeon, K. Choi, S.W. Min, H. Kwon, and S. Im, "Analysis of Self-Heating Effect on Short Channel Amorphous InGaZnO Thin-Film Transistors", *IEEE Electron Device Lett.,* vol. 36, no. 5, pp. 472-474, 2015.

http://dx.doi.org/10.1109/LED.2015.2411742

[45]   J. Shao, W-C. Su, T-C. Chang, H-C. Chen, K-J. Zhou, I-N. Lu, Y-F. Tu, Y-S. Shih, T-M. Tsai, C-H. Lien, J. Yang, and Q. Zhang, "Abnormal threshold voltage shift caused by trapped holes under hot-carrier stress in a-IGZO TFTs", *J. Phys. D Appl. Phys.,* vol. 53, no. 8, p. 085104, 2020.

http://dx.doi.org/10.1088/1361-6463/ab5999

[46]   G.F. Chen, T-C. Chang, H-M. Chen, B-W. Chen, H-C. Chen, C-Y. Li, Y-H. Tai, Y-J. Hung, K-J. Chang, K-C. Cheng, C-S. Huang, K-K. Chen, H-H. Lu, and Y-H. Lin, "Abnormal Dual Channel Formation Induced by Hydrogen Diffusion From SiN x Interlayer Dielectric in Top Gate a-InGaZnO Transistors", *IEEE Electron Device Lett.,* vol. 38, no. 3, pp. 334-337, 2017.

http://dx.doi.org/10.1109/LED.2017.2657546

[47]   A.T. Hatzopoulos, D.H. Tassis, N.A. Hastas, C.A. Dimitriadis, and G. Kamarinos, "An analytical hot-carrier induced degradation model in polysilicon TFTs", *IEEE Trans. Electron Dev.,* vol. 52, no. 10, pp. 2182-2187, 2005.

http://dx.doi.org/10.1109/TED.2005.856178

[48]   M. Zhang. Degradation of n-type polysilicon thin film transistors under dynamic electrical stress. Master's thesis, 2010.

[49]   A. Valletta, L. Mariucci, and G. Fortunato, "Hot-carrier-induced degradation of LDD polysilicon TFTs", *IEEE Trans. Electron Dev.,* vol. 53, no. 1, pp. 43-50, 2006.

http://dx.doi.org/10.1109/TED.2005.860656

[50]   J.J. Chen, T-C. Chang, H-C. Chen, K-J. Zhou, C-W. Kuo, W-C. Wu, H-C. Li, M-C. Tai, Y-F. Tu, Y-L. Tsai, P-Y. Wu, and S.M. Sze, "Enhancing Hot-Carrier Reliability of Dual-Gate Low-Temperature Polysilicon TFTs by Increasing Lightly Doped Drain Length", *IEEE Electron Device Lett.,* vol. 41, no. 10, pp. 1524-1527, 2020.

http://dx.doi.org/10.1109/LED.2020.3018196

[51]   Yong-Sang Kim, and Min-Koo Han, "Degradation due to electrical stress of poly-Si thin film transistors with various LDD lengths", *IEEE Electron Device Lett.,* vol. 16, no. 6, pp. 245-247, 1995.

http://dx.doi.org/10.1109/55.790723

[52]   M. Zhang, W. Zhou, R. Chen, S. Zhao, M. Wong, and H.S. Kwok, "P-6: Static Reliability of Bridged-Grain Poly-Si TFTs", *Dig. Tech. Pap.,* vol. 45, no. 1, pp. 964-967, 2014.

http://dx.doi.org/10.1002/j.2168-0159.2014.tb00250.x

[53] M. Zhang, W. Zhou, R. Chen, M. Wong, and H.S. Kwok, "Driving"-Stress-Induced Degradation in Polycrystalline Silicon Thin-Film Transistors and Its Suppression by a Bridged-Grain Structure", *IEEE Electron Device Lett.,* vol. 38, no. 1, pp. 52-55, 2017.

http://dx.doi.org/10.1109/LED.2016.2626481

[54] M. Zhang, H. Lin, S. Deng *et al.,* "High-Performance Polycrystalline Silicon Thin-Film Transistors without Source/Drain Doping by Utilizing Anisotropic Conductivity of Bridged-Grain Lines", Advanced Electronic Materials, vol. 6, no. 2, 2020.

http://dx.doi.org/10.1002/aelm.201900961

[55] G. Zhu, M. Zhang, Z. Jiang, J. Huang, Y. Huang, S. Deng, L. Lu, M. Wong, and H-S. Kwok, "Significant Degradation Reduction in Metal Oxide Thin-Film Transistors via the Interaction of Ionized Oxygen Vacancy Redistribution, Self-Heating Effect, and Hot Carrier Effect", *IEEE Trans. Electron Dev.,* vol. 70, no. 8, pp. 4198-4205, 2023.

http://dx.doi.org/10.1109/TED.2023.3283940

[56] M. Zhang, Z. Jiang, S. Deng, Z. Chen, X. Ma, C-H. Tien, L-C. Chen, M. Wong, and H-S. Kwok, "Hot Carrier Degradation Accompanied by Recovery in InSnZnO Thin-Film Transistors", *IEEE Electron Device Lett.,* vol. 44, no. 7, pp. 1124-1127, 2023.

http://dx.doi.org/10.1109/LED.2023.3277823

[57] J. Zha, "Self-Heating Stress Induced Degradation of a-IGZO TFTs and Modeling of Drain Offset a-IGZO TFTs", Master's Dissertation from Socchow University, 2021.

http://dx.doi.org/10.27351/d.cnki.gszhu.2020.001285

[58] H. Wang, M. Wang, Z. Yang, H. Hao, and M. Wong, "Stress Power Dependent Self-Heating Degradation of Metal-Induced Laterally Crystallized n-Type Polycrystalline Silicon Thin-Film Transistors", *IEEE Trans. Electron Dev.,* vol. 54, no. 12, pp. 3276-3284, 2007.

http://dx.doi.org/10.1109/TED.2007.908907

[59] S. Hashimoto, Y. Uraoka, T. Fuyuki, and Y. Morita, "Suppression of Self-Heating in Low-Temperature Polycrystalline Silicon Thin-Film Transitors", *Jpn. J. Appl. Phys.,* vol. 46, no. 4R, p. 1387, 2007.

http://dx.doi.org/10.1143/JJAP.46.1387

[60] M. Zhang, Z. Xia, W. Zhou, and R. Chen, "Man Wong, and Hoi-Sing Kwok, "Degradation behaviors and degradation mechanisms of bridged grain polycrystalline silicon thin film transistors under DC biass tresses,"", *Yejing Yu Xianshi,* vol. 30, no. 02, pp. 187-193, 2015.

http://dx.doi.org/10.3788/YJYXS20153002.0187

[61] M. Kimura, S. Inoue, T. Shimoda, S.W-B. Tam, O.K.B. Lui, P. Migliorato, and R. Nozawa, "Extraction of trap states in laser-crystallized polycrystalline-silicon thin-film transistors and analysis of degradation by self-heating", *J. Appl. Phys.,* vol. 91, no. 6, pp. 3855-3858, 2002.

http://dx.doi.org/10.1063/1.1446238

[62] M. Zhang, M. Wang, and H. Wang, "Degradation of metal-induced laterally crystallized n-type poly-Si thin-film transistors under dynamic voltage stress", *in 2008 15th International Symposium on the Physical and Failure Analysis of Integrated Circuits,* pp. 1-4, 2008.

http://dx.doi.org/10.1109/IPFA.2008.4588214

[63] L. Mariucci, P. Gaucci, A. Valletta, A. Pecora, L. Maiolo, M. Cuscuna, and G. Fortunato, "Edge Effects in Self-Heating-Related Instabilities in p-Channel Polycrystalline-Silicon Thin-Film Transistors", *IEEE Electron Device Lett.,* vol. 32, no. 12, pp. 1707-1709, 2011.
http://dx.doi.org/10.1109/LED.2011.2169040

[64] M. Du, J. Zhao, D. Zhang, H. Wang, Q. Shan, and M. Wang, "Roles of Gate Voltage and Stress Power in Self-Heating Degradation of a-InGaZnO Thin-Film Transistors", *IEEE Trans. Electron Dev.,* vol. 68, no. 4, pp. 1644-1648, 2021.
http://dx.doi.org/10.1109/TED.2021.3055751

[65] Y. Zhou, F. Liu, H. Yang, X. Zhou, G. Li, M. Zhang, R. Chen, S. Zhang, and L. Lu, "Competition between heating and cooling during dynamic self-heating degradation of amorphous InGaZnO thin-film transistors", *Solid-State Electron.,* vol. 195, p. 108393, 2022.
http://dx.doi.org/10.1016/j.sse.2022.108393

[66] K. Ito, "24-1: Invited Paper: Development of High Quality IGZO-TFT with Same On-Current as LTPS", *SID Symposium Digest of Technical Papers,* vol. 51, 2020pp. 343-346
http://dx.doi.org/10.1002/sdtp.13874

[67] T. Wu, C-W. Sher, Y. Lin, C-F. Lee, S. Liang, Y. Lu, S-W. Huang Chen, W. Guo, H-C. Kuo, and Z. Chen, "Mini-LED and Micro-LED: Promising Candidates for the Next Generation Display Technology", *Appl. Sci. (Basel),* vol. 8, no. 9, p. 1557, 2018.
http://dx.doi.org/10.3390/app8091557

[68] S.M. Lee, W.J. Cho, and J.T. Park, "Device Instability Under High Gate and Drain Biases in InGaZnO Thin Film Transistors", *IEEE Trans. Device Mater. Reliab.,* vol. 14, no. 1, pp. 471-476, 2014.
http://dx.doi.org/10.1109/TDMR.2013.2278990

CHAPTER 5

Alternating Current Voltage Stress-Induced Degradation in Thin-Film Transistors

Abstract. This chapter investigates the degradation induced by alternating current (AC) voltage stress in thin-film transistors (TFTs). It provides a detailed examination of AC voltage stress, discussing its impact on silicon-based TFTs and metal oxide TFTs. The chapter aims to summarize and analyze the influence of these degradation mechanisms on the performance and reliability of TFT devices.

Keywords: Alternating current (AC), AC hot carrier (HC) effect, Degradation models, Metal oxide TFTs, Polycrystalline silicon TFTs, Waveform elements.

5.1. INTRODUCTION

Chapter 4 provided a detailed discussion on the degradation of thin-film transistors (TFTs) under direct current (DC) stress. However, as mentioned in Section 1.4.1, the actual operating state of a TFT is subject to a superposition of various changing signals. Therefore, the study of alternating current (AC) reliability is more aligned with the actual working conditions of TFTs compared to DC reliability research.

Current research methods for investigating the AC reliability of TFTs primarily involve applying periodic pulse stress to either the gate or the drain and observing the resulting degradation in the device. For Si-based TFTs, the application of AC stress typically leads to changes in the device's on-state current (I_{on}), while for metal oxide (MO) TFTs, AC stress usually results in a shift in the device's threshold voltage (V_{th}).

The mechanisms proposed to explain the above degradation phenomena largely focus on the AC hot carrier (HC) effect. Under the application of AC stress, a large number of HCs are generated within the device due to the influence of a strong electric field, leading to device degradation. This chapter presents the degradation conditions of Si-based TFTs and MO TFTs under single-ended AC stress and the related degradation mechanism.

Meng Zhang & Mingxiang Wang

5.2.   ALTERNATING   CURRENT   DEGRADATION   MODEL   IN POLYCRYSTALLINE THIN-FILM TRANSISTORS

5.2.1. Degradation Behavior of Polycrystalline Thin-Film Transistors under Alternating Current Stress

As illustrated in Fig. (**5.1**), the degradation of the transfer characteristics of a polycrystalline silicon (poly-Si) TFT under AC gate stress is shown for different drain voltages of –0.1 V and –5 V. The AC stress conditions were applied with a rising time (t_r) of 1 ns, a falling time (t_f) of 50 ns, a duty cycle of 50%, a peak voltage (V_{peak}) of 10 V, and a base voltage (V_{base}) of –10 V. It is evident that the device exhibits clear signs of HC degradation, characterized by a significant reduction in I_{on}, while the subthreshold region remains largely unchanged.

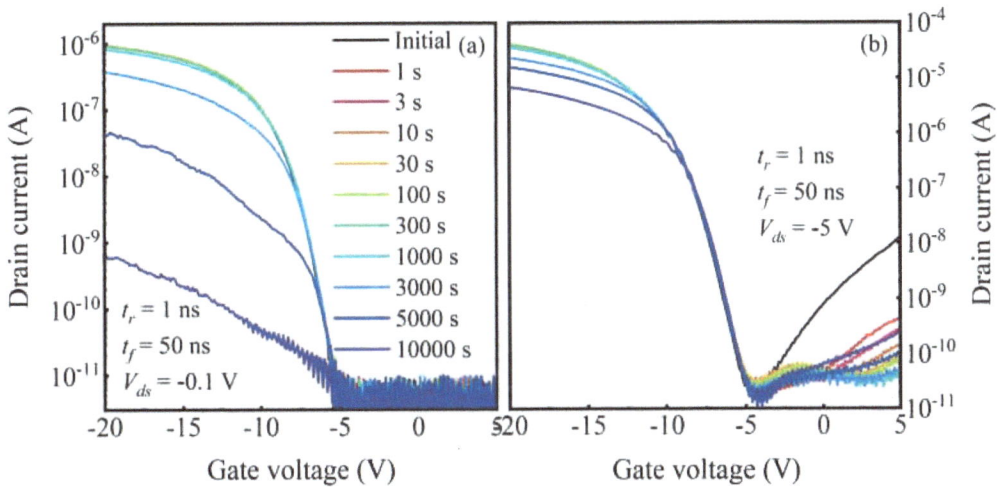

Fig. (5.1). Transfer characteristic curves with time under V_g stress for V_{ds} of –0.1 V and 5 V.

Upon comparing the transfer characteristics curves at different drain-source voltage (V_{ds}), it is apparent that the degree of degradation at –5 V is notably less than that at –0.1 V. Moreover, the device at –5 V shows a distinct phenomenon of degradation followed by an increase in the off-state current (I_{off}). These degradation phenomena are typical for poly-Si TFTs, and the extent of degradation can vary depending on the conditions of the applied AC stress [1].

5.2.2. Waveform Elements Dependence of Alternating-Current Degradation

Fig. (**5.2**) typically illustrates the pulse stress applied to TFTs [1], which can be categorized based on the location of application into AC gate stress [1-10] and AC drain stress [11-14]. Depending on the pulse shape, pulse stress can generally be divided into several components: t_r, t_f, peak time (t_{peak}), base time (t_{base}), and V_{peak} and V_{base} [1-14].

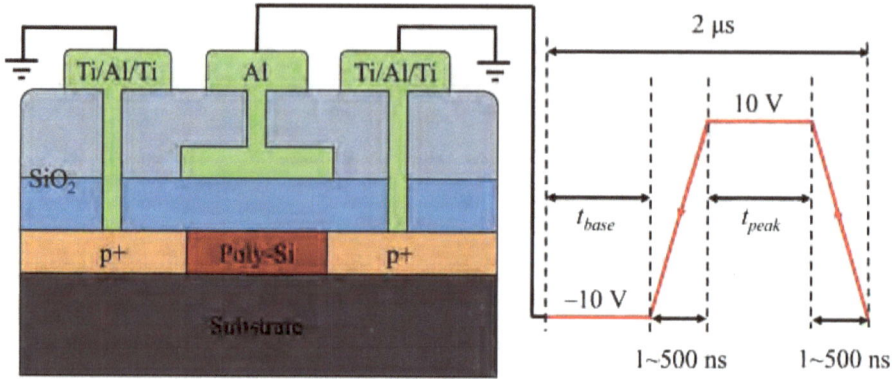

Fig. (5.2). The schematic diagram of poly-Si TFTs and the waveform of the gate voltage pulse [1].

Usually, the degradation caused by gate and drain pulse stresses are similar, but gate stress often results in less degradation due to the blocking effect of the gate insulator (GI) layer under the same voltage conditions. Variations in these pulse conditions can influence the extent of device degradation to a certain degree.

5.2.2.1. Peak Voltage and Base Voltage

As two critical parameters that determine the pulse amplitude, the magnitude of V_{peak} and V_{base} will directly affect the degradation of TFTs under AC stress. The difference between V_{peak} and V_{base}, *i.e.*, the pulse amplitude, will cause greater I_{on} degradation when larger, provided all other conditions are equal. As shown in Fig **(5.3)**, for a fixed V_{base} of 0 V and varying V_{peak}, the decay is intensified with an increased pulse magnitude. Across various amplitudes, a uniform degradation gradient of 1.0 is consistently noted, indicat

ing the potential involvement of a reaction-limited mechanism for defect generation. If the scale of the ordinate is altered, it can be seen that the device experiences a region of initial I_{on} increase, with larger voltage amplitudes leading to higher increases in I_{on}, although these increases decline more rapidly [4, 11].

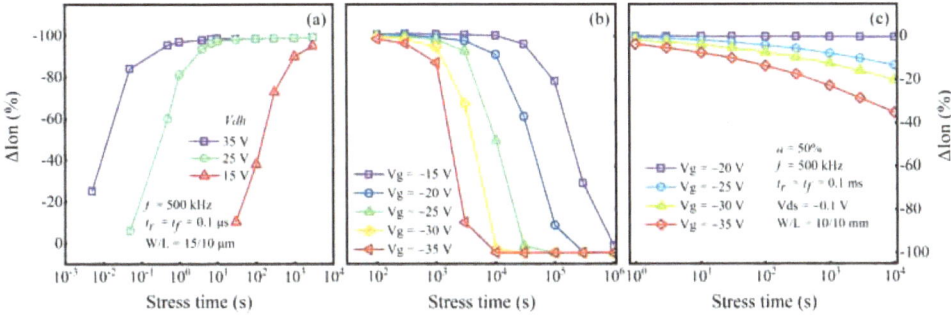

Fig. (5.3). Device degradation at different pulse amplitudes [4, 11].

Further subdivision of the AC stress can be made based on the voltage range, taking gate AC stress as an example. When the low voltage of the AC stress is greater than V_{th}, meaning the entire AC pulse falls within the device's on-state, this kind of stress can be referred to as on-state AC stress [3]. When the V_{base} of the AC stress is less than V_{th}, and V_{peak} is greater than V_{th}, the device operates in both on and off states; this kind of stress can be referred to as on-off-state AC stress [1, 5-7]. When the high voltage of the AC stress is less than V_{th}, the device operates in the off-state; this kind of stress can be referred to as off-state AC stress [2, 8-10].

Fig. (5.1) illustrates the degradation of the device under on-off-state AC stress [1], while Fig. (5.4) depicts the degradation under off-state AC stress [2]. It is noticeable that the devices demonstrate analogous degradation traits, with the I_{on} degrading while the subthreshold region remains relatively unchanged.

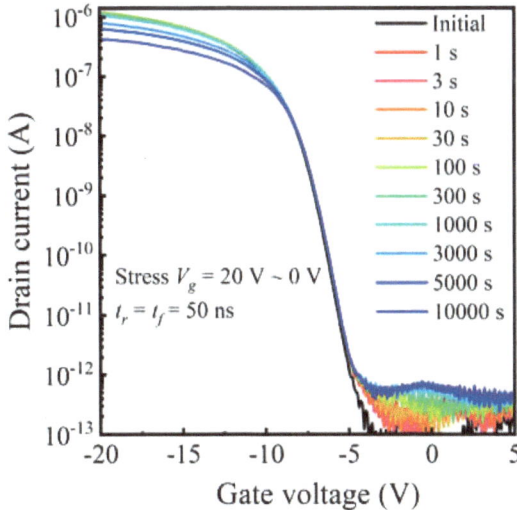

Fig. (5.4). Device degradation under off-state pulse stresses.

However, if the pulse is entirely within the on-state region, the degradation of the device will be noticeably different. As shown in Fig. (**5.5**), the device experiences a certain shift in the V_{th}. Additionally, some studies have found that the device under on-state AC stress does not undergo significant degradation.

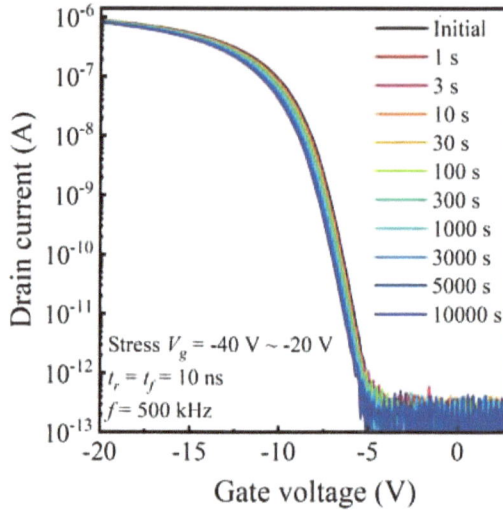

Fig. (5.5). Device degradation under on-state pulse stresses.

5.2.2.2. Rising Time and Falling Time

In the practical application of TFTs, the presence of t_r and t_f represents the rate of signal change, with shorter t_r and t_f indicating a faster change rate. As shown in Fig. (**5.6a**), the application of AC stress from 15 V to 0 V on the drain of an N-type TFT leads to noticeable changes in the device's transfer characteristics curve. A clear HC degradation can be observed (refer to Fig. **5.6**), with a decrease in I_{on} and no change in the subthreshold region. Fig. (**5.6b**) shows the change in the transfer characteristics curve's I_{off}, which gradually decreases with increasing stress time [11]. To explore the impact of t_r and t_f on device degradation during stress, as shown in Fig. (**5.7**), fixing t_r at 1 µs and varying t_f reveal no significant change in device degradation, indicating that degradation is independent of t_f. Conversely, fixing t_f at 1 µs and varying t_r show that smaller t_r leads to more severe degradation, suggesting that degradation is related only to t_r and not to t_f. [11].

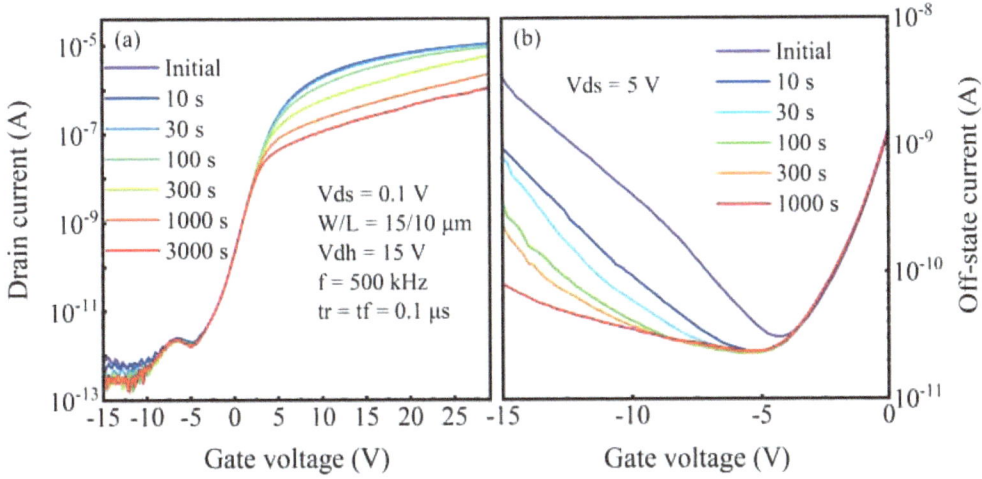

Fig. (5.6). Transfer characteristic curves of n-type LTPS TFTs measured at V_{ds} = 0.1 V under drain stress conditions as a function of time. The inset shows the I_{off} measured at V_{ds} = 5 V [11].

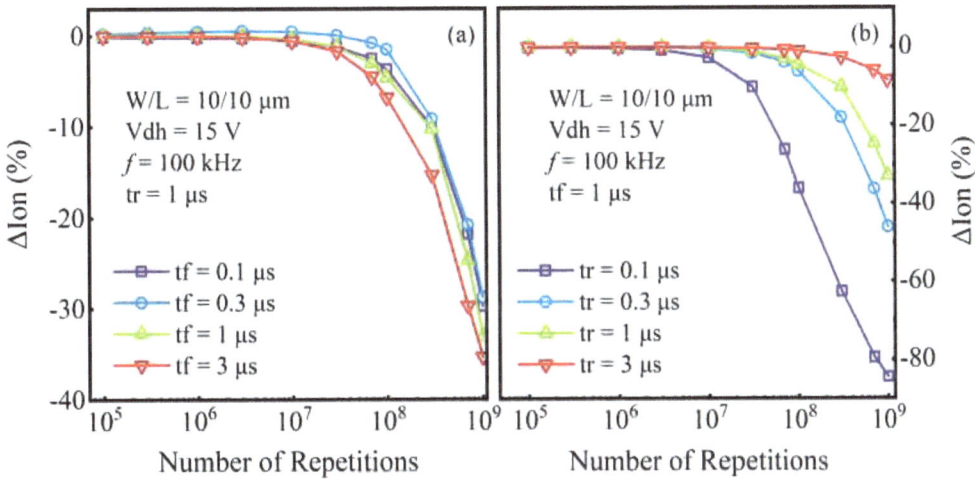

Fig. (5.7). Effect of different t_f vs. t_f on I_{on} degradation for (a) fixed t_r = 1 μs and (b) t_f = 1 μs [11].

In the case of P-type poly-Si devices, the circumstances are markedly different. As depicted in Fig. (**5.8**), the application of pulse stress to the drain exhibits this contrast; the device degradation almost overlaps when t_r is fixed and t_f is varied. However, when t_r is fixed and t_f is varied, the degradation of the device increases significantly as t_f decreases. It is evident that for P-type devices under AC stress at the drain, degradation is mainly related to t_f [12].

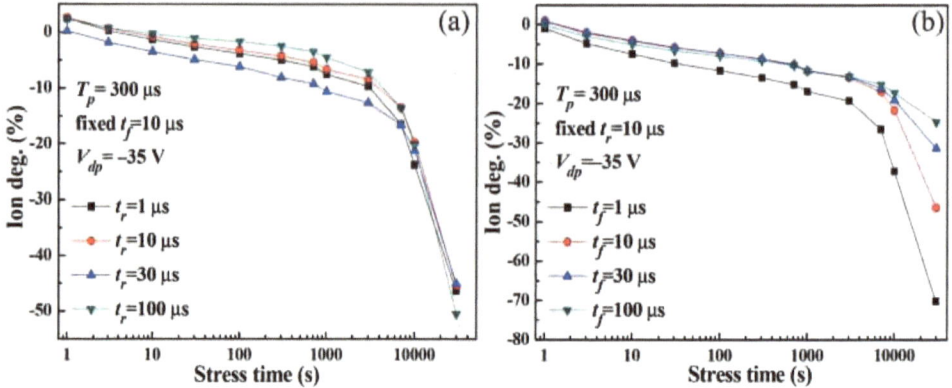

Fig. (5.8). (a) Effect of varying t_r on I_{on} degradation for fixed t_f = 10 μs (b) Effect of varying t_f on I_{on} degradation for fixed t_r = 1 μm [12].

When the application of stress is shifted from the drain to the gate, the situation is reversed compared to when stress is applied to the drain. In the case of N-type devices, the degradation correlates with t_f, whereas for P-type devices, it correlates with t_r.

In P-type semiconductors, the application of stress to the gate predominantly induces a shift in the channel's state, moving it from an on-state to an off-state, with t_r being the key parameter influencing this transition. For N-type devices, the application of stress to the gate predominantly induces a shift in the channel's state, moving it from an on-state to an off-state, with t_f being the key parameter influencing this transition. When stress is applied to the gate, for P-type devices, the electrodes are typically highly doped P-type silicon, and the electrodes and active layer can be considered as two PN junctions, with t_f causing the PN junction to switch from forward bias to reverse bias. Similarly, for N-type devices, the electrodes are typically highly doped N-type silicon, and t_r also causes the PN junction to switch from forward bias to reverse bias [12].

5.2.2.3. Peak Time and Base Time

In actual pulse signals, t_{peak} and t_{base} represent the lengths of each signal segment, and since signal changes are almost entirely random, the durations of t_{peak} and t_{base} also vary randomly. In the application of AC pulse stress, the lengths of t_{peak} and t_{base} directly affect the pulse period. As shown in Fig. (**5.9**), which illustrates the degradation of I_{on} under different periods with fixed t_r and t_f, the lengths of t_{peak} and t_{base} increase or decrease synchronously. As depicted in Fig. (**5.9a**), using stress time as the horizontal axis, it can be observed that the shorter the period, the more pronounced the device degradation. However, as mentioned in Section 5.2.2.2, the degradation of the device is directly related to the rise or fall times of the pulse, and different period lengths will inevitably result in different numbers of rise and fall times within the same duration. Therefore, measuring the degree of device degradation by time alone is unfair [15].

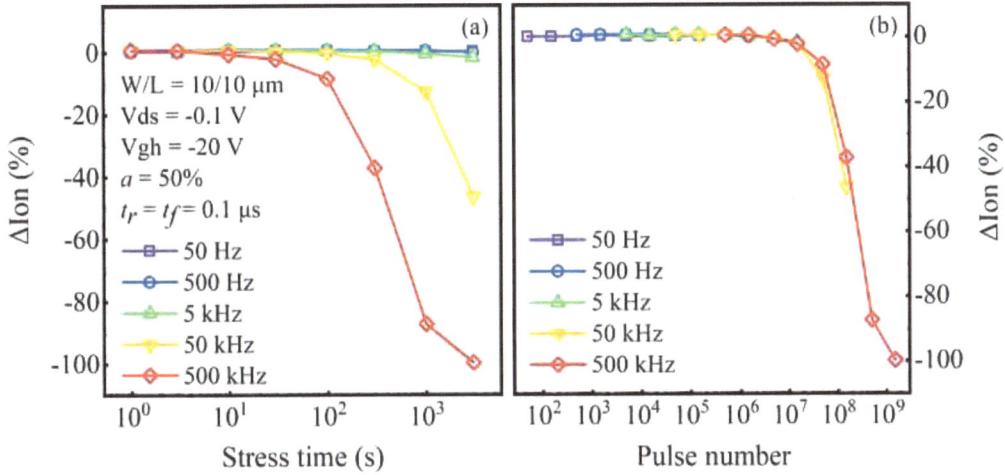

Fig. (5.9). Variation of (a) ΔI_{on} with time (b) ΔI_{on} with number of pulses for different AC pulse periods [15].

Thus, it is necessary to process the horizontal axis, as shown in Fig. (**5.9b**), which presents the number of pulses obtained by dividing the stress time by the period length. It is evident that the I_{on} degradation under different periods almost coincides. The conclusion drawn indicates that the degradation of the I_{on} is solely associated with the count of pulses, with a greater number of pulses causing more significant device degradation. The lengths of t_{peak} and t_{base} have little impact on the degree of device degradation.

The above experiment can generally conclude that the lengths of t_{peak} and t_{base} do not affect the magnitude of AC degradation. However, since the lengths of t_{peak} and t_{base} change synchronously, to further investigate the impact of t_{peak} and t_{base} on the AC stress reliability of poly-Si TFTs, Fig. (**5.10**) shows the degradation of devices under pulse stress with different duty cycles. In this case, the pulse frequency is fixed at 30 kHz, avoiding the issue of different device degradation due to varying numbers of pulses. It can be observed that under different duty cycles, the device degradation almost completely coincides [15].

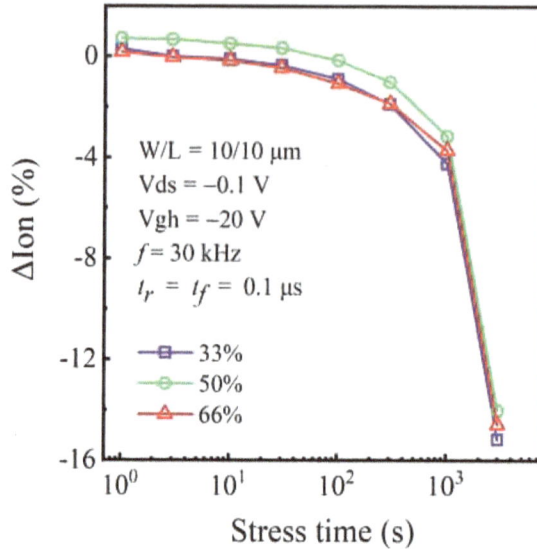

Fig. (5.10). Variation of ΔI_{on} with time for different AC duty cycles [15].

Based on the above test results, it is evident that t_{peak} and t_{base} have little to no significant impact on the AC reliability of devices. However, it is important to note that the experiments mentioned above are based on pulses with relatively long periods and slow rise and fall times, primarily on the order of microseconds. When the pulse change rate becomes faster, the degradation of the device may exhibit some differences, which are further explained in Section 6.4 of this book regarding ultrafast pulse stress.

5.2.3. Unified Alternating-Current Degradation Model for Thin-Film Transistors

Compared to DC reliability studies, AC reliability research is more aligned with the actual working environment of TFTs, as these devices are often subjected to complex and variable voltage conditions over long periods. Since 2002, a comprehensive investigation into the AC dependability of poly-Si TFTs has been spearheaded by Yukiharu Uraoka's group at the Nara Institute of Science and Technology [16]. Their findings indicated that AC pulse stress applied to the gate caused significant I_{on} degradation in the devices, which closely resembled the degradation observed in DC HC effects. Furthermore, the research revealed a significant correlation between degradation and the transient durations of pulse rise and fall. The impact was found to be more closely linked to the total pulse count instead of pulse frequency [17].

Since then, several teams have conducted a series of research works on the AC reliability of poly-Si TFTs. Notable among these are Professor Yukiharu Uraoka's team from Nara Institute of Science and Technology, Dr. Yoshiaki Toyota's team from Hitachi Ltd. Central Research Laboratory [9, 18], Professor Yaxiang Dai's team from National Chiao Tung University [7, 19], and Professor Mingxiang Wang's from Soochow University [11, 13] and Prof. Meng Zhang's teams from Shenzhen University [6, 14, 15, 20-23].

Professor Uraoka's team used a picosecond emission microscope to observe the E_x and carrier distribution in the drain and channel, offering insights into HC degradation [16, 17]. Dr. Toyota's team considered the capture and emission of carriers by traps in the channel to explain HC degradation [18]. Professor Dai's team employed slicing techniques to divide the entire TFT into many smaller series-connected TFTs and used an equivalent circuit model along with the circuit simulation software SPECTRE to calculate the transient voltage distribution at different nodes under dynamic stress, providing an explanation for HC degradation. Later, Professor Wang and Prof. Zhang's teams proposed an AC reliability model for non-equilibrium PN junctions, which effectively explains the degradation mechanisms of devices under various AC voltage stress conditions.

5.2.3.1. Simulation of Alternating-Current Voltage Stress

The development of the non-equilibrium PN junction AC reliability model is predicated on the simulation of internal electrical parameters within the device. Prior analyses have revealed that device degradation is predominantly characterized by HC effects, which are directly associated with the electric field and current density within the device. The subsequent simulations focus on N-type devices subjected to AC stress at the drain as an illustrative example [13].

The collective density of states includes two exponential tails (a donor-like valence band and an acceptor-like conduction band) as well as two Gaussian-shaped deep bands (with one having acceptor-like properties and the other donor-like traits). The settings for the poly-Si model with ongoing defects are aligned with the default parameters of the simulation tool. The mobilities for both electrons and holes are established at a common value of 70 cm^2/Vs. Crucial elements, specifically the transient electric field and current density in the lateral direction (denoted as E_x and J_x), are detected at the boundary of the channel and the drain, as well as during the transient periods of pulse transition (either t_r or t_f).

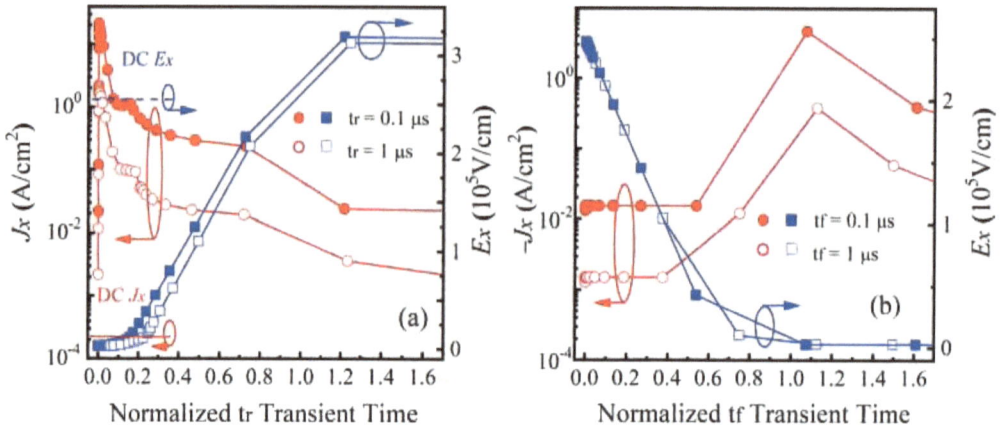

Fig. (5.11). Variation of ΔI_{on} with time for different AC duty ratios [13].

In Fig. (**5.11**), the values of E_x and J_x are illustrated with respect to the transient moment, scaled by the width of the transition edge for Figs. (**11a** and **11b**) using t_r and t_f, respectively. During the t_r transition in Fig. (**5.11a**), as V_d rises from 0 to V_{peak}, E_x soars to a notable level of 3.2×10^5 V/cm. Simultaneously, the lateral current density J_x exhibits a sharp increase and then a slow decrease. In the final stage of the t_r transition, where E_x has surpassed 10^5 V/cm, a lingering "tail" current is still detectable. These carriers are potentially affected by the high field, becoming HCs, with J_x tracking the direction of E_x. A shorter t_r results in a slightly higher E_x and a substantially larger J_x. This leads to the creation of more HCs and increased degradation, reflecting the strong t_r dependence in observed dynamic HC degradation. Conversely, under a DC stress of 15 V, J_x nearly reaches zero, even with a high electric field, thus preventing HC degradation. Additionally, under dynamic V_d stress, an increase in the vertical electric field (E_y) is also observed during the t_r transition, staying at a high level during t_{peak}, which is favorable for hole trapping and leads to a reduction in I_{off} [13].

In Fig. (**5.11b**), throughout the t_f transition phase (as V_d transitions from V_{peak} to 0), E_x rapidly diminishes, exhibiting a comparable pattern across various t_f durations. Additionally, a J_x value is detected, which is approximately 100 times smaller than during the t_r transition. Notably, this J_x flows in the opposite direction to E_x, indicating it is not a drift current influenced by the drain electric field. Consequently, this type of current does not generate HCs. This accounts for the observed t_f independence in dynamic HC degradation [13].

5.2.3.2. Nonequilibrium-Junction Degradation Model

A degradation model centered on the nonequilibrium junction is put forth, taking into account the trap-associated carrier emission and recombination processes. In the poly-Si film, a significant number of traps are present at the grain boundaries and within the grains themselves. As illustrated in Fig. (**5.12a**), during steady-state operation with the drain electrode set to zero potential, certain traps in the intrinsic channel near the n+ drain diffusion require hole emission for the creation of the drain junction's depletion zone. Upon the application of pulsed V_d, which swiftly transitions from 0 to V_{peak} during the t_r interval, the junction becomes more intensely reverse-biased, necessitating the expansion of the depletion region into the channel, as illustrated in Fig. (**5.12b**). Nonetheless, the time for carrier emission, τ_t, is contingent upon the energy level of the traps E_t by $1/\tau_t = v_{th}\sigma_h n_i \exp\left[(E_i - E_t)/kT\right]$, where v_{th}, σ_h, n_i, and E_i are the thermal speed of the carriers, the cross-sectional area for hole capture, the concentration of carriers inherent to the material, and the level of the intrinsic Fermi energy, in that order. Regarding the deep traps ($E_t \approx E_i$), the emission time τ_t might be considerably lengthy, potentially spanning the *ms* range ($\sigma_h = 10^{-14}$ cm^2); this duration is significantly more extended than the approximate microsecond-scale t_r transient. Consequently, during the t_r transient phase, ionization is limited to shallow traps (for $t_r = 1$ μs, $E_t > E_i + 0.16$ eV). Hence, the depletion zone illustrated in Fig. (**5.12b**) needs to be markedly broadened into the channel through hole emission from shallow traps only. At the same time, an electric field is established across this zone. Following this, holes that are released from deeper traps keep neutralizing the positive space charge in the n+ drain diffusion area, denoted by the yellow spots in Fig. (**5.12b**). At this juncture, the electric field E_x adjacent to the channel/drain boundary is sufficiently intense to propel certain emitted carriers into HCs, thereby generating defects in proximity to the drain junction, which leads to the observed HC degradation [13].

According to the suggested model, the principal aspects of dynamic V_d HC degradation can be elucidated. With a reduced t_r duration, E_x is capable of attaining a more elevated magnitude. Concurrently, only carriers trapped in even less depth in energy levels are liberated within a briefer V_d phase transition; a more substantial portion of deep-trap emission will take place subsequent to the establishment of a higher E_x, resulting in exacerbated degradation. Throughout the t_f phase transition, as V_d transitions from V_{peak} to 0, it is expected that the drain junction will return to the equilibrium state shown in Fig. (**5.12a**). The depletion zone, which extended into the channel, should resume its original position, as demonstrated in Fig. (**5.12c**). Carrier recombination at the outer boundary of the depletion zone neutralizes the traps that were ionized during the t_r phase. The E_x in this region is exceedingly low, precluding the generation of HCs. Consequently, the degradation is not influenced by the t_f duration [13].

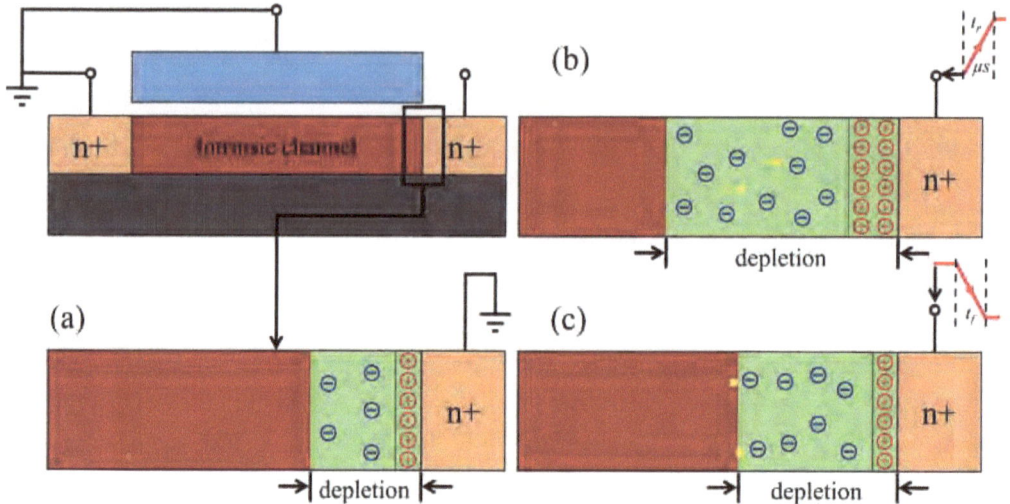

Fig. (5.12). Schematic diagram of the device degradation model. (a) Adjacent to the n+ drain, the intrinsic channel region experiences an expansion of the drain junction's steady state as the drain electrode reaches zero potential. (b) During the t_r phase, carriers that are emitted from deep traps transform into HCs. (c) Carriers emitted from outside the depletion region recombine during t_f. [13].

5.3. ALTERNATING CURRENT DEGRADATION MODELS IN METAL OXIDE THIN-FILM TRANSISTORS

Section 5.2 discussed the degradation phenomena of poly-Si TFTs under pulse stress and the impact of different pulse stresses on degradation, proposing a reliability model for the non-equilibrium junction between the source/drain and the active layer. Similarly, MO TFTs are also subject to pulse stress in high-refresh-rate pixel circuits. The following text will detail the degradation phenomena of MO TFTs under pulse stress and the pulse parameters that affect the extent of device degradation under pulse stress.

5.3.1. Degradation Behavior under Alternating Current Stress of Metal Oxide Thin-Film Transistors

In poly-Si TFTs, the degradation under pulse stress is primarily characterized by a decrease in the I_{on} of the transfer characteristic curve. For MO TFTs, the degradation phenomena under pulse stress will be discussed mainly through single-ended pulse stress, considering the types of pulse stress that may be applied to the devices in actual circuits [24-26].

Fig. (5.13) compares the changes in the transfer characteristic curves of MO TFTs under DC stress at the gate and pulse stress at the gate. When a DC positive voltage of 20 V is applied to the gate, the transfer characteristic curve gradually shifts positively with increasing stress time due to the capture of electrons by interface defects at the interface between the GI and the active layer. When the device is subjected to an AC pulse stress of 0 to 20 V with $t_r = t_f = 100$ ns and the source/drain grounded, the transfer characteristic curve significantly shifts positively with increasing stress time, and the shift is larger than that under DC +20V stress. This indicates that in addition to the degradation caused by DC, the AC mechanism also plays a role in the device's degradation [27].

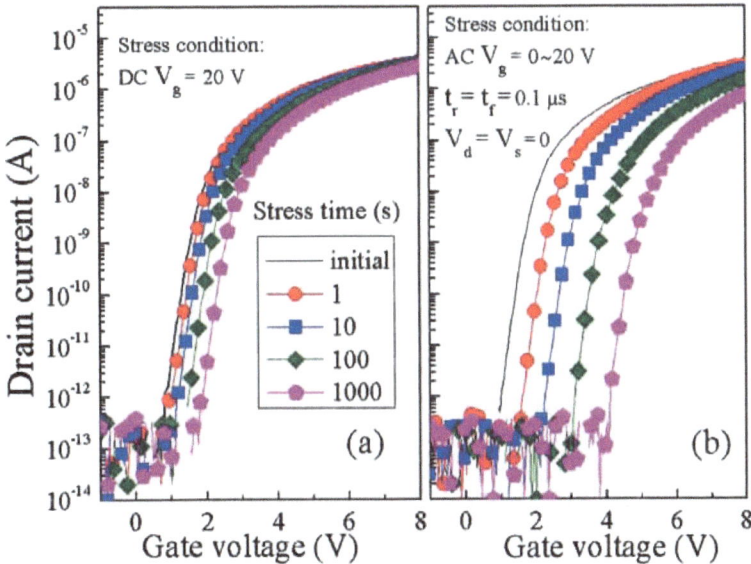

Fig. (5.13). Trend of transfer characteristics versus stress time for a-IGZO TFTs under (a) DC 20V and (b) gate pulse conditions. The data were all measured at $V_{ds} = 5$ V [27].

To elucidate which part of the pulse governs the degree of device degradation under AC stress, the durations of t_r and t_f were varied individually, and the corresponding degradation was observed. As illustrated in Fig. (5.14), when t_f is held constant and t_r is altered, the shift in the threshold voltage (V_{th}) of the transfer characteristic curve does not exhibit significant changes. This indicates that the AC effect of the device under gate AC stress is not modulated by t_r [27]. Subsequently, by fixing the pulse's t_r and varying t_f, it was found that as t_f increases, the shift in the V_{th} of the transfer characteristic curve gradually decreases. This suggests that the AC effect of the device under gate pulse stress is modulated by t_f [27]. To further investigate the specific mechanisms of modulation, TCAD simulations were conducted.

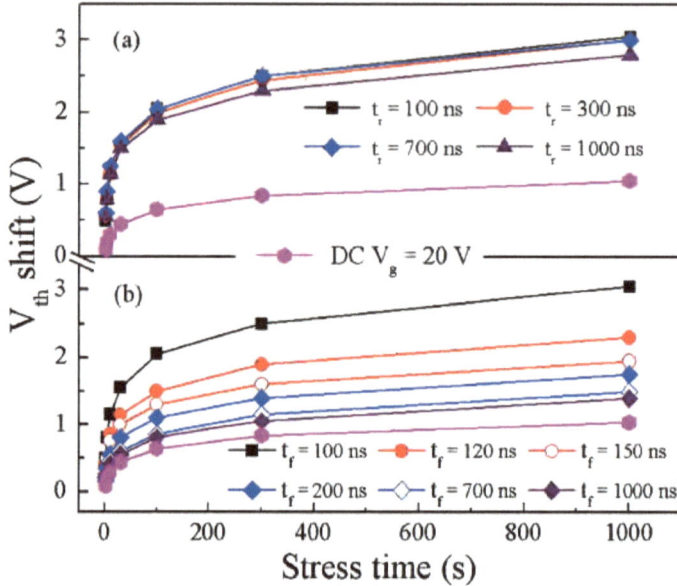

Fig. (5.14). Offset of V_{th} versus stress time for rising edge (a) and falling edge (b) conditions for different pulses. All gate pulse amplitudes are 0 ~20 V. [27].

As shown in Fig. (**5.15**), an IGZO TFT was constructed in the TCAD simulation software ATLAS, and the lateral electric field (E_x) distribution within the active layer at the end of the gate pulse's falling edge was extracted. It was observed that the E_x is primarily concentrated near the source and drain electrodes, leading to the inference that this region is the primary area for HC generation [28].

A node within the active layer near the source and drain was selected to simulate the variation of the E_x throughout the entire pulse process. When the gate bias is zero, the channel is in thermal equilibrium, and the E_x is also zero. During the rising edge of the pulse, there is a slight increase in the horizontal electric field at the channel surface, which then quickly returns to zero. This is clearly due to the presence of a horizontal electric field between the gate and source/drain when the channel is in a depleted state as the gate voltage begins to rise [28]. However, as the gate voltage continues to increase, electrons accumulate at the channel surface, forming an electron accumulation layer, allowing the source and drain to conduct and the electric field to rapidly decrease to zero. However, during the falling edge of the pulse, electrons return to the source and drain and the channel becomes depleted, causing the horizontal electric field to rapidly increase in the opposite direction and then slowly decrease. The slow decrease is likely due to the superposition of the electric field at the junction and the emission of electrons from defects (Fig. **5.16**) [5].

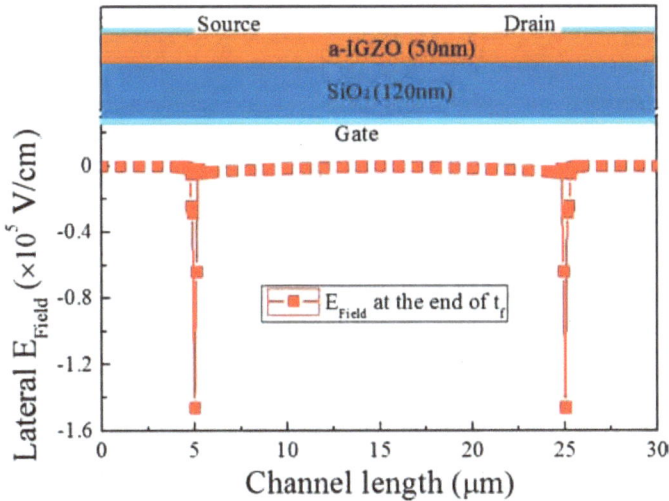

Fig. (5.15). Distribution of the peak of the horizontal electric field transient along the channel surface for a gate pulse falling edge of 0.1 μs [28].

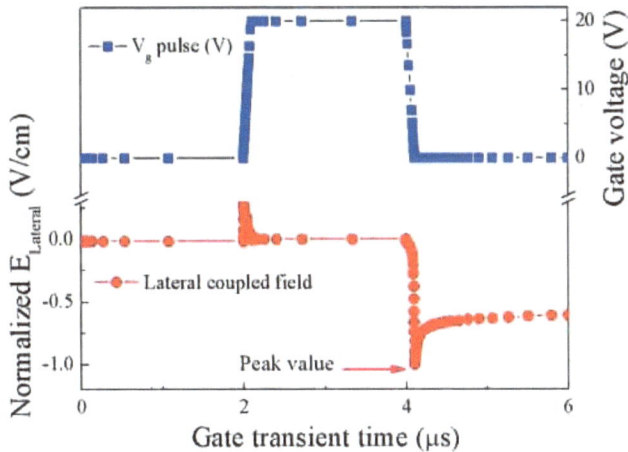

Fig. (5.16). Variation of gate pulse stress and horizontal transient electric field at the channel surface (near source/drain) [28].

Based on the simulation results, an AC HC degradation model for IGZO TFTs under gate pulse stress was proposed. According to the experimental and simulation results presented above, it is believed that the degradation mechanism of a-IGZO TFT devices under gate pulse stress occurs during the transient process of the falling edge, generating a certain number of HCs [4]. Some of these HCs can overcome the interface barrier and be captured by the GI or interface near the source/drain, creating defects at the interface [4]. When V_g is at a high level, electrons accumulate at the channel surface, forming an accumulation layer, and the interface traps between a-IGZO and the GI are also filled. When V_g begins to decrease, to maintain charge balance, the number of accumulated electrons in the channel will decrease, and the excess electrons will flow through the source/drain electrodes to the ground. However, if the falling edge of the pulse is short enough, the gate voltage will quickly drop below the flatband voltage, causing the channel to become depleted. However, some electrons captured by defect states, especially those captured by deep-level defects, require a certain amount of time to be emitted from the defects [4]. They cannot return to the source/drain electrode in time. At this point, the non-equilibrium junction formed between the source/drain electrode and the active layer is reverse-biased, and the voltage of the gate mainly falls on the depleted region of the non-equilibrium junction formed between the source/drain electrode and the active layer.

Consequently, a transient lateral E_x is formed between the channel and the source/drain electrode, with the peak value of the electric field being close to the source/drain electrode [4]. As a result, electrons emitted from defects in the channel are exposed to the transient E_x and absorb energy from the field, being accelerated into HCs. Some of these HCs may gain sufficient energy from the field to overcome the interface barrier and become injected into the GI, forming localized trap charges near the source/drain electrode, as shown in Fig (**5.17**). For steeper falling edges, the channel-coupled voltage (V_{ch}) is larger, resulting in a higher transient electric field E_x, leading to greater degradation [27].

The reliability of IGZO TFTs under drain pulse stress is similar to those under gate pulse stress, exhibiting a noticeable positive shift in the transfer curve and an increase in *SS*. As t_r increases, the degradation of the transfer characteristic curve gradually decreases, which is consistent with the degradation model mentioned earlier.

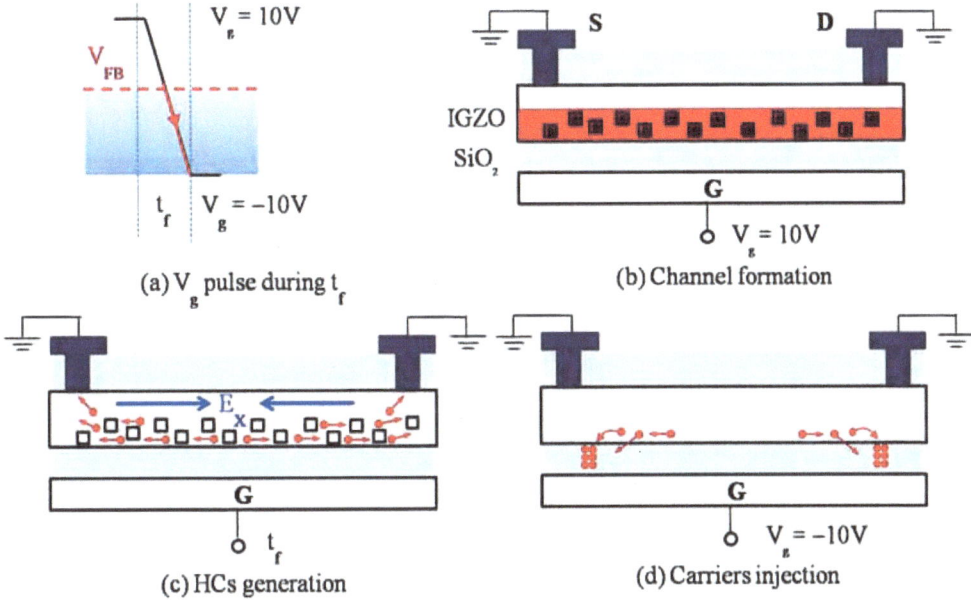

Fig. (5.17). Modelling of dynamic HC degradation due to a-IGZO TFT gate pulse conditions. (a) The gate pulse V_g switches back and forth rapidly from 10 V to –10 V; (b) At V_g = 10 V, the channel is in an electron accumulation state and the defects in the semiconductor layer are filled with electrons; (c) At the falling edge, the electrons remaining in the channel are in a horizontal electric field and absorb energy from the field to become HCs; (d) HCs are injected into the gate near the source drain (e) HCs are injected into the gate insulation layer near the source-drain or are trapped by nearby interface defects [27].

For the reliability of ITZO TFTs under dynamic drain voltage stress, similar degradation phenomena are observed. As shown in Fig (**5.18b**), the transfer characteristic curve significantly shifts positively with increasing stress time. However, it is worth noting that, unlike IGZO TFTs, the magnitude of the degradation is not only related to t_r but also to t_f. Through simulation, it is found that the degradation tuned by t_r is attributed to the AC HC mechanism; steeper t_r causes a larger E_x and induces a more severe AC HC degradation. The degradation tuned by t_f is attributed to the HC recovery mechanism; a steeper t_f causes a smaller E_y and induces a more severe AC HC degradation. The associated degradation model is shown in Fig. (**5.19**) [29].

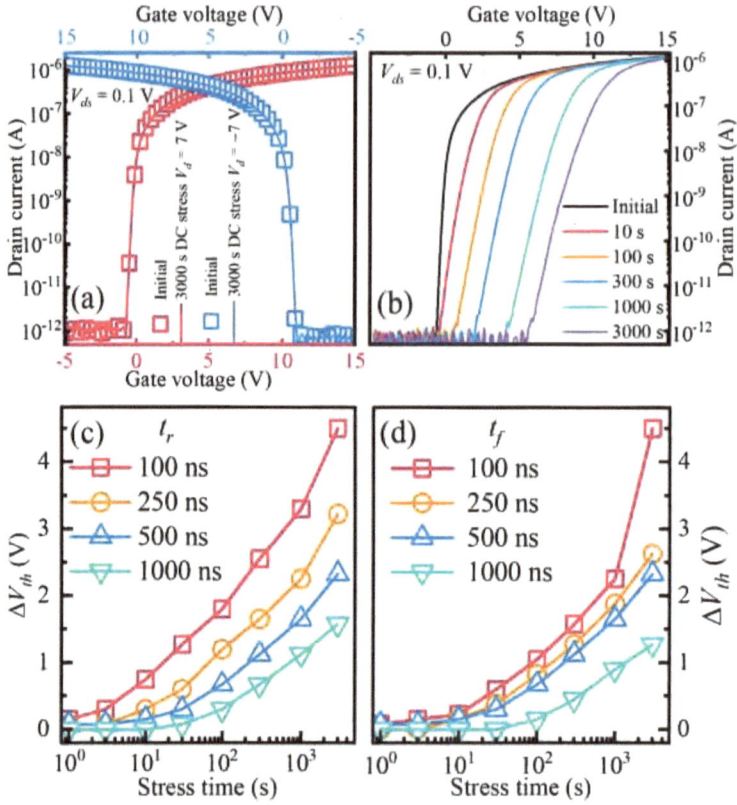

Fig. (5.18). Time evolution of transfer curves of ITZO TFT under (a) DC stress $V_d = 7$ V and DC stress $V_d = -7$ V and (b) AC V_d stress with $t_r = t_f = 100$ ns and $f = 500$ kHz. V_{th} shift is dependent on stress time under AC V_d stress with (c) various t_{rs} and fixed $t_f = 500$ ns and (d) various t_{fs} and fixed $t_r = 500$ ns. The f is fixed at 250 kHz.

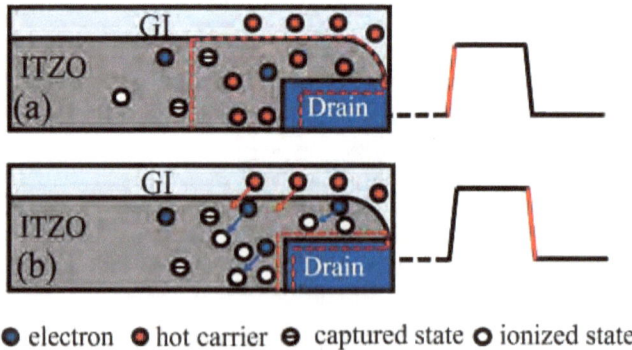

● electron   ● hot carrier   ⊖ captured state   ○ ionized state

Fig. (5.19). Schematic diagram of the degradation model in ITZO TFTs during (a) t_r and (b) t_f.

5.3.2. Dependence of Alternating Current Degradation on Waveform Elements

The previous discussion addressed the degradation phenomena of IGZO and ITZO TFTs under single-ended and double-ended pulse stress, as well as the degradation models. It also explored the impact of the rise and fall edges of the pulses on the AC HC effect degradation. Currently, there is a lack of research on the influence of high and low-level voltage magnitudes and their durations on device degradation after pulse stress in MO TFTs. Most discussions still focus on direct current mechanisms. Based on the study in Section 5.2.2 of poly-Si's varying pulse parameters and their effects on AC stress degradation, it is hypothesized that MO TFTs may exhibit similar phenomena, and the recovery effect in ITZO TFTs remains a significant factor.

5.3.3. Comparison between Polycrystalline Silicon Thin-Film Transistors and Metal Oxide Thin-Film Transistors

According to the discussions in Sections 5.2.2 and 5.3.2, the degradation patterns under pulse stress for both poly-Si and MO TFTs can be explained by the non-equilibrium junction model between the source/drain and the active layer (refer to Sections 5.2.2 and 5.3.2).

Under steady-state conditions with the drain electrode grounded, there is a certain amount of traps in the intrinsic channel region near the n+ drain diffusion that need to be ionized through hole emission to form the depleted region of the drain junction. When a pulse V_d is applied, V_d rapidly swings from 0 to V_{peak} during the t_r transient, causing the junction to become more reverse-biased and the depleted region to extend into the channel, as shown in Fig. (**5.12b**). As a result, the depleted region needs to extend far into the channel by emitting holes only from shallow traps. Concurrently, an electric field is established in the depleted region. Subsequently, holes emitted from deeper traps continue to balance the positive space charge in the n^+ drain diffusion region. At this time, the E_x near the channel/leak edge is high enough to accelerate some emitted carriers into HCs, generating defects near the drain end and leading to observed HC degradation.

Based on the proposed model, the main features of dynamic V_d HC degradation can be explained. For shorter tr, E_x can reach higher levels. At the same time, only those traps with shallower energy levels can emit within shorter V_d transitions. After a higher E_x is established, more deep trap emissions will occur, leading to more severe degradation. During the t_f transient, V_d swings from V_{peak} to 0; the drain junction should recover to the steady state. The extended depleted region entering the channel should retract. The traps ionized during the t_r transient are neutralized from the outer edge of the depleted region through carrier recombination. During this process, E_x is very low and cannot generate HCs. Therefore, the degree of degradation is unrelated to t_f. The model above can be seen in Section 5.2. [27, 30, 31]

There are two main differences in reliability between MO TFTs and poly-Si TFTs under dynamic stress:

1. The degradation phenomena are different. [27, 30, 31]

2. The modulation effect of changing pulse parameters on degradation is different. [6, 7]

As shown in Fig. (**5.20**), The dynamic HC degradation in poly-Si TFTs primarily manifests as a decrease in the I_{on} of the transfer curve (refer to Section 5.2.1), while the dynamic HC degradation in MO TFTs mainly manifests as a positive shift in the transfer curve, an increase in SS, and a decrease in I_{on} (refer to Section 5.3.1).

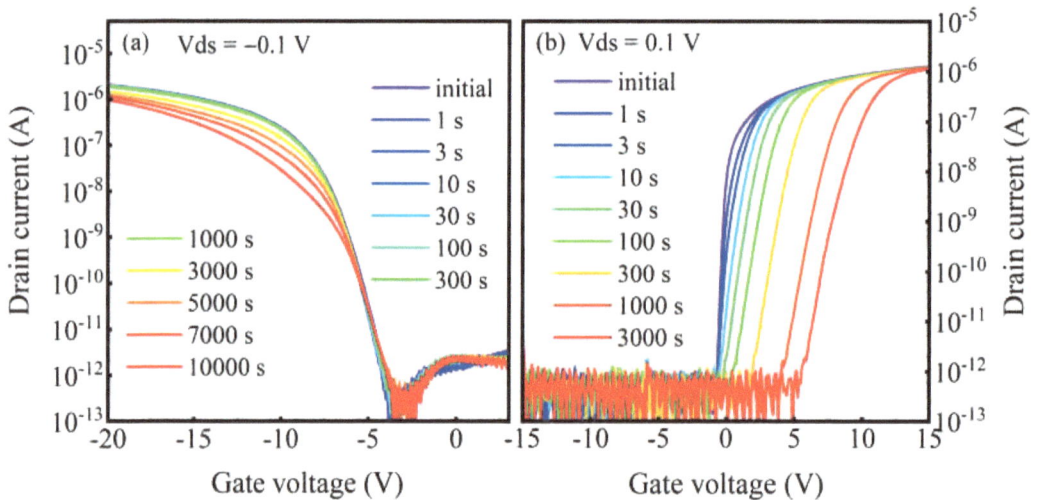

Fig. (5.20a). Typical AC stress degradation in poly-Si TFT and (b) typical AC stress degradation in MO TFT.

Furthermore, the degree of degradation in p-type poly-Si TFTs under pulse stress is only regulated by the rise edge (gate pulse) or the fall edge (drain pulse) (refer to Section 5.2.2). However, for the MO TFTs under drain pulse stress, degradation was found to be not only related to t_r. Detailed investigations must be carried out in the future.

CONCLUSION

Table 1. Degradation of TFTs under single AC stress.

Type of TFTs	Degradation Phenomenon	Type of TFTs	Electrode	Increasing parameters				
				t_r	t_f	t_{peak}	t_{base}	f
Poly-Si TFTs	Decreasing of I_{on}	P type	Gate	-	↓	-	-	↑
			Drain	↓	-	-	-	↑
		N type	Gate	↓	-	-	-	↑
			Drain	-	↓	-	-	↑
MO TFTs	Positive shift of V_{th} and degradation of SS	N type	Gate	↓	↓	-	-	↑
			Drain	↓	↓	-	-	↑

* "↓" represents the decrease of degradation, "↑" represents the increase of degradation and "-"represents no effect on the degradation.

This chapter conducts a thorough investigation into the effects of AC voltage stress on the performance of TFTs, encompassing both poly-Si and MO TFTs. As shown in the above Table **1**, it begins with an examination of the degradation behaviors exhibited by poly-Si TFTs under AC stress, exploring how various waveform elements influence these degradation patterns. A unified degradation model is presented, which integrates the findings from multiple experiments to better characterize the degradation mechanisms specific to TFTs subjected to AC voltage. The chapter further extends this analysis to MO TFTs, detailing their unique degradation behaviors under similar AC conditions and highlighting the differences between these two types of transistors. By elucidating the intricate mechanisms that drive performance degradation under AC stress, this chapter provides valuable insights that pave the way for enhanced stability and reliability of TFTs in AC-driven applications. Ultimately, the findings presented contribute significantly to the ongoing development of more robust and durable TFT technologies for advanced electronic devices, ensuring their optimal performance in diverse operating environments.

REFERENCES

[1]   M. Zhang, "Dynamic Hot Carrier Degradation Behavior of Polycrystalline Silicon Thin-Film Transistors under Gate Voltage Pulse Stress with Fast Transition Time", *2023 IEEE International Symposium on the Physical and Failure Analysis of Integrated Circuits (IPFA)*, pp. 1-5, 2023.

http://dx.doi.org/10.1109/IPFA58228.2023.10249065

[2]   H. Wang, M. Wang, and M. Zhang, "Dynamic HC-induced degradation in n-type poly-Si thin film transistors under off-state gate pulse voltage", *Proceedings of the 20th IEEE International Symposium on the Physical and Failure Analysis of Integrated Circuits (IPFA)*, pp. 381-384, 2013.

http://dx.doi.org/10.1109/IPFA.2013.6599186

[3]   M. Zhang, Z. Xia, W. Zhou, R. Chen, and M. Wong, "Significant Reduction of Dynamic Negative Bias Stress-Induced Degradation in Bridged-Grain Poly-Si TFTs", *IEEE Electron Device Lett.*, vol. 36, no. 2, pp. 141-143, 2015.

http://dx.doi.org/10.1109/LED.2014.2377040

[4]   M. Zhang, Z. Xia, W. Zhou, R. Chen, M. Wong, and H.S. Kwok, "Dynamic-Gate-Stress-Induced Degradation in Bridged-Grain Polycrystalline Silicon Thin-Film Transistors", *IEEE Trans. Electron Dev.*, vol. 63, no. 10, pp. 3964-3970, 2016.

http://dx.doi.org/10.1109/TED.2016.2601218

[5]   M. Zhang, "Degradation Behaviors of Driving Thin-Film Transistors in Active-Matrix Organic Light-Emitting Diode Displays", *2018 IEEE International Conference on Electron Devices and Solid State Circuits (EDSSC)*, pp. 1-3, 2018.

http://dx.doi.org/10.1109/EDSSC.2018.8487157

[6]   M. Zhang, Z. Jiang, L. Lu, M. Wong, and H.S. Kwok, "Analysis of Degradation Mechanism in Poly-Si TFTs Under Dynamic Gate Voltage Stress With Short Pulse Width Duration", *IEEE Electron Device Lett.*, vol. 45, no. 2, pp. 204-207, 2024.

http://dx.doi.org/10.1109/LED.2023.3345282

[7]   Y.H. Tai, S.C. Huang, and C.K. Chen, "Analysis of Poly-Si TFT Degradation Under Gate Pulse Stress Using the Slicing Model", *IEEE Electron Device Lett.*, vol. 27, no. 12, pp. 981-983, 2006.

http://dx.doi.org/10.1109/LED.2006.886416

[8]   D. Zhang, M. Wang, and X. Lu, "Two-Stage Degradation of p-Type Polycrystalline Silicon Thin-Film Transistors Under Dynamic Positive Bias Temperature Stress", *IEEE Trans. Electron Dev.*, vol. 61, no. 11, pp. 3751-3756, 2014.

http://dx.doi.org/10.1109/TED.2014.2359299

[9]   Y. Toyota, T. Shiba, and M. Ohkura, "Effects of the timing of AC stress on device degradation produced by trap states in low-temperature polycrystalline-silicon TFTs", *IEEE Trans. Electron Dev.*, vol. 52, no. 8, pp. 1766-1771, 2005.

http://dx.doi.org/10.1109/TED.2005.852726

[10]   L. Chen, M. Wang, D. Zhang, and H. Wang, "Gate Voltage Pulse Rising Edge Dependent Dynamic Hot Carrier Degradation in Poly-Si Thin-Film Transistors", *IEEE Electron Device Lett.,* vol. 42, no. 11, pp. 1615-1618, 2021.

http://dx.doi.org/10.1109/LED.2021.3110916

[11]   M. Zhang, and M. Wang, "An investigation of drain pulse induced hot carrier degradation in n-type low temperature polycrystalline silicon thin film transistors", *Microelectron. Reliab.,* vol. 50, no. 5, pp. 713-716, 2010.

http://dx.doi.org/10.1016/j.microrel.2010.01.024

[12]   X. Lu, M. Wang, M. Zhang, and M. Wong, "Negative drain pulse stress induced two-stage degradation of P-channel poly-Si thin-film transistors", *18th IEEE International Symposium on the Physical and Failure Analysis of Integrated Circuits (IPFA),* pp. 1-4, 2011.

http://dx.doi.org/10.1109/IPFA.2011.5992756

[13]   M. Zhang, M. Wang, X. Lu, M. Wong, and H.S. Kwok, "Analysis of Degradation Mechanisms in Low-Temperature Polycrystalline Silicon Thin-Film Transistors under Dynamic Drain Stress", *IEEE Trans. Electron Dev.,* vol. 59, no. 6, pp. 1730-1737, 2012.

http://dx.doi.org/10.1109/TED.2012.2189218

[14]   Y. Zeng, Y. Yan, Z. Jiang, Z. Xia, M. Zhang, M. Wong, J.J. Liou, and H-S. Kwok, "Reliability of Poly-Si TFTs Under Voltage Pulse With Fast Transition Time", *IEEE Electron Device Lett.,* vol. 42, no. 12, pp. 1782-1785, 2021.

http://dx.doi.org/10.1109/LED.2021.3124755

[15]   Y. Wang, Z. Jiang, L. Zou, and M. Zhang, "Analysis of degradation mechanism in polycrystalline silicon thin-film transistors under dynamic off-state stress with fast transition time", *Physica Scripta,* vol. 99, no. 12, 2024.

http://dx.doi.org/10.1088/1402-4896/ad896a

[16]   Y. Uraoka, H. Yano, T. Hatayama, and T. Fuyuki, "Comprehensive study on reliability of low-temperature poly-Si thin-film transistors under dynamic complimentary metal-oxide semiconductor operations," Jpn. J. Appl. Phys. Part 1 - Regul", *Jpn. J. Appl. Phys.,* vol. 41, no. Part 1, No. 4B, pp. 2414-2418, 2002.

http://dx.doi.org/10.1143/JJAP.41.2414

[17]   Y. Uraoka, N. Hirai, H. Yano, T. Hatayama, and T. Fuyuki, "Hot carrier analysis in low-temperature poly-Si TFTs using picosecond emission microscope", *IEEE Trans. Electron Dev.,* vol. 51, no. 1, pp. 28-35, 2004.

http://dx.doi.org/10.1109/TED.2003.820937

[18]   Y. Toyota, T. Shiba, and M. Ohkura, "A new model for device degradation in low-temperature N-channel polycrystalline silicon TFTs under AC stress", *IEEE Trans. Electron Dev.,* vol. 51, no. 6, pp. 927-933, 2004.

http://dx.doi.org/10.1109/TED.2004.828163

[19]   Ya-Hsiang Tai, Shih-Che Huang, and Po-Ting Chen, "Degradation Mechanism of Poly-Si TFTs Dynamically Operated in OFF Region", *IEEE Electron Device Lett.,* vol. 30, no. 3, pp. 231-233, 2009.

http://dx.doi.org/10.1109/LED.2008.2010784

[20]   M. Zhang, W. Zhou, R. Chen, M. Wong, and H.S. Kwok, ""Driving"-Stress-Induced Degradation in Polycrystalline Silicon Thin-Film Transistors and Its Suppression by a Bridged-Grain Structure", *IEEE Electron Device Lett.*, vol. 38, no. 1, pp. 52-55, 2017.

http://dx.doi.org/10.1109/LED.2016.2626481

[21]   M. Zhang, Y. Yan, G. Li, S. Deng, W. Zhou, R. Chen, M. Wong, and H-S. Kwok, "OFF-State-Stress-Induced Instability in Switching Polycrystalline Silicon Thin-Film Transistors and Its Improvement by a Bridged-Grain Structure", *IEEE Electron Device Lett.*, vol. 39, no. 11, pp. 1684-1687, 2018.

http://dx.doi.org/10.1109/LED.2018.2872350

[22]   M. Zhang, X. Ma, S. Deng, W. Zhou, Y. Yan, M. Wong, and H-S. Kwok, "Degradation Induced by Forward Synchronized Stress in Poly-Si TFTs and Its Reduction by a Bridged-Grain Structure", *IEEE Electron Device Lett.*, vol. 40, no. 9, pp. 1467-1470, 2019.

http://dx.doi.org/10.1109/LED.2019.2931007

[23]   "Reversely-Synchronized-Stress-Induced Degradation in Polycrystalline Silicon Thin-Film Transistors and Its Suppression by a Bridged-Grain Structure | IEEE Journals & Magazine | IEEE Xplore", 2024. Available from: https://ieeexplore.ieee.org/document/9125948

[24]   T. Chen, C. Wang, G. Yang, Q. Lou, Q. Lin, S. Zhang, and H. Zhou, "Monolithic Integration of Perovskite Photoabsorbers with IGZO Thin-Film Transistor Backplane for Phototransistor-Based Image Sensor", *Adv. Mater. Technol.*, vol. 8, no. 1, p. 2200679, 2023.

http://dx.doi.org/10.1002/admt.202200679

[25]   C. Liu, L. Zhou, Z-H. Zhou, F-F. Li, X-Q. Wei, M. Xu, L. Wang, W-J. Wu, and J-B. Peng, "A high-reliability scan driver integrated circuit by MO TFTs for a foldable display", *Flex. Print. Electron.*, vol. 8, no. 1, p. 015015, 2023.

http://dx.doi.org/10.1088/2058-8585/acbbdd

[26]   S. Mao, J. Li, A. Guo, T. Zhao, and J. Zhang, "An Active Multielectrode Array for Collecting Surface Electromyogram Signals Using a-IGZO TFT Technology on Polyimide Substrate", *IEEE Trans. Electron Dev.*, vol. 67, no. 4, pp. 1613-1618, 2020.

http://dx.doi.org/10.1109/TED.2020.2974971

[27]   H. Wang, M. Wang, and Q. Shan, "Dynamic degradation of a-InGaZnO thin-film transistors under pulsed gate voltage stress", *Appl. Phys. Lett.*, vol. 106, no. 13, p. 133506, 2015.

http://dx.doi.org/10.1063/1.4916825

[28]   H. Wang, *Degradation of thin film transistor devices under Dynamic stress.* Ph.D., Soochow University, 2016.

[29]   G. Zhu, Z. Chen, M. Zhang, L. Lu, S. Deng, M. Wong, and H-S. Kwok, "Reliability of indium-tin-zinc-oxide thin-film transistors under dynamic drain voltage stress", *Appl. Phys. Lett.*, vol. 125, no. 2, p. 023505, 2024.

http://dx.doi.org/10.1063/5.0213509

[30]   P.-H. Chen, Y.-Z. Zheng, T.-H. Yeh, and T.-Y. Nieh, "Investigating DC and AC degradation behaviors to P-type low temperature polycrystalline silicon thin film transistor with fin-like structure", Journal of Physics D: *Applied Physics*, vol. 56, no. 43, 2023.

http://dx.doi.org/10.1088/1361-6463/ace835

[31]   M. Zhang, M. Wang, and H. Wang, "Degradation of metal-induced laterally crystallized n-type poly-Si thin-film transistors under dynamic voltage stress", *in 2008 15th International Symposium on the Physical and Failure Analysis of Integrated Circuits,* singapore: IEEE, pp. 1-4, 2008.

http://dx.doi.org/10.1109/IPFA.2008.4588214

CHAPTER 6

Circuit-Level Stress-Induced Degradation in Thin-Film Transistors

Abstract. This chapter discusses the degradation of Thin-Film Transistors (TFTs) under circuit-level stress, which is crucial for understanding their performance and reliability in practical applications. It covers AC degradation under DC bias, bipolar AC degradation, and ultra-fast AC degradation, providing insights into the impact of different stress conditions on the TFTs' performance and reliability.

Keywords: AC degradation, Bipolar AC degradation, Circuit-level stress, DC bias, Performance, Reliability, Ultra-fast AC degradation.

6.1. INTRODUCTION

Chapter 5 provided a detailed discussion on the degradation of thin-film transistors (TFT) under alternating current (AC) stress and its causes, with the non-equilibrium PN junction model effectively explaining the basic degradation patterns. However, Chapter 5 primarily focused on the degradation under single-ended AC stress. In the actual operating state of TFTs, multiple electrodes of the device are simultaneously exposed to various stress conditions. As mentioned in Section 1.4.1, a switch TFT may be exposed to an AC state at the gate and drain simultaneously or a direct current (DC) state at the gate with an AC state at the drain; for a driver TFT, the drain is typically connected to V_{DD} for extended periods, while the gate is directly exposed to various AC environments.

Therefore, based on the actual working conditions of TFTs, the degradation of TFTs can be categorized into AC stress reliability under DC bias and dual-end AC stress reliability.

With the current pursuit of high refresh rates and high-resolution displays, TFTs will be exposed to increasingly faster AC frequencies, transitioning from the μs range to the ns range. Consequently, to reflect the actual working conditions of TFTs more accurately, further research is needed on the degradation of TFTs under faster pulse stress.

Meng Zhang & Mingxiang Wang

6.2. ALTERNATING CURRENT DEGRADATION UNDER DIRECT CURRENT BIAS

This section primarily examines the degradation phenomena and mechanisms of TFT devices when one end is subjected to AC stress while the other end is under DC bias [1-4]. This mode of stress is more aligned with the actual forces exerted on devices in real-world scenarios and is better suited for circuit-level reliability studies. Consequently, an understanding of the fundamental Active Matrix Organic Light Emitting Diode (AMOLED) circuitry is required. An AMOLED display circuit typically comprises two TFTs and a capacitor [5-7]. In practical applications, both the gate and drain of these two TFTs are subjected to complex stresses. For instance, they may experience simultaneous AC stress [8-11], or one end may be under AC stress while the other end is under DC stress [1-4], as depicted in Fig. (**6.1b**). Therefore, studying the degradation of TFT devices under stress at both ends is essential. The stress conditions illustrated in Fig. (**6.1c**) are designed to simulate the stresses encountered by TFTs in real-world applications.

The condition for driving stress [1, 2] is that the gate terminal of the TFT is subjected to an AC voltage of positive and negative 11.5 V in magnitude, with adjustable rise time, high-level duration, fall time, and low-level duration. At the same time, the source terminal of the TFT is subjected to a positive DC voltage, as shown in Fig. (**6.1c**). Furthermore, in an alternative experimental setup, a DC voltage is applied to the gate terminal of the TFT, while an AC voltage, oscillating between positive and negative 10 V, is imposed on the drain terminal. This configuration, illustrated in Fig. (**6.1d**), is known as"OFF-state stress."

The performance deterioration of polycrystalline silicon (poly-Si) TFTs under the described "driving" stress is examined and characterized. It is evident that this type of stress-induced degradation is mainly attributed to a dynamic hot carrier (HC) effect. The severity of the degradation is exacerbated by a shorter t_r, leading to more pronounced dynamic HC effects.

Furthermore, the analysis of poly-Si TFTs under OFF-state stress conditions [3, 4] indicates that the degradation is chiefly governed by the dynamic HC effect. This effect is predominantly determined by the pulse t_f and appears to be largely unaffected by the pulse t_r.

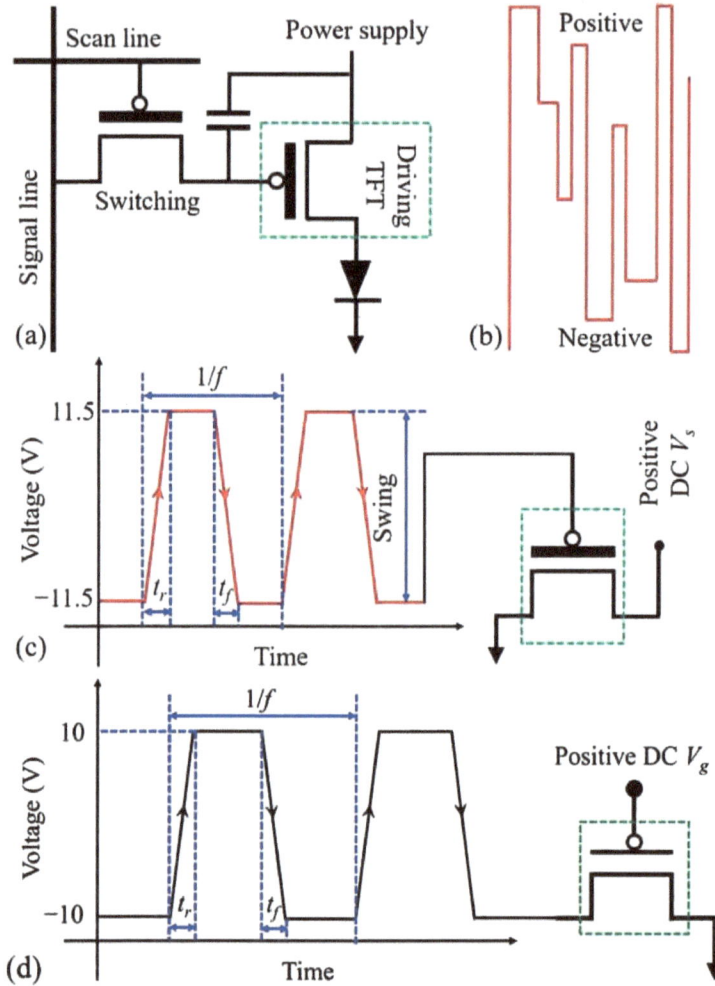

Fig. (6.1). (a) Schematic of 2T1C circuit of a pixel in an AMOLED display. (b) Practical voltage pattern applied to the gate electrode of the driving TFT. (c) "Driving" stress conditions for a driving TFT. (d) OFF-state stress conditions for a switching TFT [2, 4].

6.2.1. "Driving"-Stress-Induced Degradation

Fig. (**6.2a**) illustrates the temporal evolution of the transfer characteristics in a conventional poly-Si TFT under "driving" stress. This stress condition is characterized by an AC gate stress superimposed on a DC V_s of 10 V. Observations indicate that both the I_{on} and gate-induced drain leakage (GIDL) current [12] experience a continuous decline with increasing stress duration, while the subthreshold swing (*SS*) remains relatively unaffected. Additionally, a recovery in

I_{on} is noted at elevated V_{sd}, which can be attributed to the source-induced reduction of the potential barrier caused by stress-induced traps. These degradation characteristics are consistent with the hallmarks of typical HC degradation [13, 14], suggesting that the degradation induced by "driving" stress is associated with HC mechanisms.

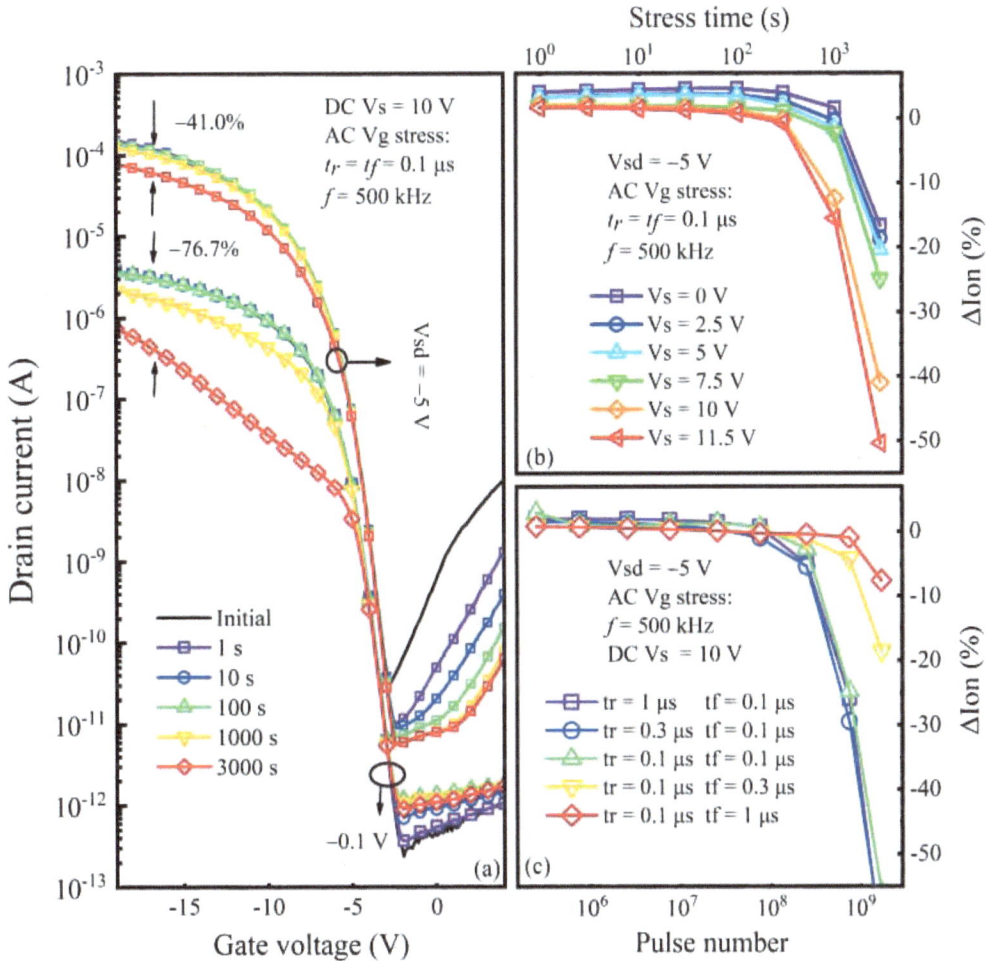

Fig. (6.2). (a) Degeneration phenomenon. (b) Degradation of I_{on} at different DC Vs. (c) Degradation of I_{on} at different AC V_g [2].

To ascertain the predominant parameter of "driving" stress affecting device degradation, a series of stress tests were conducted under AC gate stress with varying DC V_s, f, t_r, and t_f. Fig. (**6.2b**) presents the degradation of I_{on} as a function of stress time under AC gate stress with different positive DC V_s conditions in standard poly-Si TFTs. The I_{on} degradation displays a two-stage behavior. Initially, I_{on} experiences a modest increase with stress time, potentially due to electron trapping or injection into the gate oxide. Subsequently, I_{on} undergoes a significant decrease, likely driven by the dynamic HC effect. It is observed that a higher positive DC V_s leads to a reduced I_{on} increase in the initial stage and an exacerbated I_{on} decrease in the latter stage. Moreover, at a DC V_s of 0 V, the degradation after 3000 seconds of stress reaches −36.6% (V_g swing = 23 V, V_{sd} = −0.1 V). Comparatively, a similar stress magnitude results in −65.5% degradation in solution-based metal-induced lateral crystallization TFTs (with V_g swing = 20 V) and −90.0% degradation in excimer laser annealing TFTs (with V_g swing = 30 V). The pronounced degradation is primarily attributed to the dynamic HC effect [8-11].

To delve into the degradation mechanism engendered by "driving" stress in poly-Si TFTs, transient simulations were executed using the Silvaco ATLAS platform. Fig. (**6.3a**) illustrates the temporal variation of E_x at the channel/source and channel/drain boundaries under AC gate stress, subjected to varying DC V_s in standard poly-Si TFTs. The positive x-axis is oriented from the source towards the drain. Observations reveal that a higher DC V_s diminishes the magnitude of $|E_x|$ at the source side of the channel, whereas it amplifies E_x at the drain side. Given that HC degradation is predominantly influenced by the peak electric field, it is inferred that the dynamic HC degradation during the second stage of "driving" stress predominantly occurs at the drain side. An elevated DC V_s intensifies E_x in the channel near the drain, consequently leading to a more pronounced degradation of I_{on}, as depicted in Fig. (**6.2b**).

Fig. (**6.3b** and **6.3c**) depict the E_x and the lateral current density (J_x) as a function of normalized transient time at the channel/drain boundary. The positive DC V_s is consistently set at 10 V. In Fig. (**6.3b**), during the t_r interval, the E_x profiles for varying t_r values escalate rapidly to a high electric field, while J_x diminishes gradually. This environment exposes carriers to the elevated E_x, potentially transforming them into HCs. A reduced t_r yields a more substantial E_x and J_x, culminating in significant I_{on} degradation during the second stage, as illustrated in Fig. (**6.2c**). In the t_f phase, demonstrated in Fig. (**6.3c**), E_x declines over time. Initially, despite high E_x, Jx direction is counter to that of E_x, as indicated in the inset of Fig. (**6.3c**), suggesting it is not a drift current driven by E_x. Under these conditions, HC generation is negligible. As t_f progresses, J_x aligns with the direction of E_x. However, by this point, the E_x has diminished to a level insufficient to accelerate carriers into HCs. Consequently, the dynamic HC degradation during the second stage under "driving" stress is found to be independent of t_f, as shown in Fig. (**6.2c**).

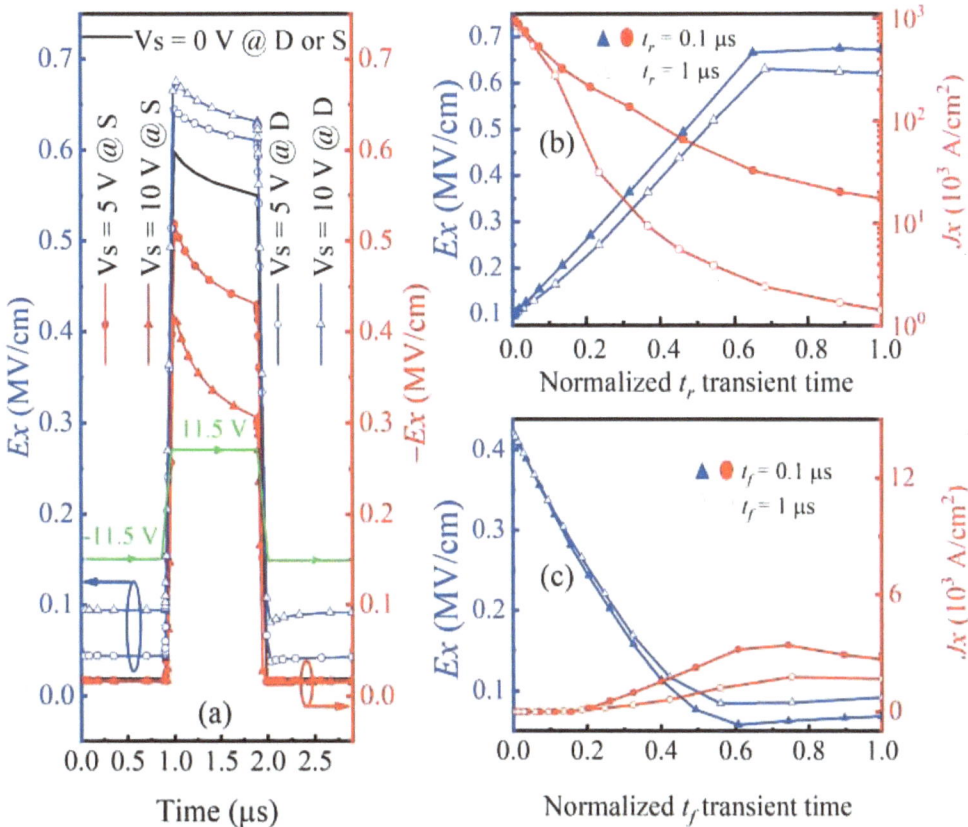

Fig. (6.3). (a) Change of E_x with stress time. (b) Change of E_x with stress t_r. (c) Change of E_x with stress t_f [2].

To elucidate the degradation mechanism at play during the second stage, the non-equilibrium junction degradation model has been utilized and further developed. Under "driving" stress, which comprises an AC gate stress and a positive DC V_s, the initial phase of t_r (−11.5 V to 11.5 V) results in a forward-biased source junction and a reverse-biased drain junction. The magnitude of the $|E_x|$ at the drain side of the channel exceeds that at the source side. An increased positive DC V_s further biases the source junction forward and the drain junction reverse, leading to a suppression of $|E_x|$ at the source side and an amplification at the drain side, as depicted in Fig. (**6.3a**). As t_r progresses, the initially forward-biased source junction transitions to reverse bias, while the reverse-biased drain junction becomes increasingly reverse-biased. The depletion regions surrounding both junctions in the channel must expand by releasing carriers associated with traps. For the drain junction, the pre-existing depletion region and the high electric field enable the emitted carriers to acquire sufficient energy to transform into HCs. Conversely, for the source junction, only carriers emitted later from deep trap states encounter a substantial E_x across the depletion region, empowering them to become HCs. This accounts for the predominance of HC degradation at the drain junction under "driving" stress. During t_f transition, the expanded depletion region in the channel is expected to retract. Traps ionized during the tr phase are neutralized from the outer edge of the depletion region through carrier recombination. The electric field in this region remains too low to facilitate the generation of HCs. Consequently, the degradation process is found to be independent of t_f.

The non-equilibrium junction degradation model has been further advanced to account for the observed device degradation, which is predominantly governed by the dynamic HC effect.

6.2.2. OFF-State Stress-Induced Degradation

Fig. (**6.4a**) displays the temporal evolution of the transfer characteristics in standard poly-Si TFTs under OFF-state stress. With a positive DC V_g of 15 V applied to ensure channel deactivation, a continuous decline in the I_{on} is observed over stress time, while the subthreshold region appears to be largely unaffected. This behavior aligns with the typical patterns associated with HC degradation. It is particularly noteworthy that under DC stress conditions with V_g set to 15 V and a V_d of ±10V, only minimal degradation is detected, as indicated by the black lines with open symbols in Fig. (**6.4b**). This observation implies that the DC component of the OFF-state stress may not substantially contribute to device degradation. On the contrary, the degradation is suspected to be associated with the AC component of the OFF-state stress. Fig. (**6.4b**) further investigates the impact of the positive DC V_g during OFF-state stress, revealing a two-stage degradation pattern: an initial slight increase in I_{on}, followed by a significant decrease in the subsequent stage, akin to earlier findings [8, 15].

The increase and subsequent decrease in I_{on} under stress are exacerbated by a higher positive DC V_g, with the transition point occurring more rapidly at elevated V_g levels. The initial phase of degradation is potentially dominated by electron trapping or injection into the gate oxide, while the dynamic HC effect is suspected to govern the second phase. Moreover, OFF-state stress is found to induce more severe degradation in comparison to pure drain pulse stress, particularly when the DC V_g is set to 0 V. The influence of f on standard poly-Si TFTs has been examined, although not displayed here. The degradation appears to be independent of frequency when considered in relation to pulse number, suggesting a link to the pulse transient time rather than frequency. Fig. (**6.4c**) illustrates the degradation of I_{on} as a function of pulse transient time in conventional poly-Si TFTs. For a constant t_f, the change in I_{on} (ΔI_{on}) exhibits no significant variation with different t_r. However, a reduced tf results in increased Ion degradation for a given t_r, with the most pronounced degradation occurring at the shortest t_f. This degradation pattern aligns with HC degradation caused by AC drain stress [8] and contrasts with AC gate stress-induced HC degradation [16], where shorter t_f and t_r result in more severe degradation. These observations indicate that OFF-state-induced degradation is predominantly driven by dynamic HC effects that manifest during the t_f period.

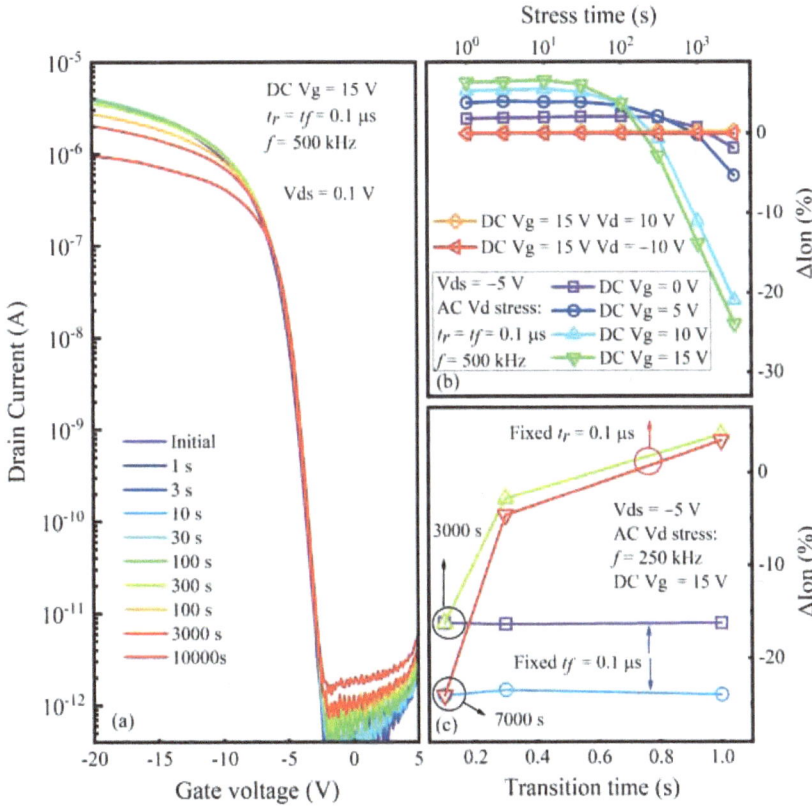

Fig. (**6.4**). (a) Time evolution of transfer characteristics of normal poly-Si TFTs under OFF-state stress. (b) ΔI_{on} as a function of stress time under OFF-state stress with various positive DC stress V_g in normal poly-Si TFTs. Open squares and circles represent ΔIon for stress conditions with DC $V_g =$ 15 V/$V_d =$ 10 V and DC $V_g =$ 15 V/DC $V_d =-$10 V. (c) ΔI_{on} as a function of transient time in normal poly-Si TFTs [4].

To delve into the mechanism behind OFF-state stress-induced degradation in poly-Si TFTs, transient simulations were executed using the Silvaco ATLAS platform. Focusing on the parameters strongly associated with HC effects, the study concentrated on E_x and J_x. These parameters were tracked 15 nm beneath the oxide/channel interface in proximity to the drain junction, as indicated by the red dot in Fig. (**6.5b**), with the positive x-direction extending from the drain towards the source. The influence of the positive DC stress V_g on E_x during various pulse durations was initially examined. Fig. (**6.5a** and **6.5b**) illustrate that an increased positive V_g amplifies the magnitude of E_x during t_r, t_f, and periods of low V_d. This suggests that higher V_g intensifies HC degradation, as depicted in Fig. (**6.4b**). Transient simulations were subsequently conducted during t_r and t_f. The outcomes, presented in Fig. (**6.5c** and **6.5d**), portray E_x and J_x in relation to different transient times, normalized by the transition edge width (t_r for Fig. **6.5c** and t_f for Fig. **6.5d**). During t_r (Fig, **6.5c**), despite an initially high $|E_x|$, J_x moves in the reverse direction, precluding HC generation, and thus no degradation is observed, as shown in Fig. (**6.4c**). For a V_d of 10 V and a DC V_g of 15 V, both E_x and J_x (solid lines) approach zero, resulting in negligible degradation (Fig. **6.4b**). Conversely, during t_f (Fig.**6.5d**), J_x is directed in accordance with $|E_x|$. As $|E_x|$ escalates, J_x sustains a high level, facilitating HC formation and ensuing device degradation. A reduced t_f yields a higher $|J_x|$, culminating in more pronounced HC degradation, as indicated in Fig. (**6.4c**). For DC $V_d = -10$ V and DC $V_g = 15$ V, both $|E_x|$ and $|J_x|$ (solid lines) are minimal, with no degradation detected (Fig. **6.4b**).

To clarify the dynamic HC degradation mechanism under OFF-state stress, the non-equilibrium junction degradation model has been employed and refined. During the OFF-state stress period, as the V_d transitions from –10 V to 10 V during t_r, the drain junction becomes forward-biased. This transition leads to a reduction in the depletion region width in the channel, as ionized traps at the edge of this region are neutralized through carrier recombination, resulting in a minimal E_x. As a consequence, HC generation is prevented, and no device degradation is observed, as depicted in Fig. (**6.4c**). In contrast, during t_f, when V_d switches from 10 V to –10 V, the drain junction becomes reverse-biased. The previously shrunken depletion region rapidly expands by emitting carriers associated with shallow traps, followed by those associated with deep traps. These carriers, once energized by a high E_x, transform into HCs. A shorter tf results in a larger proportion of deep-trap-related carriers being emitted under a high E_x, leading to more severe device degradation, as illustrated in Fig. (**6.4c**). An increased positive V_g further intensifies the reverse bias of the drain junction, leading to a higher E_x and more pronounced device degradation, as shown in Fig. (**6.4b**). Under DC stress conditions (DC $V_g = 15$ V and DC $V_d = \pm 10V$), no transient effects are observed, and thus, no degradation is recorded, as noted in Fig. (**6.4b**).

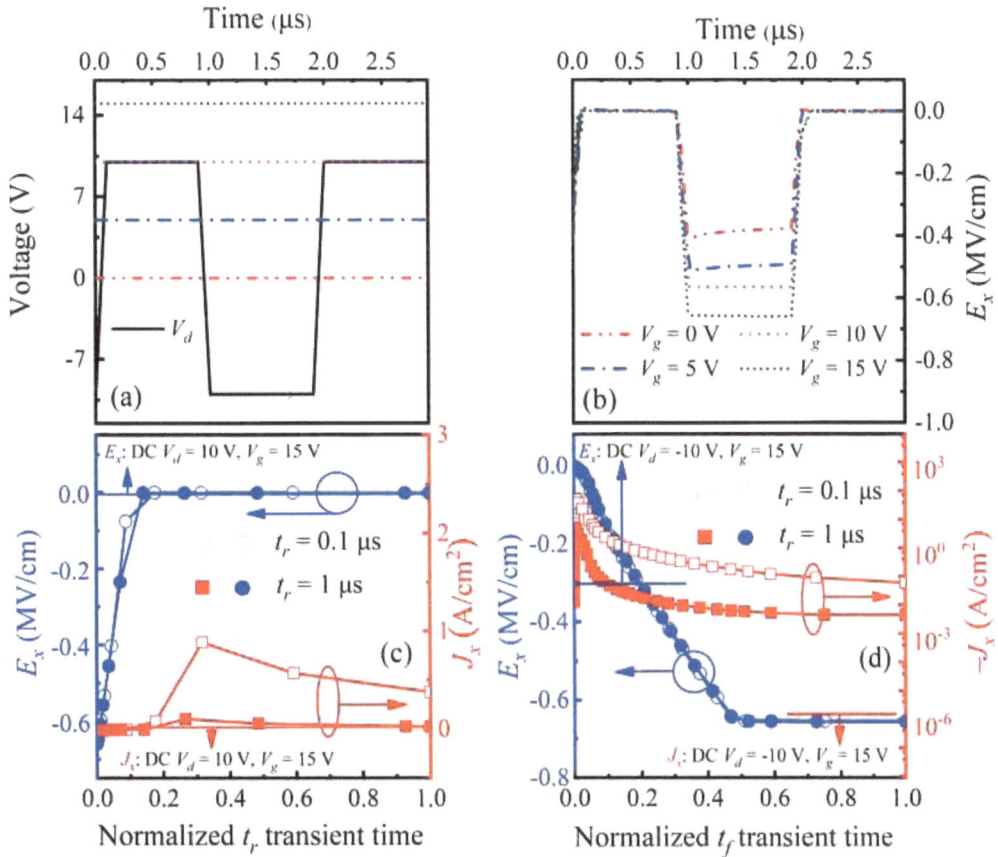

Fig. (6.5). (a) Various OFF-state stresses applied to the normal poly-Si TFTs. (b) Simulated E_x in the channel at the drain side on the pulse time for various positive DC stress V_{gs}. $t_r = t_f = 0.1$ μs. (c) E_x and J_x at the channel/drain edge on normalized t_r transient time. The blue solid line E_x and the red solid line represent J_x under DC $V_d = 10$ V and DC $V_g = 15$ V. (d) E_x and J_x at the channel/drain edge on normalized t_f transient time. The blue solid line and red line are E_x and J_x under DC $V_d = -10$ V and DC $V_g = 15$ V, respectively [4].

6.2.3. Conclusion

This section systematically investigates the degradation of practical stress conditions in circuit applications, along with their degradation mechanisms. Simulation analysis enhances the understanding of previous degradation mechanisms. Driving stress and OFF-state stress are common stress conditions that

TFT devices easily encounter in practice, with their reliability significantly influenced by the t_r and t_f, respectively. Through these two studies, the reliability of stress at both ends of TFT devices can be improved.

6.3 BIPOLAR ALTERNATING CURRENT DEGRADATION

In the previous section, device degradation under one-end DC and one-end AC was briefly introduced. However, in actual work, TFT may also encounter a situation where the level of the two electrodes rises at the same time or the level of the two electrodes moves in the opposite direction at the same time. We refer to this degradation under voltage stress as bipolar alternating current degradation. In the following introduction, we will introduce the bipolar alternating current degradation of polysilicon and metal oxide TFTs, respectively.

6.3.1 Synchronous Alternating Current Degradation

TFTs functioning in saturation within driver circuits frequently encounter concurrent pulses of V_g and V_d pulses, which can be called forward synchronized stress (FSS). Such synchronized voltage stress data may cause HC degradation. When the stress V_g and V_d are larger, the self-heating (SH) degradation may happen at the same time.

For N-channel MILC TFTs experiencing synchronized voltage stress, the width (W) varies from 10 to 30 μm, while the length (L) is consistently 10 μm. Fig. (**6.6**) shows the pulse signals applied to the gate and drain terminals. V_{g_h} and V_{d_h} denote the peak voltage levels for the respective V_g and V_d pulses. The baseline voltage for both pulses is zeroed. By fine-tuning V_{g_h} and V_{d_h}, the stressed devices are sustained in a saturated state, and the peak power density under stress (P_{max}) is unified at approximately 0.24 mW/μm² across various devices. Typically, under a DC bias with this power density, SH degradation would take place. Assuming P_{max} is constant, the SH effect is anticipated to be consistent across devices experiencing a variety of AC stress scenarios. Additional pulse parameters encompass the duty ratio (α), f, t_r, t_f, and the total number of pulses (N). For reference, supplementary stress evaluations under increased thermal conditions, at half the P_{max}, or outside saturation are also executed [17]. Assuming a constant P_{max}, the SH impact is anticipated to be similar across devices experiencing diverse AC stress scenarios. Supplementary pulse characteristics comprise the α, f, t_r, t_f, and N. Furthermore, comparative stress evaluations are executed under conditions of heightened temperature, a P_{max} reduced by half, or in non-saturated states. Device degradation is ascertained by scrutinizing shifts in device parameters, encompassing the I_{on}, delineated in a transfer curve at V_d = 0.1 V and V_g = 12 V. The threshold voltage (V_{th}) is ascertained from the crossing point of the linearly extrapolated transfer curve when the V_d is set to 0.1 V, and the I_{off} [18] is defined when V_d = 5 V and V_g = −12 V [19].

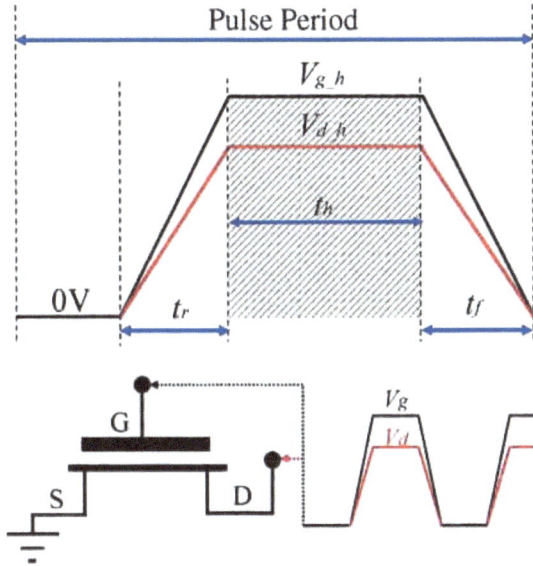

Fig. (6.6). A synchronized voltage pulse signal is imposed on the gate side and drain side terminals of the TFT. The pulse characteristics, namely V_{g_h} and V_{d_h}, represent the peak voltage levels for the gate and drain, while t_r and t_f denote the respective rise and fall times [19].

Fig. (**6.7**) illustrates the I_{on} degradation's reliance on the duration of stress across different α settings, with a constant frequency of 100 Hz and equal t_r and t_f of 0.1 μs. Initially, device performance either remains stable or experiences marginal enhancement, which is then followed by substantial degradation. Notably, this degradation pattern is consistent with that observed under DC SH conditions [17]. The turnaround phenomenon is due to the floating-body impact, which arises from the emission of H$^+$ following the breakdown of fragile bonds in the early stages. Subsequently, devices under higher α values demonstrate a more pronounced ΔI_{on}. This outcome is expected, as within an equivalent stress duration, a larger α equates to an extended cumulative duty period, during which the device endures SH stress. Therefore, Fig. (**6.8**) shows that when the change in the ΔI_{on} is plotted against duty cycle time rather than overall stress duration, the lines for different α values almost overlap. The inset further demonstrates that degradation is unaffected by α as long as the duty cycle time is steady. A comparable trend is noticed for the V_{th} degradation. Hence, with low-*f* pulses, the device's degradation correlates with the effective duty cycle time instead of the stress duration. This correlation strongly suggests that the predominant degradation mechanism under low-*f* stress is the SH effect [19].

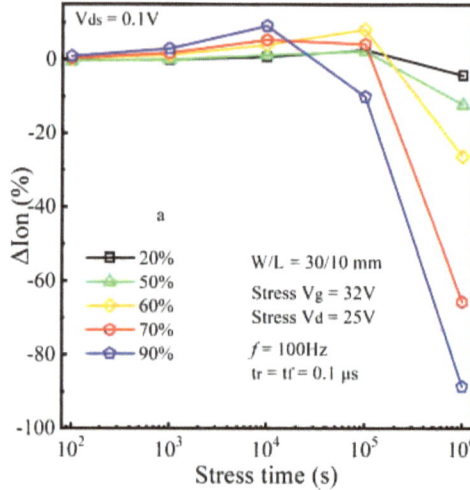

Fig. (6.7). The correlation of I_{on} current degradation with the duration of stress across various α values is examined, with the applied stress frequency being 100 Hz.

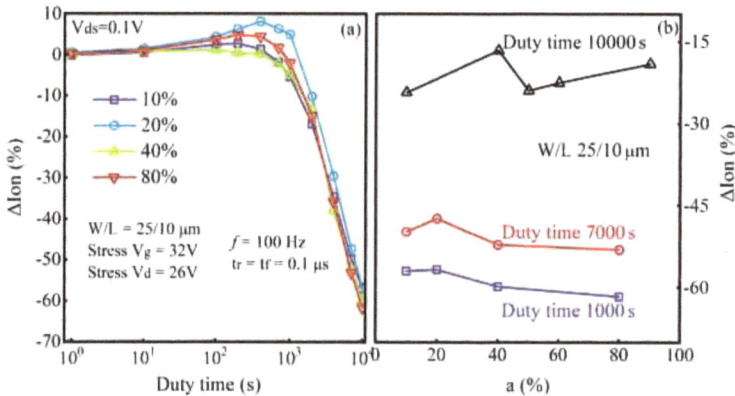

Fig. (6.8). The relationship between I_{on} degradation and stress time under different α. The inserted figure shows the influence of different α on ΔI_{on} when the applied stress is 1000s. The stress frequency is 100Hz.

In Fig. (**6.9**) and the inset, the variation of ΔI_{on} with stress time for different *f*s is plotted using logarithmic and linear scales. The α is consistently maintained at 50%, with both t_r and t_f fixed at 0.1 μs for all tests. The degradation of the I_{on} current shows distinct differences between low and high *f*s. The inset reveals that the two-stage degradation at low *f*s (below 20 kHz) mimics the DC SH degradation observed in MILC TFTs, while high *f*s (above 20 kHz) exhibit a straightforward degradation pattern. Furthermore, the slope of I_{on} degradation at low fs is roughly 0.88, which is consistent with DC SH degradation [17]. However, at high fs, the degradation slope is considerably reduced to about 0.24. Thirdly, at low fs, all degradation trajectories adhere to a uniform pattern, irrespective of the f. Nonetheless, at high fs, a more pronounced device degradation is encountered with increased f pulses. The markedly distinct degradation patterns strongly suggest the involvement of an additional degradation mechanism beyond SH under high-f stress [19].

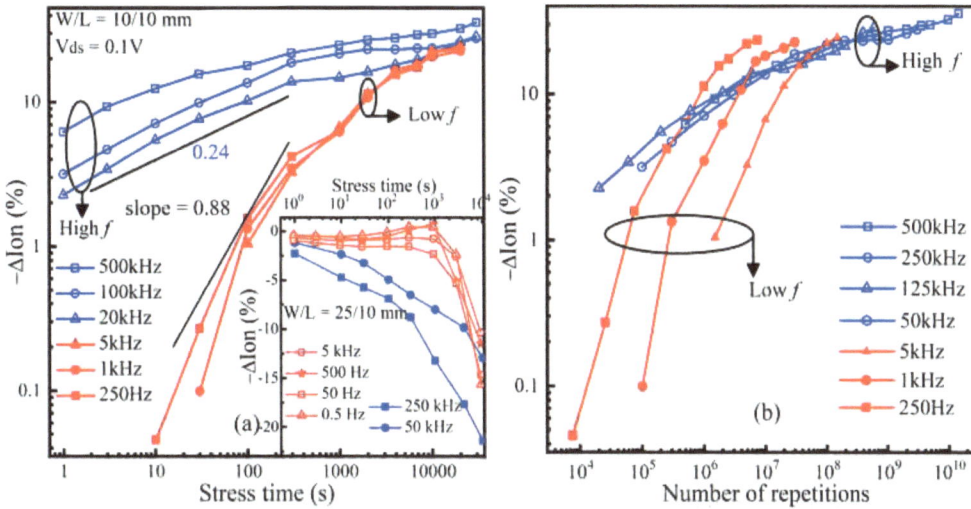

Fig. (6.9). (a) The variation of ΔI_{on} with stress time, depicted on a logarithmic scale for a range of stress fs (with V_g fixed at 32 V and V_d at 29 V), is presented. The inset illustrates the degradation of I_{on} in a linear scale for different stress fs (V_g at 32 V and V_d at 26 V). Additionally, part (b) exhibits the relationship between ΔI_{on} and pulse count on a logarithmic scale for assorted pulse frequencies [19].

As per our prior finite element analysis, a temporal constant indicative of the channel's thermal increase during SH events denoted as τ_r is contingent upon the device's structural attributes and the thermal characteristics of the material [20]. The estimated value is approximately 0.6 μs for devices under stress (with a width, W, of 10 μm). When subjected to low-f stress, the channel temperature dynamics of the device can synchronize with the voltage pulse, given that the pulse duration is significantly greater than τ_r, leading to continuous SH impact as V_{g_h} and V_{d_h} are in effect. This rationale clarifies the predominance of SH as the degradation mechanism under low-f stress conditions. With an identical α, the SH exposure time remains consistent across various low-f stresses, resulting in the superposition of degradation curves. In the case of high-f stress, such as at 500 kHz, τ_r approaches the t_{peak}, suggesting a reduced degradation potential due to inadequate SH conditions as that in [21, 22]. Nonetheless, within the scope of this research, device degradation intensifies with elevated stress fs. With a constant α, t_r, t_f, and for a specified stress duration, increased stress f correlates with an increased count of pulse cycles, yet the cumulative SH exposure time remains roughly equivalent. Consequently, the exacerbated degradation at higher fs suggests the operation of an alternative degradation pathway, distinct from SH, potentially associated with the transient phases of pulse switching, t_r or t_f [23-29]. In fact, Fig. (**6.9**) illustrates that when ΔI_{on} at assorted fs is regraphed against pulse count N, the trajectories for all high-f stress conditions align closely, suggesting a degradation mechanism associated with the transients t_r or t_f. Conversely, under low-f stress, more pronounced device degradation occurs at lower fs due to an extended SH duration [19].

To discern the effects of stress-induced defects, the transfer characteristics of the device before and after the pulse stress are juxtaposed, as shown in Fig. (**6.10**). The parameters α, t_r/t_f, and stress duration are set at 50%, 0.1 µs, and 30000 seconds, respectively. The comparison of transfer curves before and after stress at a low f of 5 kHz is depicted at drain-source voltages V_{ds} of 0.1 V and 5 V in both normal and reverse modes in Fig. (**6.10a**). At the outset, the transfer curves exhibit symmetry in both operational modes. Before the stress, the ON-state transfer curves maintain symmetry at V_{ds} = 0.1 V and exhibit minimal alteration at V_{ds} = 5 V, indicating a distribution that is close to even for the SH-generated shallow traps. In the normal mode, the degradation of I_{on} current (ΔI_{on}) is 24.3% at V_{ds} = 0.1 V, escalating to 39.5% at V_{ds} = 5V, as demonstrated by the pointed indicators, signifying no resumption of I_{on} degradation. In contrast, within the OFF-state sector, a pronounced difference in the GIDL current I_{off} between the conventional and reverse modes implies a selective formation of deep leakage states adjacent to the drain. Subsequently, the transfer properties subsequent to high-f stress at 500 kHz are examined at V_{ds} of 0.1 and 5 V under both operational modes in Fig. (**6.10b**). Post this high-f stress, the transfer curves preserve their symmetry at V_{ds} of 0.1 V yet become asymmetric at V_{ds} of 5 V. A clear recovery in I_{on} degradation is detected in the standard mode, with the degradation percentage decreasing from 32.3% at V_{ds} of 0.1 V to 20.7% at V_{ds} of 5 V, as signified by the arrows. Commonly, in HC degradation scenarios, I_{on} at higher V_{ds} is less affected by degradation [30]. The recovery of I_{on} current at elevated V_{ds} values can be attributed to the reduction in the potential barrier caused by stress-induced traps in the vicinity of the drain, which is influenced by the drain-induced barrier-lowering effect [30, 31]. Consequently, traps generated under high-f stress are likely concentrated near the drain boundary, which explains the recovery in ΔI_{on} at elevated V_{ds}. Notably, while the ΔI_{on} recovery is a common characteristic of HC degradation, the rise in I_{off} in the OFF-state region deviates from this pattern. Traditionally, within the context of DC HC degradation processes, the standard mode I_{off} has been observed to decrease as a result of the reduced electric field strength at the drain caused by the injection of hot holes near the drain region. The observed increase in I_{off} in this case suggests the formation of deep traps that contribute to leakage near the drain in the absence of HC injection [19].

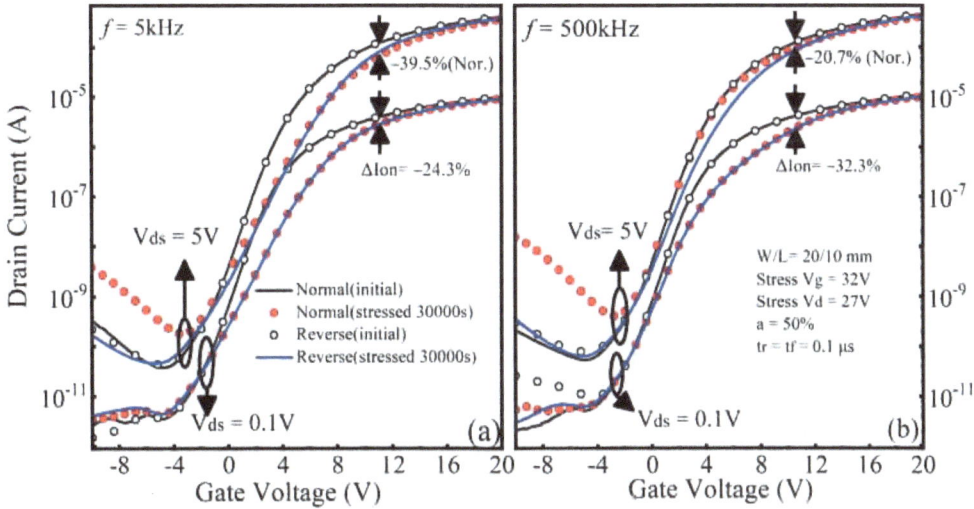

Fig. (6.10). Comparison of initial transfer curves under normal and reverse sweep at $V_{ds} = 0.1$ V and 5 V and transfer curves after stress at (a) a low frequency of 5 kHz and (b) a high frequency of 500 kHz. The stress time is 30000s.

To examine the degradation in the OFF-state, Fig. (**6.11a**) illustrates the dependence of ΔI_{off} on stress time across various pulse fs, with α consistently set at 50% and t_r/t_f at 0.1 μs. Under higher f stresses, ΔI_{off} exhibits a greater magnitude, adhering to a uniform degradation slope. Given that a higher f corresponds to an increased pulse count N, the link between ΔI_{off} and f suggests a connection between trap formation and the edges of pulse transitions. In contrast to the I_{on} degradation depicted in Fig. (**6.11**), the degradation of I_{off} does not display a significant disparity between low and high f regions. Fig. (**6.11b**) presents ΔI_{off} regraphed against N, revealing a remarkable universal trend where all I_{off} degradation curves align, regardless of f, following the equation $\Delta I_{off} = 0.23$. This straightforward universal trend is well-suited for forecasting I_{off} degradation under diverse pulse stress conditions. It should also be highlighted that the I_{off} degradation slope closely mirrors that of I_{on} degradation at high fs, indicating a shared degradation mechanism. This mechanism is evidently tied to the transitions during pulse switching (t_r and t_f), as opposed to the accumulated duty time linked to the SH mechanism [19].

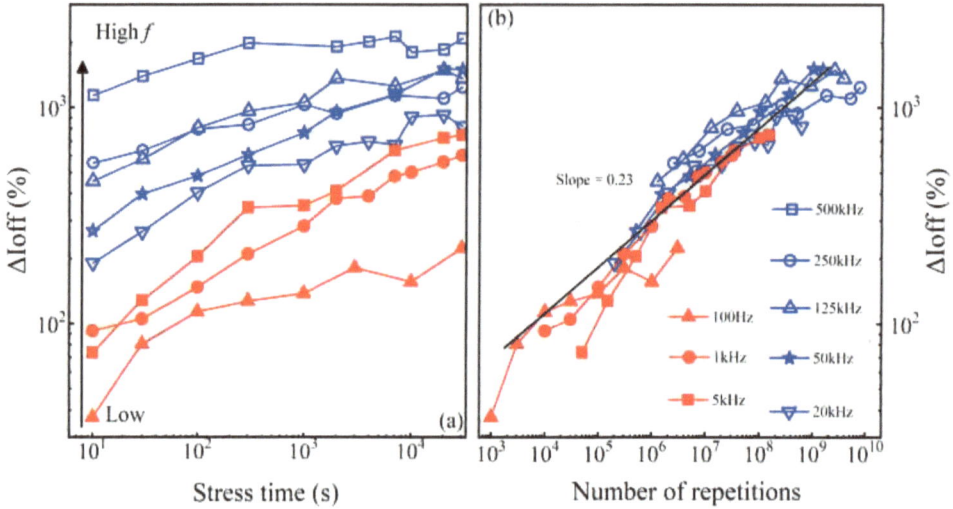

Fig. (6.11). ΔI_{off} degradation in logarithmic format under different f stresses depending on (a) stress time and (b) pulse number.

To explore the degradation mechanism, Fig. (**6.12** and **6.13**), respectively, depict the N-dependence of I_{on} and I_{off} degradations under high f stress with modified t_f or t_r. In Fig. (**6.12**), with t_r set at 1 µs and t_f ranging from 0.1 to 4 µs, it becomes evident that t_f exerts minimal influence on device degradation. This finding diverges from prior research on dynamic stresses, where t_f was a significant determinant of device degradation. With a reduced t_f, a heightened quantity of electrons that are trapped experience a stronger temporary transient coupling electric field (E_c) [28, 29, 32] and become HCs. In Fig. (**6.13**), the t_f is stabilized at 1 µs, and the t_r is incremented from 0.1 to 4 µs. Both the variations in the ΔI_{on} and ΔI_{off} exhibit a pronounced reliance on t_r, with more severe degradation observed under conditions of reduced t_r. Consistently, across both figures, the degradation gradients for I_{on} and I_{off} are closely aligned, suggesting a commonality in the degradation processes. The distinct sensitivities of device degradation to t_r versus t_f indicate that the degradation process is more closely associated with t_r. A recent study has highlighted that the transition of the V_g during both t_r and t_f phases generates a coupling E_c, which contributes to the degradation of the device performance that is contingent on both t_r and t_f [19, 29].

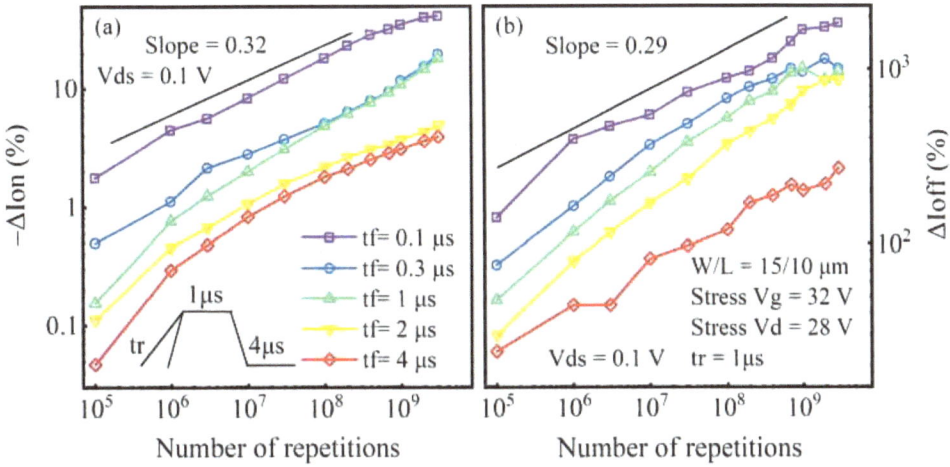

Fig. (6.12). Fixed $t_r = 1$ µs, the relationship between (a) I_{on} and (b) I_{off} degradation and the number of pulses under different t_f.

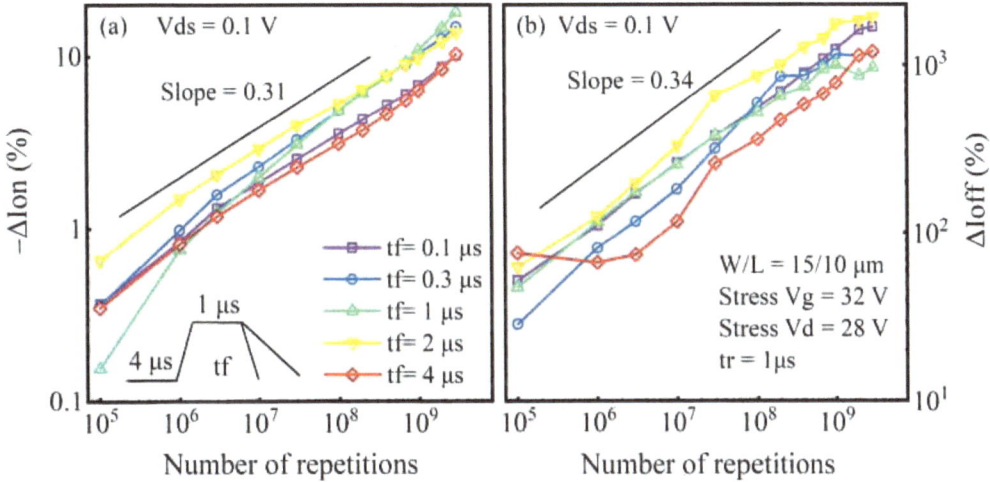

Fig. (6.13). Fixed $t_f = 1$ µs, the relationship between (a) I_{on} and (b) I_{off} degradation and the number of pulses under different t_r.

To elucidate the degradation mechanism further, the device undergoes stress at increased substrate temperatures, as depicted in Fig. (**6.14**). Under low-f stress (100 Hz, Fig. (**6.14a**), the device exhibits exacerbated degradation at higher temperatures, predominantly governed by the SH mechanism, which intensifies with temperature elevation. In contrast, high-f stress (164 kHz, Fig. (**6.14b**) demonstrates a converse outcome, with I_{on} degradation diminishing at elevated temperatures. Additionally, the SH degradation's initial turnaround phenomenon, indicated by the positive ΔI_{on} at 80°C, is also observed. With high-f stress, the known reduction in HC impact ionization rate at higher temperatures suggests that an increase in substrate temperature mitigates the HC mechanism while augmenting the SH mechanism. Consequently, the pronounced suppression of I_{on} degradation, coupled with the resurgence of the turnaround effect at higher temperatures, unmistakably points to a t_r-related HC mechanism as the operative degradation process [19].

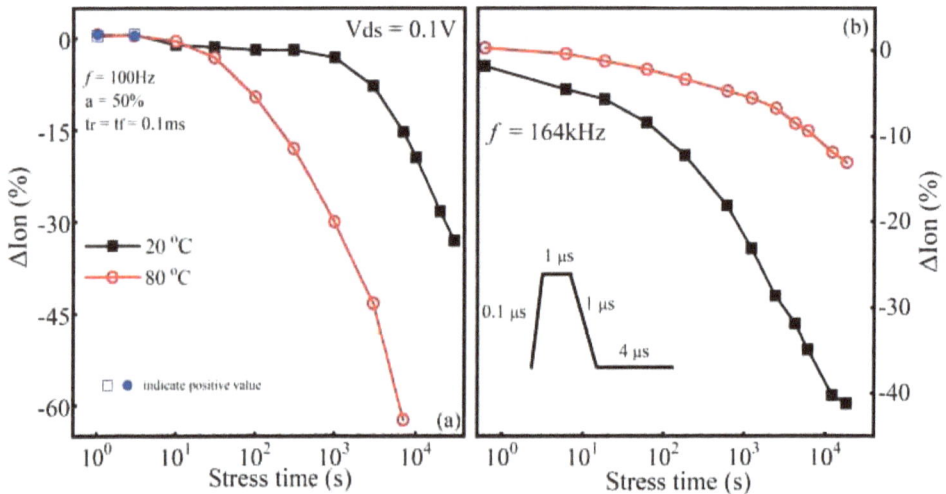

Fig. (6.14). Comparison of I_{on} degradation under different temperatures (20 °C, and 80 °C) for (a) low f = 100 Hz stress and (b) high f = 163 kHz stress (V_g = 32 V, V_d = 28 V, W/L = 15/10 µm).

A model delineating degradation under synchronized voltage pulses has been introduced. In the ON-state, the SH mechanism predominates under conditions of low-f stress, while a high-f stress scenario is characterized by a t_r-related HC mechanism. Additionally, this HC mechanism is also responsible for OFF-state degradation, regardless of f. Notably, significant degradation is contingent upon the device being in a state of saturation. Following a study [33], the drain saturation voltage (V_{Dsat}) of a stressed TFT can be estimated by $V_{Dsat} = V_{gt}(1+1/\xi - \sqrt{1+1/\xi^2})$, $\xi = \delta S V_{gt}/(V_{gt}-\delta V_d)^2$ and $V_{gt} = V_g - V_t$, and the inversion voltage of the device within silicon grains, which is approximately −1 V, has been deduced from CV measurements [34]. The parameter S, associated with grain boundary defects, is deduced to be approximately 13.2 V for devices under stress, while δ remains nearly invariant at around 0. Utilizing the equation, the relationship between V_{Dsat} and V_g during a pulse cycle is determined, exemplified by a standard synchronized voltage stress scenario with V_{g_h} at 32 V and V_{d_h} at 28 V, as depicted in Fig. (**6.15a**). The persistently higher V_d compared to V_{Dsat} signifies that the device is saturated, ensuring the channel's drain side remains depleted. Consequently, the E_c in the depleted region escalates with V_g variations and peaks during shorter transition times of t_r/t_f. For the stressed devices, characterized by a width (W) of 15 μm, the time constants for the channel temperature's increase (τ_r) and decrease (τ_f) are approximately 0.65 μs and 1.4 μs, respectively. The transit time for carriers is approximated by the formula $t_{transit} = L/[\mu(V_d/L)]$, given a channel length (L) of 10 μm, a carrier mobility (μ) of 50 cm²/Vs, and a calculated transit time ($t_{transit}$) of 4 ns using V_d of 5 V, indicating that mobile carriers can keep pace with the pulse switching transients [19].

Fig. (**6.15a**) delineates a voltage pulse within a single period, partitioned into four distinct regions. Region I (from 0 to t_1) marks the onset of the rise phase, devoid of degradation. Upon V_g surpassing V_{th}, the TFT enters saturation, instigating E_c from V_g oscillations. Yet, E_c remains insufficient for HC generation until t_1. Region II (from t1 to t2) encompasses the majority of the rise phase, where HC degradation initiates, as illustrated in Fig. (**6.15b**). The TFT, saturated and with a depletion region at the drain, witnesses a growing E_c capable of generating HCs. Electrons swiftly traverse the channel within a $t_{transit}$ of less than 4 ns, subsequently encountering the intense E_c in the depletion region, transforming into HCs and generating defects near the drain, devoid of HC injection. Concurrently, the synchronous escalation of V_g and V_d renders SH prominent, screening the HC effect by t_2. Region III (from t_2 to t_3) is characterized by the SH mechanism under high power density, as shown in Fig. (**6.15c**), where channel temperature escalates post a τ_r of 0.65 μs. Even as E_c peaks at the rise phase's conclusion, it is overshadowed by the SH effect. In Region IV (from t_3 to t_4), despite the power density's transition, an elevated channel temperature persists, governed by τ_f of 1.4 μs, surpassing t_f at 1 μs, thereby obscuring the t_f-related HC mechanism, as depicted in Fig. (**6.15d**). Should t_f exceed τ_f, the HC mechanism might be anticipated at the fall phase's terminus, yet it remains inconsequential given the diminished E_c at elongated t_f. Consequently, the degradation observed is independent of t_f. At abbreviated t_r (for instance, t_r less than τ_r), the HC mechanism not only engenders a more potent E_c but also extends its influence over a larger portion of t_r and potentially the t_{peak}'s inception, where channel temperature remains low, culminating in exacerbated degradation [19].

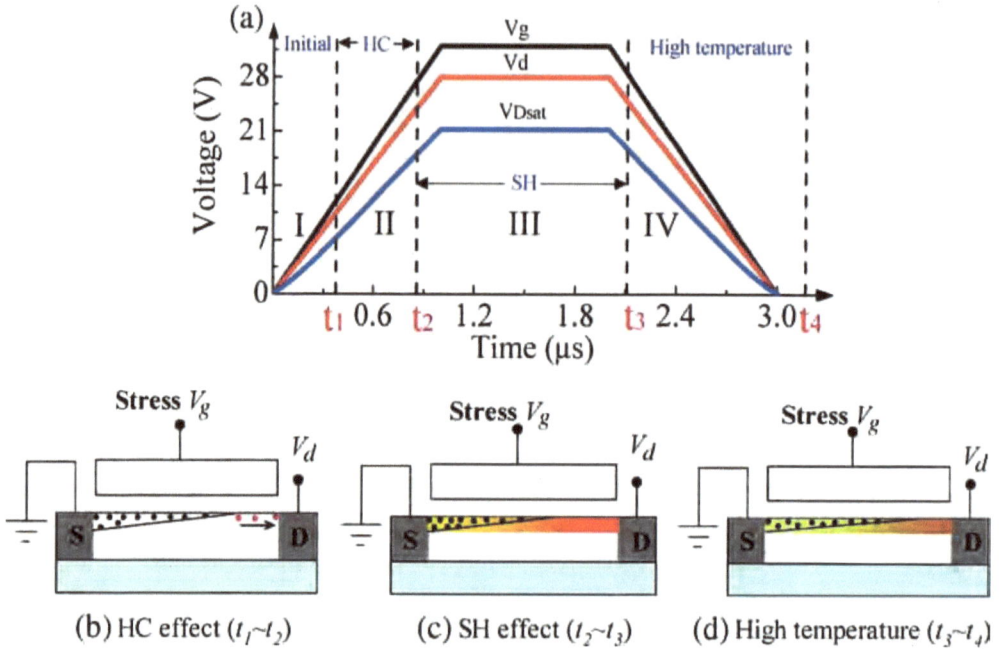

Fig. (6.15). Schematic diagram of device degradation model. (a) Changes in V_{Dsat} with V_g during synchronous pulse cycle (b) HC degradation in the second region (c) SH degradation in the third region (d) High-temperature degradation in the fourth region [19].

To validate the degradation model, the device undergoes stress at a significantly decreased P_{max} to mitigate the SH effect. In Fig. (**6.16**), P_{max} is adjusted from 0.24 to 0.14 mW/μm^2 by reducing V_{g_h} and V_{d_h} to 25 V and 20 V, respectively. With the attenuated SH effect, the channel temperature during the falling edge (Region IV) declines, prompting the re-emergence of t_f-dependent HC degradation, as anticipated by the model. This is evidenced in Fig. (**6.16a**), where increased degradation is noted for reduced t_f. Fig. (**6.16b**) corroborates that the t_r-related HC mechanism persists, with greater degradation for shorter t_r. Saturation of the device is integral to the degradation model, necessitating drain depletion to generate E_c. In Fig. (**6.17**), P_{max} is similarly lowered to approximately 0.14 mW/μm^2, but this time by decreasing V_{d_h} to 15 V while maintaining V_{g_h} at 32 V. The calculation of V_{Dsat} reveals that V_d consistently falls below V_{Dsat}, as depicted in the inset, indicating a conductive inversion channel throughout most of the stress pulse, precluding E_c induction at t_r or t_f. Consequently, in Fig. (**6.17**), the I_{on} degradation observed is minimal and detached from both t_r and t_f, with the exception of minor SH degradation, confirming that device saturation during synchronized voltage pulses is essential for the dynamic HC mechanism.

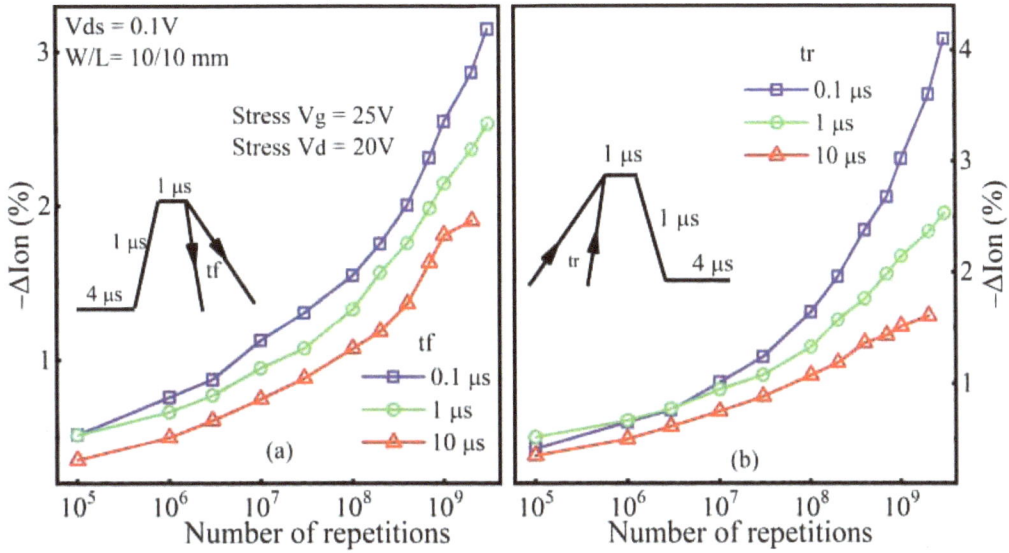

Fig. (6.16). (a) Fixed t_r to 1 μs. Different t_f and (b) fixed t_f of 1 μs. I_{on} degradation dependent on pulse number under different t_r [19].

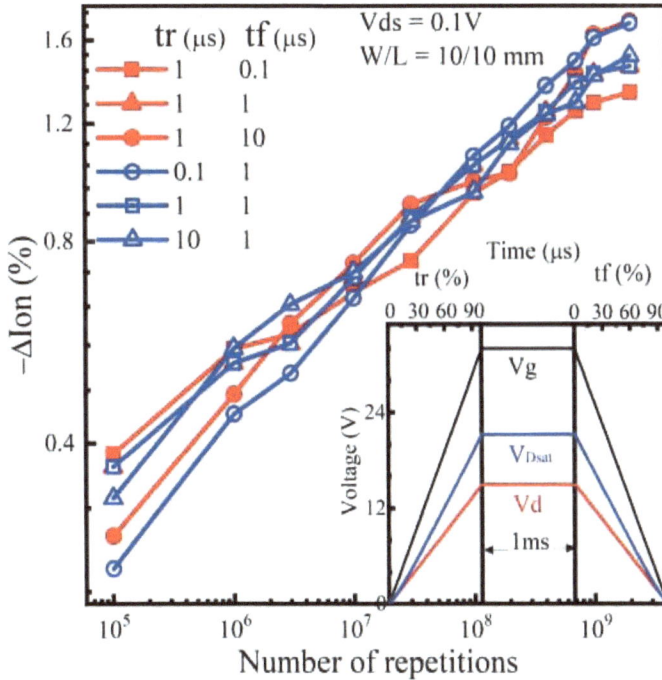

Fig. (6.17). The degradation of Ion depends on the number of pulses under different tr and tf. Insertion chart showing the variation of VDsat with Vg under pulse [19].

When the voltage amplitude of FSS stress becomes low and the device works under the saturation region, HC degradation may occur. Fig. (**6.18**) illustrates the gate and drain voltage pulses, which respectively oscillate between −15 and 15 volts and −10 and 10 volts. Given that V_g and V_d pulses are synchronized in the same direction, their respective duty cycles (α), *fs*, t_r, and t_f (denoted as α*g*, f_g, t_{r-g}, t_{f-g} for V_g, and α*d*, f_d, t_{r-d}, t_{f-d} for V_d) match. Both duty cycles, α*g* and α*d*, are configured at a 50% level. The parameters that are subject to variation encompass the frequencies f_g and f_d, as well as the rise and fall times t_{r-g}, t_{r-d}, t_{f-g}, and t_{f-d}. To streamline notation, *f*, t_r and t_f collectively denote the frequencies and the timing parameters of both pulses. The research utilizes top-gate self-aligned poly-Si TFTs with a consistent L and W of 10 micrometers. The transient analysis is executed through the Silvaco ATLAS simulation tool [35].

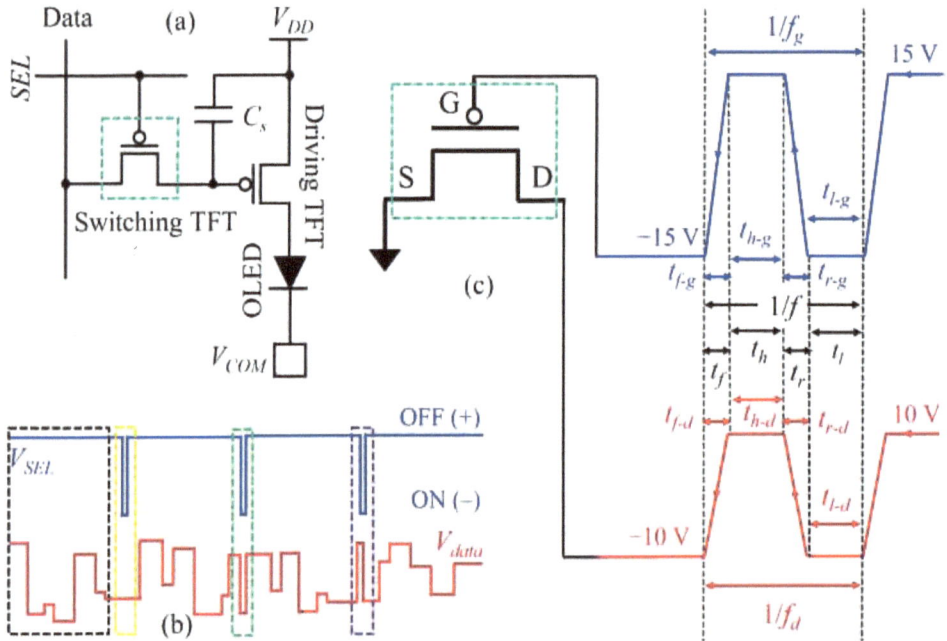

Fig. (6.18). (a) 2T1C pixel circuit diagram of AMOLED (b) actual voltage mode applied to the gate and drain of switching TFT (c) schematic diagram of stress waveform of switching TFT [35].

The impact of FSS on the transfer characteristics of a standard poly-Si TFT is scrutinized, as depicted in Fig. (**6.19a**). The parameters of *f*, t_r and t_f are all configured at 500 kHz and 0.1 microseconds, respectively. It is evident that FSS leads to a significant decline in device performance, with the I_{on} experiencing a reduction of approximately 100% after a 1000-second stress period. In contrast, under pure V_g stress with a range of -15 to 15 V and pure V_d stress with a range of -10 to 10 V, the I_{on} only diminishes by approximately 9% and 1%, respectively, over the identical stress duration. Additionally, the patterns of degradation observed under FSS closely mirror those observed in HC stress scenarios [30, 32, 36-38], where the I_{on} experiences a decline, and the subthreshold area remains unaffected by the degradation. [35].

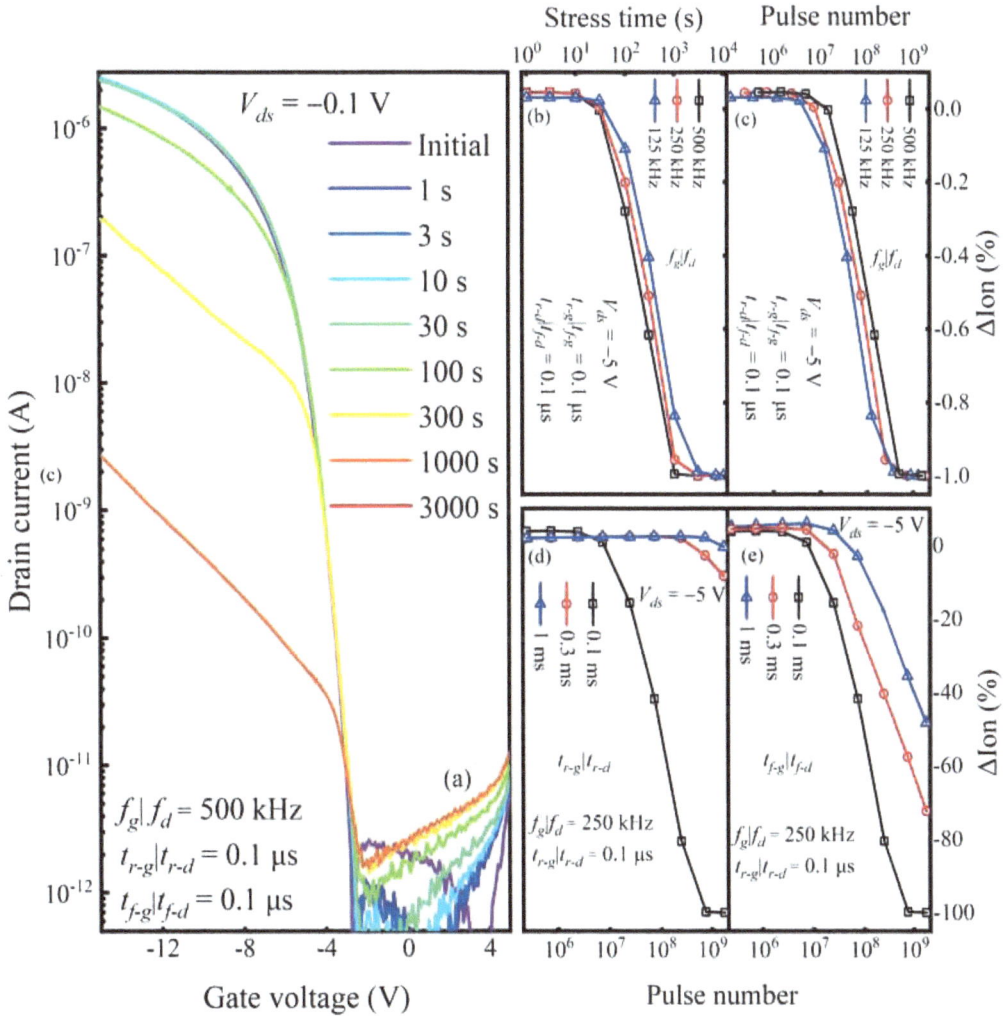

Fig. (6.19). (a) The transfer characteristic curve of normal poly-Si TFT under FSS depends on time. (b) Different f values ΔI_{on}'s dependence on stress time. Different (c) f, (d) t_r, and (e) t_f ΔI_{on}'s dependence on pulse count [35].

In order to ascertain the predominant factor of FSS that leads to device deterioration, a sequential analysis of the impacts of f, t_r and t_f is conducted. Fig. (**6.19b**) illustrates the variation in I_{on} degradation with stress duration across different fs, with t_r and t_f consistently maintained at 0.1 μs. It is evident that an increased f correlates with more pronounced device degradation over an equivalent stress period. A contrasting observation is made when the degradation data is juxtaposed against pulse count rather than stress duration, as depicted in Fig. (**6.19c**); a lower f results in greater device degradation for an identical pulse count, suggesting a correlation with the f of the pulses. This finding contrasts with the degradation patterns observed under isolated dynamic V_g stress [39] or isolated dynamic V_d stress [40], where the degradation does not depend on the f. Additionally, the influences of t_r and t_f are scrutinized in Fig. (**6.19d** and **6.19e**), respectively, with the f fixed at 250 kHz. Notably, both a reduced rise time and a reduced fall time are associated with increased device degradation, a behavior that diverges from the degradation observed under pure dynamic V_g or V_d stress, where the degradation is solely attributed to either the rise time or the fall time [35].

To elucidate the intrinsic mechanisms behind the degradation caused by FSS in poly-Si TFTs, a time-variant simulation was conducted. Given the striking resemblance between FSS-induced degradation and HC degradation, a meticulous examination was undertaken of two key parameters, E_x and J_x, within the context of poly-Si TFTs.

Fig. (**6.20a** and **6.20b**) display the temporal evolution of E_x and J_x across various fs at the channel's source and drain terminals. It is evident that the magnitude of E_x and J_x peaks and troughs remain consistent across different fs. Consequently, at lower fs, the duration of effective stress is extended for an equivalent number of pulses, leading to an exacerbation of HC-induced degradation, as illustrated in Figs. (**6.19c**, **6.15d** and **6.15e**) delineate how E_x and J_x vary with normalized time for distinct t_r and t_f. During the rise time, E_x escalates while J_x diminishes at the drain terminal, as shown in Fig. (**6.20d**). Despite a reduction in E_x, a substantial J_x persists, indicating that carriers are still susceptible to the influence of a strong electric field, thus transforming into HCs. A reduced t_r is associated with amplified E_x and J_x, culminating in intensified device degradation. On the source side, depicted in Fig. (**6.20e**), the scenario is analogous; a shorter t_r results in an increase in the absolute values of E_x and J_x, further aggravating the degradation. Accordingly, during the t_r, HC degradation is evident at both the source and drain terminals, with a more pronounced effect for shorter t_rs, as indicated in Fig. (**6.19d**). In the case of the t_f, at the drain terminal, even though E_x declines, it retains a significant value, enabling carriers to acquire sufficient energy to become HCs. A shorter t_f is linked to an increase in E_x and J_x. Conversely, on the source side, J_x starts to move in the reverse direction to E_x, precluding the formation of HCs [41]. Thus, it is inferred that HC-induced damage is confined to the drain side during the t_f and is intensified by a reduced t_f, as demonstrated in Fig. (**6.19e**).

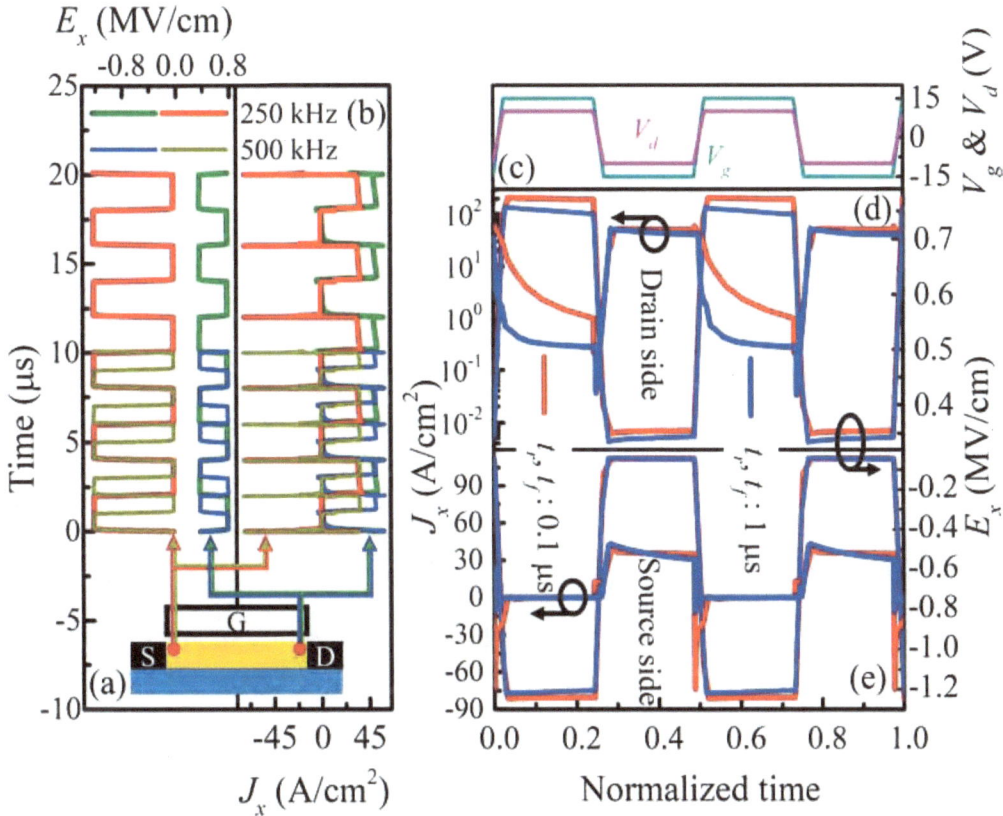

Fig. (6.20). The (a) E_x and (b) J_x under different f simulated in a standard poly-Si TFT depend on time, with the channel surfaces on the source region with red and brown lines and drain region with blue and olive lines near the gate oxide layer, $t_r = t_f = 0.1$ μs. (c) Apply FSS stress to normal poly-Si TFT. The time-dependent relationship between E_x and J_x at different t_r and t_f near the gate oxide layer on the channel surface of the (d) drain side and (e) source side [35].

To elucidate the mechanisms of degradation induced by FSS, an advanced version of the nonequilibrium PN junction degradation model has been utilized. As the t_r commences, the junctions on both the drain and source terminals are subjected to reverse biasing. The channel's depletion zone enlarges, initiating the release of carriers associated with shallow traps, succeeded by the emission of carriers from deeper traps. Carriers emitted in the latter phase from deep traps encounter the developed depletion zone, transforming into HCs, which leads to device deterioration. A reduced t_r results in a greater emission of deep-trap carriers subsequent to the establishment of a potent E_x, intensifying the degradation (as illustrated in Fig. (**6.19d**). In the t_f, on the source side, the junction contracts due to carrier recombination at the depletion region's periphery, where the electric field is minimal, precluding the formation of HCs. Conversely, on the drain side, the depletion region persists despite its reduction, as the V_d consistently exceeds the overdrive voltage during channel activation. The electric field in this region remains significant (as depicted in Fig. **6.20d**), enabling carriers to acquire sufficient energy and turn into HCs, thus causing further device degradation. A swifter t_f is linked to an amplified coupling E_x, which in turn results in more pronounced device degradation (as shown in Fig. **6.19e**).

6.3.2 Reverse synchronous Alternating-Current Degradation

The previously mentioned "driving" stress, which includes the stress applied during the OFF-state phase and the forward synchronized stress known as FSS, aims to mimic the real-world operating conditions of TFTs in pixel matrices for AM displays. However, the stress experienced in the OFF-state and the FSS might not entirely represent the full range of operational dynamics encountered by TFTs during switching processes [42].

In this research, we introduce a reversely synchronized stress (RSS) to more accurately replicate the operational scenarios of switching TFTs. The experimental findings indicate that RSS leads to significant deterioration in poly-Si TFTs, primarily driven by an AC HC impact. This degradation model, when integrated with transient simulation analysis, is both introduced and deliberated.

As discussed in ref [35], the operational scenarios for switching TFTs in AM displays encompass four distinct cases: OFF-state stress, a V_g pulse in conjunction with a DC V_d, FSS, and RSS. Fig. (**6.21a**) illustrates the RSS scenario, where the V_g pulse transitions from a base voltage (V_{base}) to a peak voltage (V_{peak}) concurrently with the V_d pulse moving from V_{peak} to V_{base}. The frequencies of the Vg and V_d pulses are denoted by f_g and f_d, respectively. The terms t_{r-g} and t_{f-g} represent the rise and fall times of the V_g pulse, while t_{r-d} and t_{f-d} indicate the rise and fall times of the V_d pulse.

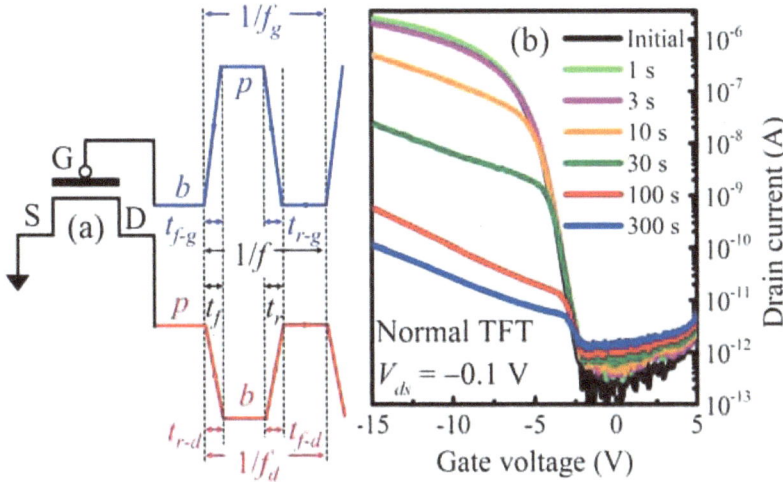

Fig. (6.21). (a) RSS conditions for switching TFT. (b) The time evolution of the transmission curve of normal TFT under RSS, measured at $V_{ds} = -0.1$ V [42].

To streamline the representation, the duty cycle (α) is utilized to denote the duty cycle of both the V_g and V_d pulses. The term f is used to signify the frequencies f_g and f_d. Additionally, t_r denotes the rise time for both the V_g and V_d pulses, while t_f signifies the fall time for the V_g pulse and the rise time for the V_d pulse. The baseline voltages for the V_g and V_d pulses are designated as −15 V and −10 V, respectively, and conversely, as 15 V and 10 V. The duty cycle α is established at 50%. The RSS pulses are produced using the Agilent 41501B pulse generator.

Comprehensive procedures for device fabrication are documented in other sources. The dimensions of standard TFTs, with a fixed channel length and width of 10 μm, are utilized. Degradation assessment is conducted through the examination of the relative change in the on-current (ΔI_{on}), measured at $V_g = -15$ V and $V_d = -5$ V. Transient simulations are executed using the Silvaco ATLAS tool, grounded on a persistent defect poly-Si model, which encompasses a total density of states with two exponential tail bands and two Gaussian deep bands.

Fig. (**6.21b**) displays the degradation of the transfer curve in standard TFTs under RSS, as measured at V_d of −0.1 V. The f, t_r and t_f are configured to 125 kHz and 0.1 μs, respectively. It is evident that the degradation induced by RSS is substantially greater than that caused by gate voltage pulses [29], drain voltage pulses [43], and FSS [35]. Within just 30 seconds of RSS, the I_{on} experiences nearly a complete degradation. Additionally, it is noted that as stress time progresses, I_{on} diminishes, the I_{off} escalates, and the SS remains largely constant, akin to HC degradation observed in poly-Si TFTs.

Fig. (**6.22a**) illustrates the dependence of I_{on} degradation on stress time under RSS across various fs. It is evident that RSS significantly contributes to device degradation, with higher fs exacerbating the I_{on} degradation. Upon re-plotting the data against pulse count in Fig. (**6.22b**), the data points for different fs appear to converge. However, upon closer inspection, it is observed that at lower fs, the degradation is more pronounced for the same number of pulses, akin to the pattern observed under FSS, albeit with a less pronounced dependency.

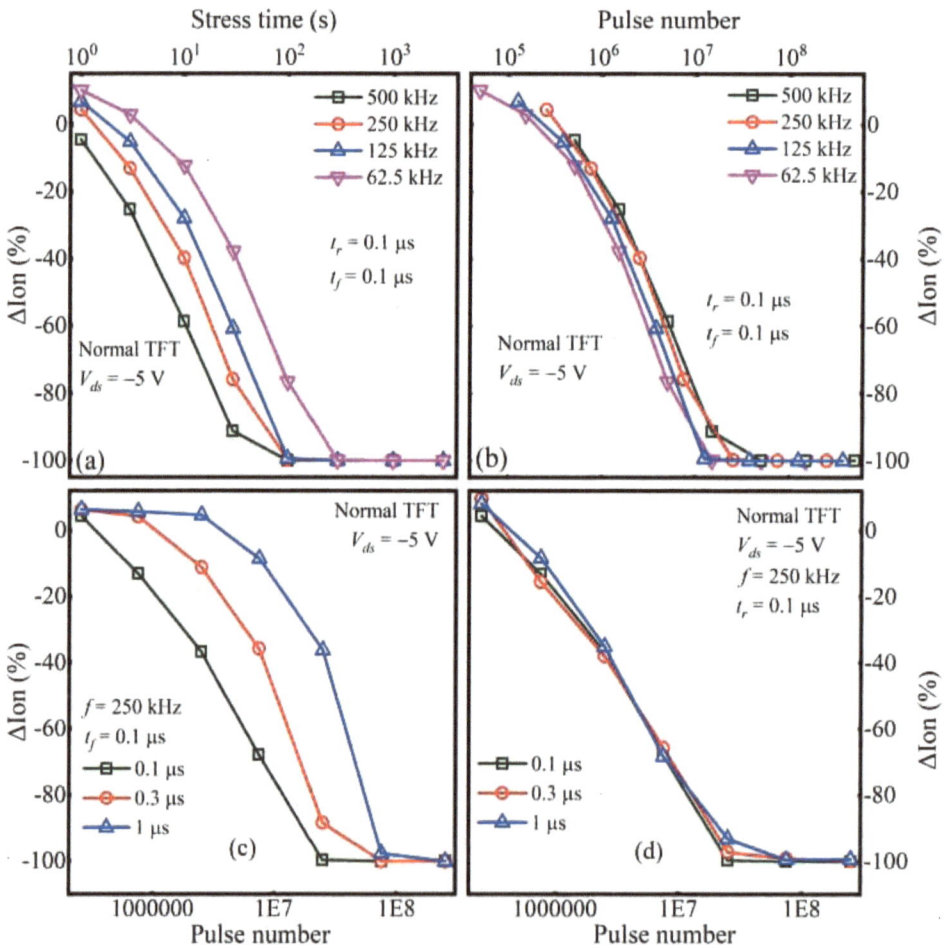

Fig. (6.22). (a) ΔI_{on} depends on the stress time under RSS and has various fs in normal TFT, measured at $V_{ds} = -5$V. ΔI_{on} depends on the number of pulses under RSS and has various (b) fs, (c) t_rs, and (d) t_fs in normal TFT, measured at $V_{ds} = -5$V [42].

Following an analysis of f, the impacts of t_r and t_f under RSS are assessed, as illustrated in Fig. (**6.22c** and **6.22d**). It is evident that degradation caused by RSS is contingent solely on t_r, with no dependence on t_f, suggesting that HCs could be produced during the t_r phase. Additionally, RSS with a reduced t_r duration results in more pronounced device degradation, as depicted in Fig. (**6.22c**). This pattern of degradation aligns with that observed in stress induced by V_d and V_g pulses, which are also solely influenced by t_r or t_f, respectively. In contrast, degradation under FSS is influenced by both transition edges.

Given that degradation caused by RSS is significantly associated with dynamic HC effects, there are two crucial parameters that play a key role [29], E_x and J_x, which are determined in relation to the RSS imposed on poly-Si TFTs.

Fig. (**6.23a**) illustrates the application of the RSS pattern to a standard TFT. Fig. (**6.23b**) displays the relationship between the E_x and J_x with respect to normalized time at the drain side, considering various t_r and t_f. It is evident that E_x escalates while J_x diminishes during the t_r phase, with J_x aligning in the direction of E_x. Upon reaching a significant value, E_x continues to influence J_x, which then absorbs energy from the field, transforming into HCs and causing device degradation. The intensity of E_x correlates with the rate of change of RSS, such that a quicker t_r phase results in a more substantial E_x. Additionally, a reduced t_r duration also leads to an increased J_x. The combination of a larger E_x and J_x during a rapid t_r phase culminates in more severe HC degradation, as depicted in Fig. (**6.22c**). Conversely, during the t_f phase, even though the absolute value of J_x ($|J_x|$) reaches a peak, E_x plummets to a minimal value (approximately 0.04 MV/cm). Moreover, J_x is directed opposite to E_x, precluding the generation of HCs during the t_f phase. Consequently, the degradation engendered by RSS is independent of t_f, as shown in Fig. (**6.22d**).

Fig. (**6.23c**) illustrates how the E_x and J_x vary with normalized time at the source side, presenting a comparable scenario. During the t_r phase, the maximum absolute value of E_x is reduced compared to the drain side, signifying a mitigated HC degradation effect at the source. In the t_f phase, even though J_x aligns with the direction of E_x, the absolute value of E_x remains minimal (approximately 0.07 MV/cm), insufficient to propel carriers into HCs. Consequently, the degradation remains unaffected by t_f, as shown in Fig. (**6.23d**). Simulation outcomes further suggest that the RSS-induced HC degradation occurs over a shorter effective stress period compared to FSS, leading to a less pronounced dependence on f for RSS compared to FSS.

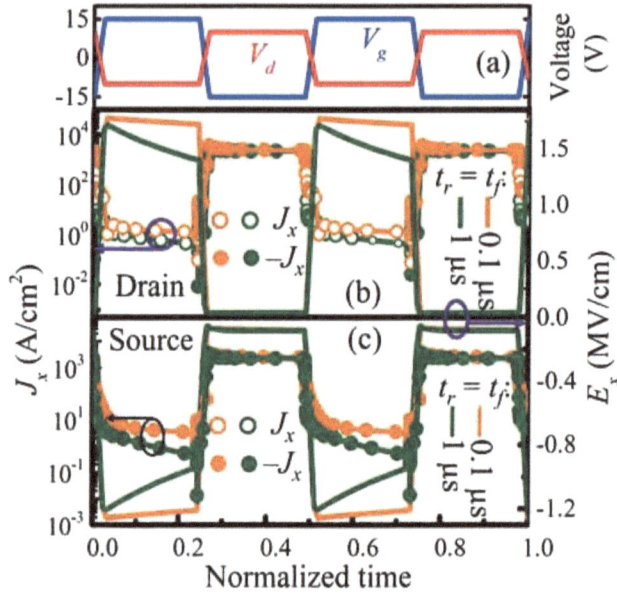

Fig. (6.23). (a) The normal TFT is subjected to RSS. The variation of the E_x and J_x near the channel/oxide interface with respect to normalized time is depicted for distinct t_r and t_f on the (b) drain side and (c) source side. The orange and olive lines correspond to the conditions of RSS with equal t_r and t_f of 0.1 μs and 1 μs, respectively. The hollow and solid circle markers denote J_x and the negative of J_x ($-J_x$), respectively [42].

To provide additional insights into the degradation mechanism under RSS, the nonequilibrium PN junction degradation model has been utilized and refined. During the t_r, the V_g and V_d collaborate to reverse-bias the drain junction, a process distinct from that under FSS, where the V_d pulse mitigates the reverse-biasing effect. The channel's depletion region forms initially through the emission of carriers from shallow traps, followed by those from deep traps. Carriers emitted later are subjected to the existing depletion region, absorb energy, and transform into HCs. A reduced t_r duration leads to a more substantial E_x and a greater emission of carriers post-high E_x establishment, as seen in Fig. (**6.23b** and **6.23c**), resulting in increased device degradation (Fig. **6.22c**). Moreover, in contrast to FSS, the E_x at the drain junction is significantly amplified by the reverse transition during t_r, exacerbating device degradation. At the source junction, the effect is primarily driven by the V_g pulse, with a more limited voltage fluctuation range compared to the drain side, thus diminishing the HC effect. During the t_f, the junctions become forward-biased, the depletion region contracts through carrier recombination at the outer edge, and the absence of HC generation ensures that degradation is t_f-independent, as illustrated in Fig. (**6.22d**).

6.3.3 Bipolar Alternating Current Degradation in Metal Oxide Thin-Film Transistors

Research has also been conducted on the effects of AC synchronous stress on MO TFTs. The degradation of MO TFTs under single-ended pulse stress is primarily attributed to the AC HC effect, which is triggered by the falling edge of the gate pulse or the rising edge of the drain pulse. The left side of Fig. (**6.24**) displays the transfer curve degradation of an IGZO TFT under DC stress of V_g = 15V and V_d = 25V. Due to the SH effect, the transfer curve exhibits an increase in V_{th}, an increase in SS, and a decrease in I_{on} after stress (refer to Section 2.5.2 for details). While maintaining the same stress amplitude at the gate and drain electrodes, the DC gate/drain bias stress was replaced with synchronized gate/drain pulses to investigate the degradation patterns of a-IGZO TFTs. The synchronized pulses had a frequency of 10 kHz and a duty cycle of 50%, and both rise and fall times were set to 0.1 µs. As shown in Fig. (**6.25**), the left and right graphs represent the changes in the device's transfer curves under different stress times for V_d of 5 V and 0.1 V, respectively [44]. It is evident that the degradation caused by synchronized pulses is somewhat mitigated compared to that induced by DC gate/drain bias. For instance, after 3000 seconds of stress, the degradation amounts for the V_{th} and I_{on} are 1.9 V and 67%, respectively, while under DC gate/drain bias conditions, the values are 3.4 V and 90%, respectively. Additionally, under synchronized pulse stress, recovery of the I_{on} degradation is observed for measurements at V_d of 0.1 V and 5 V. Similar DIBL effects are observed under both DC bias and synchronized pulse stress conditions, indicating that the DC effect plays a significant role in the degradation process under synchronized pulses, whereas the dynamic HC effect either does not occur or is of minor importance [44].

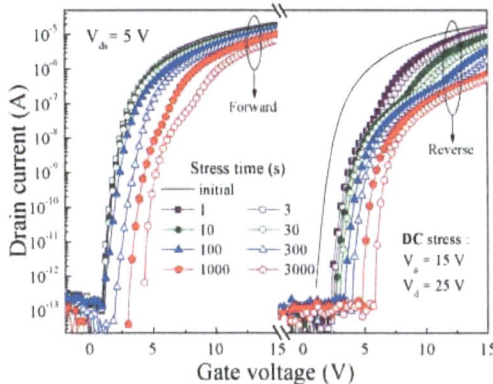

Fig. (6.24). Degradation of a-IGZO TFT transfer curves after different stress times in forward (left) and reverse (right) measurement modes. The gate/drain DC stresses are V_g = 15 V and V_d = 25 V. All the above data are measured at V_d = 5 V. [44].

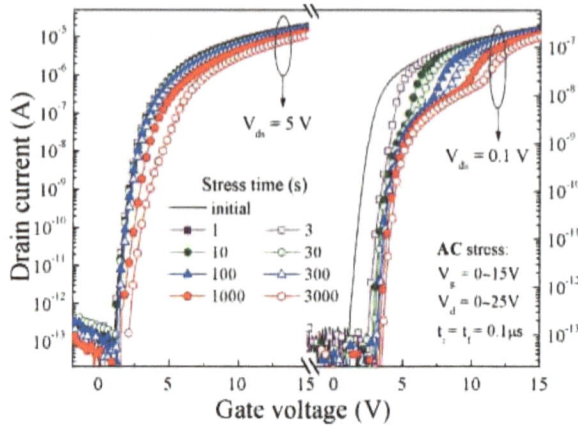

Fig. (6.25). Transfer curves of a-IGZO TFTs after different stress times under the measurement conditions of $V_d = 5$ V (left) and 0.1 V (right). The synchronous pulse stresses are $V_g = 0 \sim 15$ V, $V_d = 0 \sim 25$ V, both rising/falling edges are 0.1 μs, and the duty cycles are both 50%. [44].

To investigate the impact of the pulse rise and fall times on synchronized pulse degradation, the rise and fall times of the pulses are gradually increased from 0.1 μs to 5 μs. Fig. (**6.26**) illustrates the comparison between the changes in V_{th} and I_{on} under DC and synchronized pulse conditions with stress time. It is evident that all degradation caused by synchronized pulses is weaker than that caused by equivalent DC bias. Moreover, regardless of changes in the rise and fall times, the degradation curves for V_{th} and I_{on} follow the same trend. Based on this observation, it can be concluded that the devices are primarily influenced by DC SH under synchronized AC stress, with little to no contribution from the AC effect.

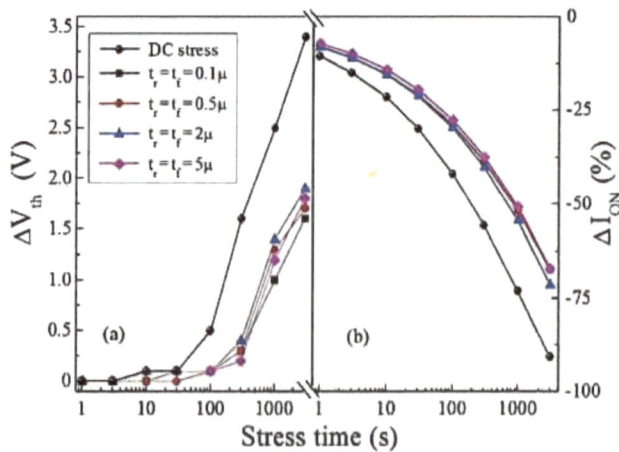

Fig. (6.26). Changes in V_{th} and open-state current as a function of stress time for different pulse rising (a) or falling (b) edges increasing from 0.1 μs to 5 μs. [44].

6.4 ULTRA-FAST ALTERNATING CURRENT DEGRADATION

In the pursuit of high-resolution and high-refresh-rate performance for screens, the operating frequency of TFT devices is gradually increasing, whether they are used in AM display technology or as SoP applications. Among the various aspects, the AC reliability of poly-Si TFTs has received thorough investigation, with dynamic V_g stress being one of the fundamental research areas. Under dynamic gate voltage stress, the AC HC effect plays a dominant role. Previous research has found that device degradation is solely related to the transition time of the voltage levels and is independent of the pulse width duration (t_d) [45-47]. However, earlier studies primarily focused on degradation under relatively slow voltage pulses, typically at 100 kHz. These low-frequency pulses, corresponding to µs-level transition times or t_d, do not accurately simulate the working conditions of TFTs in the peripheral drive circuits of high-refresh-rate, high-resolution display, and SoP applications. The degradation mechanism may change when the pulse transition time or t_d is reduced to the ns level [48]. We called the transition time of ns level in alternating-current stress ultra-fast AC stress.

As shown in Fig. (**6.27**), the devices used in this reliability measurement are top-gate coplanar structure poly-Si TFTs, with the active layer converted from amorphous silicon (a-Si) to poly-Si thin films using the solid phase crystallization (SPC) method [49]. The dimensions of channel W and channel L for the poly-Si TFTs examined in this chapter are uniformly set to 10 µm. A collective of 87 poly-Si TFTs exhibiting comparable initial traits were utilized in the endurance assessment. This chapter primarily employs gate stress to observe the degradation of poly-Si TFTs under AC stress.

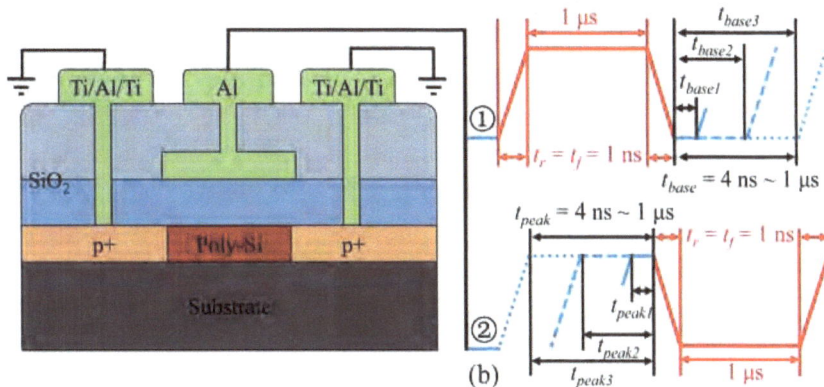

Fig. (6.27). (a) Schematic diagram of device structure and pressurization diagram (b) Changes in the pulse waveform of the device.

6.4.1 Stress Application Conditions and Degradation

To initially eliminate the DC effect, a DC stress of –12 V and 12 V was first applied to the gate while grounding the source and drain electrodes. Under DC stress of V_g = –12 V and 12 V, no degradation occurred in the devices. Based on these results, it can be concluded that no DC degradation mechanisms will be induced under AC stress with voltages less than 12 V.

The V_g pulse is composed of the t_r, t_{peak}, t_f, and t_{base}, with t_r and t_f both standardized at 1 nanosecond (ns). This research utilizes two distinct pulse waveforms to thoroughly examine the impact of t_{base} and t_{peak} on device degradation. The first waveform has a constant t_{peak} of 1 microsecond (μs) and a t_{base} that ranges from 4 ns to 1 μs. Conversely, the second waveform maintains a t_{base} of 1 μs while varying t_{peak} from 4 ns to 1 μs. Prior to stress application, each waveform type was validated using an oscilloscope through a cable to ensure its integrity. The waveforms were found to preserve their form effectively when transmitted through the cable, albeit with a reduction in amplitude of approximately 1 V in each case. For various stress conditions, the number of pulses was uniformly set to 1.5×10^9.

As mentioned in Chapter 1, the performance of each poly-Si TFT inherently varies due to the randomness of grain boundary distribution. Since the HC effect primarily occurs near the grain boundaries in the poly-Si film, degradation may differ significantly depending on the distribution of these grain boundaries. Particularly for SPC devices, which have smaller grain sizes, the degree of device degradation under the same conditions may also show noticeable differences. To address this issue, the degradation of devices was characterized under different AC pulse stresses by measuring five poly-Si TFTs for each experimental condition, as illustrated in the figure below.

To reduce the impact of device variability on reliability assessments, tests were conducted on a set of quintuple poly-Si TFTs for each stress category. Fig. (**6.28a**) displays the initial I_{on} and the I_{on} post-dynamic stress V_g across a range of t_{peak} and t_{base} values. Notably, device degradation appears to be affected by both t_{peak} and t_{base}, diverging from outcomes noted in prior studies where degradation was exclusively linked to t_r and/or t_f and unrelated to t_{peak} or t_{base}.

In the left segment of Fig. (**6.28a**), device degradation exhibits no dependence on t_{peak} for values exceeding 0.25 μs. Likewise, on the right side of Fig. (**6.28a**), for t_{base} surpassing 0.5 μs, degradation remains unaffected by t_{base}. These findings align with prior research where the t_d was within the microsecond range. Fig. (**6.28c**) presents a comparative analysis of the initial and post-stress transfer curves under dynamic stress V_g with t_{peak} and t_{base} both set at 1 µs, illustrating a classic instance of dynamic HC degradation in accordance with earlier studies.

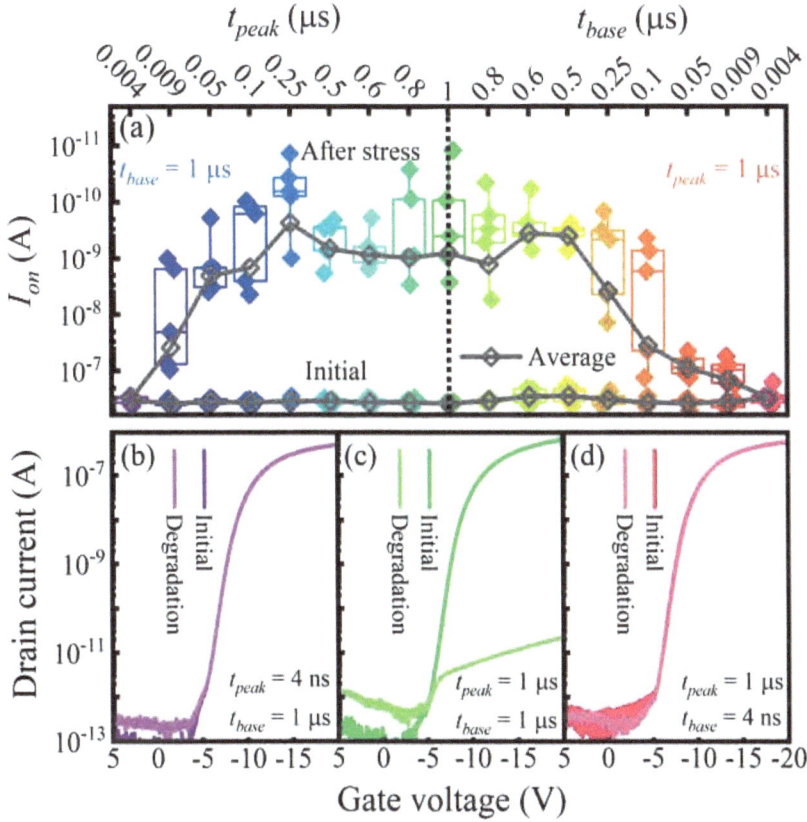

Fig. (6.28). (a) The initial I_{on} and the subsequent I_{on} following dynamic stress V_g are extracted, differentiated by varying t_{peak}s on the left and t_{base}s on the right. A comparative analysis is conducted between the initial transfer curve and the after-stress transfer curve under dynamic stress V_g for configurations (b) with t_{peak} at 4 ns and t_{base} at 1 µs, (c) both t_{peak} and t_{base} at 1 µs, and (d) t_{peak} at 1 µs with t_{base} at 4 ns [48].

Curiously, device degradation is contingent upon t_{peak} or t_{base} when t_{peak} is less than 0.25 µs or t_{base} is below 0.5 µs, as depicted in Fig. (**6.28a**). A reduced t_{peak} leads to mitigated device degradation, with t_{base} exhibiting a similar trend. Additionally, upon reducing t_{peak} or t_{base} to 4 ns, degradation is eliminated, as illustrated in Figs. (**6.28b** and **6.28d**). The assessment of output characteristic degradation reveals minimal impact under stress with brief t_d durations, whereas substantial current degradation is evident under stress with t_{peak} and t_{base} both at 1 μs, aligning with transfer curve degradation. The degradation patterns under dynamic V_g stress with abbreviated t_d markedly differ from those with extended t_d, suggesting the potential involvement of an alternative degradation mechanism and necessitating a revision of the existing model.

In prior research, the degradation patterns observed under dynamic voltage stress at lower frequencies were effectively elucidated by the nonequilibrium junction degradation model, which is predicated on pulse parameters in the microsecond range.

6.4.2 Capture and Emission Rates of Defect States

Based on the experiments and degradation results presented above, it is evident that when the device experiences waveform variations in the ns range, the internal degradation model differs from that observed under slower pulse stress. The previous non-equilibrium PN junction AC degradation model suggested that the key to device degradation lies in the slow ionization rate of deep levels within the device, leading to the exposure of ionized carriers to the high electric field in the space charge region of the PN junction [45]. This exposure accelerates the carriers into HCs, causing device degradation. From this model, it is clear that after deep levels ionize carriers, they must recapture carriers to ionize again in the next pulse. Therefore, in this chapter, the pulse stress degradation model was divided into two processes: defect ionization and defect capture. The ionization of defects is primarily composed of t_r and t_{peak}, while the capture process is mainly constituted by t_f and t_{base}. To analyze the degradation of poly-Si under ultrafast pulse stress, analyzing the capture and emission rates of defects is essential.

6.4.2.1 Emission Rate of Defects

Previous research has indicated that the emission rate of defects is related to the capture interface of the defect and the defect's position within the energy band. This rate can be expressed by the formula [45]:

$$\frac{1}{\tau_t} = \nu_{th}\sigma_h n_i \exp\left(\frac{E_i - E_t}{kT}\right) \tag{1}$$

where τ_t is the carrier emission time constant, ν_{th} is the thermal velocity of carriers, σ_h is the capture cross-section of hole defects, n_i is the intrinsic carrier concentration, E_i is the intrinsic Fermi level, and E_t is the energy level of the defect state. Based on this formula, a relationship diagram of τ_t versus E_t can be derived, as shown by the black line in Fig. (**6.29**). It can be seen that when E_t is at the position of E_i, τ_t can reach the ms range. However, according to the results shown in Fig. (**6.29**), significant degradation reduction only occurs after approximately 0.25 μs of t_d. This indicates that the formula alone is not sufficiently accurate to describe defect emission.

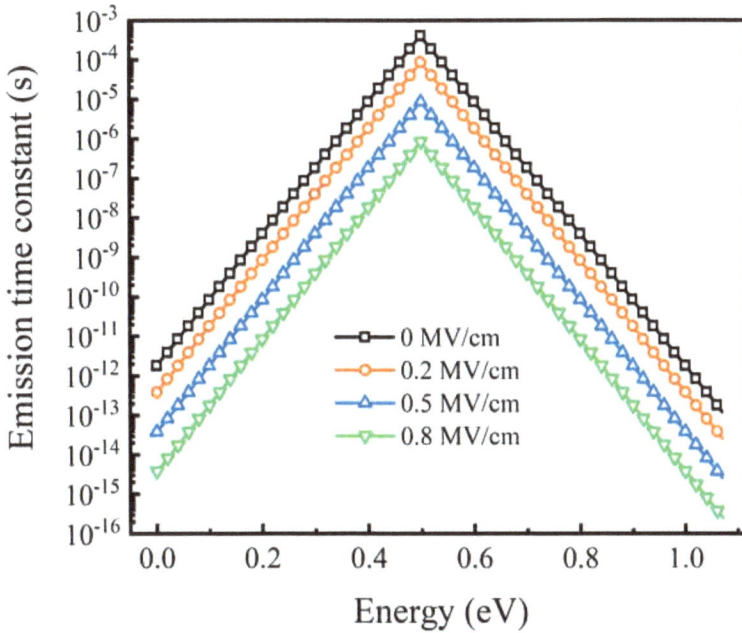

Fig. (6.29). The relationship between the position of defect states in the bandgap and τ_t under different applied electric fields.

Poole Frenkel effect in previous studies points out that the ionization rate of defect states is also related to the externally applied electric field [50]:

$$1/\tau_e(\Delta E)=1/\tau_e(0)\exp\left[(\Delta E)/kT\right] \tag{2}$$

Based on this formula, a relationship diagram of τ_t versus E_t can be further drawn, showing that as the external electric field increases, the ionization time required by defects also decreases gradually.

6.4.2.2 Capture Rate of Defects

Regarding the capture rate of defects, many previous studies have indicated that defect capture is quite fast, generally below the ns range [48, 51, 52]. However, this conflicts with the experimental results shown in the right half of Fig. (**6.28a**), where device degradation begins to significantly reduce after t_{base} is less than 0.5 μs. It is worth noting that the capture rate of defects is mainly related to the capture cross-section of the defect and the carrier concentration. This rate can be expressed as [53]:

$$1/\tau_r = v_{th}\sigma_h n_s \tag{3}$$

where τ_r is the capture time constant, and n_s is the concentration of free holes. Based on this formula, it can be inferred that when the carrier concentration reaches 10^9, the capture rate of defects may only be in the ms range. From the above analysis, it can be roughly speculated that the slow capture rate of defects may be primarily due to the low carrier concentration.

6.4.3 Ultrafast Pulse Stress Degradation Simulation

To substantiate the preceding hypothesis, TCAD simulations were conducted, concentrating on the E_x, lateral hole current density (J_x), hole concentration (h), and trantrap (tt). The study incorporated three distinct pulse configurations as depicted in Fig. (**6.30a**): Pulse 1 featuring a brief emission duration (t_{peak} of 4 ns), Pulse 2 characterized by extended t_{peak} and t_{base}, and Pulse 3 with a truncated recapture duration (t_{base} of 4 ns).

To delve into the emission process, the E_x and J_x profiles adjacent to the channel interface at the drain end are analyzed under the conditions of Pulse 1 and Pulse 2, as depicted in Figs. (**6.30b** and **6.30c**). The E_x closely mirrors the fluctuations of the applied stress signals, while the J_x undergoes a swift increase and subsequent decrease during the t_r. Relative to Pulse 2, Pulse 1 is characterized by a reduced peak value of E_x, and the J_x exhibits minimal areas of slight reduction, potentially contributing to the absence of degradation observed in Pulse 1.

Fig. (**6.30b** and **6.30c**) represent the degradation of the device under normal AC stress (Pulse 2) and during extremely short emission time (Pulse 1). It can be observed that at the initial t_r stage, there is a significant lateral current density, which quickly decreases afterward, leading to a rapid increase in the E_x. Upon magnifying both figures, it is evident that the peak J_x occurs earlier than the electric field reaching its maximum value. When the waveform is in the t_{peak} state, Pulse 1, due to its extremely short t_{peak}, experiences a rapid decrease in J_x after t_{peak} ends. In contrast, Pulse 2 maintains a smaller J_x during the subsequent t_{peak}.

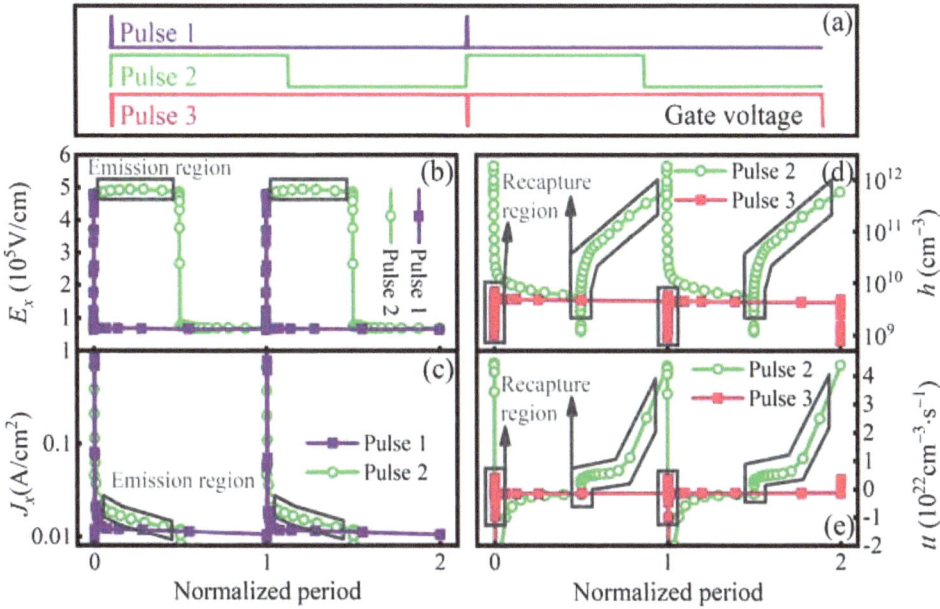

Fig. (6.30). (a) The initial I_{on} and the I_{on} following dynamic stress V_g are extracted, categorized by varying t_{peak}s on the left and t_{base}s on the right. A juxtaposition of the initial transfer curve with the curve subsequent to dynamic stress V_g is presented for (b) a t_{peak} of 4 ns and t_{base} of 1 μs, (c) t_{peak} and t_{base} both at 1 μs, and (d) t_{peak} at 1 μs with t_{base} at 4 ns.

Fig. (**6.30d** and **6.30e**) represent the degradation of the device under normal AC stress (Pulse 2) and during extremely short capture time (Pulse 3). It is evident that throughout the entire recapture process, the carriers in the channel exhibit a process of rapidly decreasing and then quickly increasing. For Pulse 2, which has a larger t_{base}, there is a noticeable increase in hole concentration within the device, and the capture rate of defect states also increases rapidly. In the case of Pulse 3, with a smaller t_{base}, the device remains mostly in the off-state, with its internal hole concentration remaining around 10^{10} cm^{-3} throughout.

6.4.4 Ultrafast Pulse Stress Degradation Mechanism

Based on the experimental results and simulation analysis presented in session 6.4.3, a degradation model was developed for poly-Si TFTs under ultrafast AC stress, as shown in Fig. (**6.31**). As discussed in Chapter 1, the active layer and the source and drain electrodes of a poly-Si TFT can be considered as two PN junctions. When pulse stress is applied to the gate, it can be viewed as applying an external voltage to the N-side of the PN junction [45]. During the t_r and t_{peak} phases of the pulse, the PN junction transitions from a forward-biased state to a reverse-biased state (Fig. **6.31a** and **6.31c**), corresponding to the emission phase described in Chapter 3.3; during the t_f and t_{base} phases, the PN junction transitions from a reverse-biased state back to a forward-biased state (Fig. **6.31b** and **6.31d**), corresponding to the capture phase described in the section 3.3.

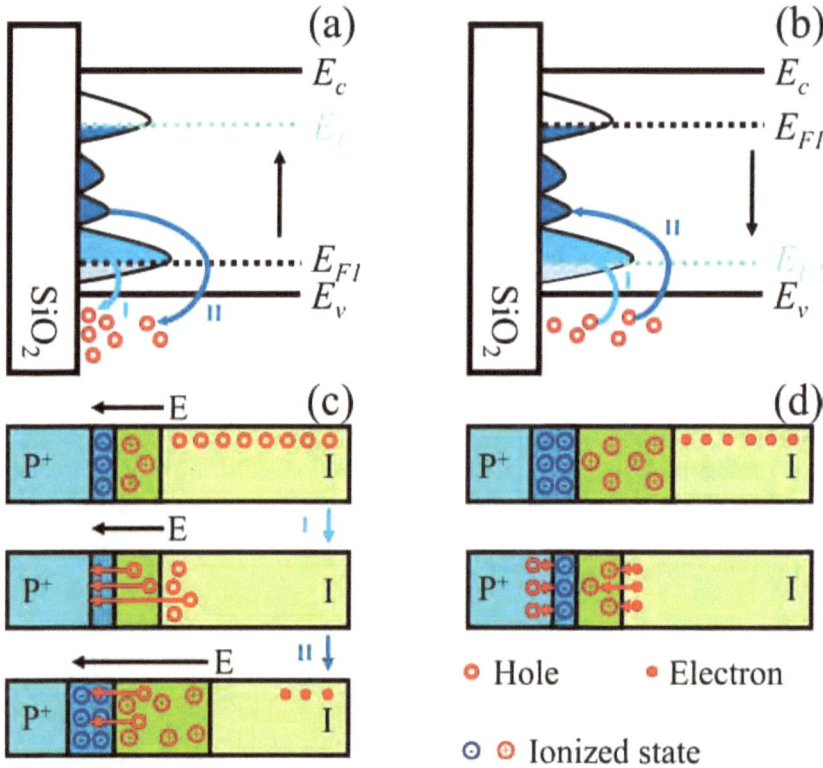

Fig. (6.31). A depiction of the energy band from the gate to the channel vicinity at the interface is presented for both (a) reverse and (b) forward modes. Figures illustrate the schematics for (c) the emission and (d) recapture processes, respectively.

At the beginning of the PN junction's inversion, the Fermi level at the device interface moves from near the conduction band toward the valence band. Concurrently, a large number of free holes from within the channel and holes ionized from potential level defects move towards the electrode from the active layer, generating a significant J_x, which is the first stage in Fig. (**6.31a** and **6.31c**). After a substantial number of potential level defects are ionized, the space charge region within the PN junction expands, and the electric field E_x increases. At this point, deep-level defects within the active layer begin to ionize, and the ionized free holes are accelerated into HCs (HC) by the larger E_x, marking the second stage in Fig. (**6.31a** and **6.31c**).

As the junction between the channel and the drain/source undergoes reversal during the transition (t_r), the Fermi level (E_F) transitions from E_{F1} to E_{F2}, as illustrated in Fig. (**6.31a**). At the outset, a significant flux of holes from the channel and those released from shallow traps move toward the electrode, generating a substantial J_x, demonstrated in Step I of Fig. (**6.31a** and **6.31c**). Post-emission from shallow traps, an E_x arises across the depletion region. The ionization of deep trap states begins, and the emitted carriers, under the large E_x, are propelled into HCs, as detailed in Step II of Fig. (**6.31a** and **6.31c**). With t_r in the microsecond range, the key HC degradation occurs throughout t_r, making it t_{peak}-independent. When t_r is at the nanosecond scale ($t_r = 1$ ns), the τ_e, although significantly curtailed by the intense electric field, is still inadequate for the depletion region's formation from deep trap state emission in such a short t_r. Hence, the main HC degradation occurs during t_{peak} rather than t_r. For t_{peak} over 250 ns, deep trap states have sufficient time to emit, rendering HC degradation independent of t_{peak}. For t_{peak} below 250 ns, the deep trap states show incomplete emission, leading to decreased HC degradation, as seen in Fig. (**5.28a**). At a t_{peak} of 4 ns, nearly all deep trap states stay un-ionized, resulting in the absence of HC degradation.

In the forward bias state (t_f), the E_F retreats, as depicted in Fig. (**6.31b**). Holes are first recaptured by the shallow traps (Step I, Fig. **6.31b**), then by the deep traps (Step II, Fig. **6.31b**). The reduction of t_{base} shortens the emission time, thereby increasing the release time (τ_r). When t_{base} is reduced to 4 ns, there are hardly any holes left for recapture (Fig. **6.30e**), and thus, minimal hole emission occurs during the next reverse transition (t_r), illustrated in Fig. (**6.30c**). As a result, HC degradation is avoided (Figs. **6.29a** & **6.29c**). For t_{base} exceeding 4 ns, traps can recapture more holes, with a longer t_{base} correlating with more recapture. This leads to more hole emission during the subsequent t_r and increased HC degradation (Fig. (**6.29a**). At t_{base} values above 500 ns, all ionized traps have sufficient time to recapture holes, making HC degradation t_{base}-independent (Fig. **6.29a**) [54].

CONCLUSION

Table 1. Degradation of Poly-TFT under different circuit-level degradation methods.

Types of TFTs	Degradation method	Increasing parameters				
		t_r	t_f	t_{peak}	t_{base}	f
	Driving stress	↓	-	-	-	↑
P-Type	Off-state stress	-	↓	-	-	↑
Poly-Si	Synchronous stress	↓	-	-	-	↑
TFTs	Reverse Synchronous stress	↓	↓	-	-	↑
	Ultrafast stress (gate)	-	↓	↓	↓	↑

* "↓" represents the decrease of degradation, "↑"represents the increase of degradation and "-"represents no effect on the degradation.

This chapter provides an in-depth examination of the various stress conditions that TFTs encounter in real-world circuit applications, emphasizing the complexities of maintaining device reliability under these scenarios. As shown in the above Table **1**, it begins with an analysis of AC degradation under DC bias, highlighting both "driving"-stress-induced and off-state-stress-induced degradation phenomena. The discussion then extends to bipolar AC degradation, differentiating between synchronous and reverse synchronous degradation mechanisms and exploring how these factors uniquely affect both polycrystalline and metal oxide TFTs. Furthermore, the chapter investigates ultra-fast AC degradation, detailing stress application conditions, defect state capture and emission rates, and simulation models that elucidate the underlying degradation mechanisms associated with ultrafast pulse stress. By synthesizing these diverse degradation phenomena, the chapter enhances our understanding of the challenges TFTs face in practical applications, ultimately contributing valuable insights that inform future research directions and the development of more resilient TFT technologies. These findings underscore the importance of addressing these stress-induced degradation mechanisms to ensure the longevity and reliability of TFTs in advanced electronic circuits.

REFERENCES

[1]   D. Zhou, M. Wang, M. Zhang, H. Hao, D. Zhang, and M. Wong, "An investigation of drain pulse induced hot carrier degradation in n-type low temperature polycrystalline silicon thin film transistors", *Microelectronics Reliability,* vol. 50, 2009no. 5, pp. 713-716
      http://dx.doi.org/10.1109/IPFA.2009.5232717

[2]   M. Zhang, W. Zhou, R. Chen, M. Wong, and H.S. Kwok, ""Driving"-Stress-Induced Degradation in Polycrystalline Silicon Thin-Film Transistors and Its Suppression by a Bridged-Grain Structure", *IEEE Electron Device Lett.,* vol. 38, no. 1, pp. 52-55, 2017.
      http://dx.doi.org/10.1109/LED.2016.2626481

[3]   M. Zhang, M. Wang, X. Lu, and M. Wong, ""Characterization of hot carrier degradation in n-type poly-Si TFTs under dynamic drain pulse Stress with DC gate bias", *in 2010 17th IEEE International Symposium on the Physical and Failure Analysis of Integrated Circuits,* pp. 1-4, 2010.
      http://dx.doi.org/10.1109/IPFA.2010.5532002

[4]   M. Zhang, Y. Yan, G. Li, S. Deng, W. Zhou, R. Chen, M. Wong, and H-S. Kwok, "OFF-State-Stress-Induced Instability in Switching Polycrystalline Silicon Thin-Film Transistors and Its Improvement by a Bridged-Grain Structure", *IEEE Electron Device Lett.,* vol. 39, no. 11, pp. 1684-1687, 2018.
      http://dx.doi.org/10.1109/LED.2018.2872350

[5]   Y. Kim, K. Chung, J. Lim, and O.K. Kwon, "A Highly Uniform Luminance and Low-Flicker Pixel Circuit and Its Driving Methods for Variable Frame Rate AMOLED Displays", *IEEE Access,* vol. 11, pp. 74301-74311, 2023.
      http://dx.doi.org/10.1109/ACCESS.2023.3296787

[6]   C.L. Lin, W.Y. Chang, and C.C. Hung, "Compensating Pixel Circuit Driving AMOLED Display With a-IGZO TFTs", *IEEE Electron Device Lett.,* vol. 34, no. 9, pp. 1166-1168, 2013.
      http://dx.doi.org/10.1109/LED.2013.2271783

[7]   C.L. Lin, P-C. Lai, L-W. Shih, C-C. Hung, P-C. Lai, T-Y. Lin, K-H. Liu, and T-H. Wang, "Compensation Pixel Circuit to Improve Image Quality for Mobile AMOLED Displays", *IEEE J. Solid-State Circuits,* vol. 54, no. 2, pp. 489-500, 2019.
      http://dx.doi.org/10.1109/JSSC.2018.2881922

[8]   M. Zhang, M. Wang, X. Lu, M. Wong, and H.S. Kwok, "Analysis of Degradation Mechanisms in Low-Temperature Polycrystalline Silicon Thin-Film Transistors under Dynamic Drain Stress", *IEEE Trans. Electron Dev.,* vol. 59, no. 6, pp. 1730-1737, 2012.
      http://dx.doi.org/10.1109/TED.2012.2189218

[9]   M. Zhang, Z. Xia, W. Zhou, R. Chen, M. Wong, and H.S. Kwok, "Dynamic-Gate-Stress-Induced Degradation in Bridged-Grain Polycrystalline Silicon Thin-Film Transistors", *IEEE Trans. Electron Dev.,* vol. 63, no. 10, pp. 3964-3970, 2016.
      http://dx.doi.org/10.1109/TED.2016.2601218

[10] C.W. Chen, T.C. Chang, P.T. Liu, H.Y. Lu, T.M. Tsai, C.F. Weng, C.W. Hu, and T.Y. Tseng, "Electrical Degradation of N-Channel Poly-Si TFT under AC Stress", *Electrochem. Solid-State Lett.,* vol. 8, no. 9, p. H69, 2005.

http://dx.doi.org/10.1149/1.1960173

[11]   Huaisheng Wang, Mingxiang Wang, and Dongli Zhang, "Suppress Dynamic Hot-Carrier Induced Degradation in Polycrystalline Si Thin-Film Transistors by Using a Substrate Terminal", *IEEE Electron Device Lett.,* vol. 35, no. 5, pp. 551-553, 2014.

http://dx.doi.org/10.1109/LED.2014.2308987

[12]   M. Zhang, Y. Song, Z. Jiang, X. Xu, M. Wong, and H-S. Kwok, "Off-State Current Degradation Behavior of Polycrystalline Silicon Thin-Film Transistors under Dynamic Drain Voltage Stress", *2023 IEEE International Symposium on the Physical and Failure Analysis of Integrated Circuits (IPFA),* pp. 1-5, 2023.

http://dx.doi.org/10.1109/IPFA58228.2023.10249125

[13]   Meng Zhang, Wei Zhou, Rongsheng Chen, Man Wong, and Hoi-Sing Kwok, "Characterization of DC-Stress-Induced Degradation in Bridged-Grain Polycrystalline Silicon Thin-Film Transistors", *IEEE Trans. Electron Dev.,* vol. 61, no. 9, pp. 3206-3212, 2014.

http://dx.doi.org/10.1109/TED.2014.2341676

[14]   M. Xue, M. Wang, Z. Zhu, D. Zhang, and M. Wong, "Degradation Behaviors of Metal-Induced Laterally Crystallized n-Type Polycrystalline Silicon Thin-Film Transistors Under DC Bias Stresses", *IEEE Trans. Electron Dev.,* vol. 54, no. 2, pp. 225-232, 2007.

http://dx.doi.org/10.1109/TED.2006.888723

[15]   X. Lu, M. Wang, M. Zhang, and M. Wong, "Negative drain pulse stress induced two-stage degradation of P-channel poly-Si thin-film transistors", *18th IEEE International Symposium on the Physical and Failure Analysis of Integrated Circuits (IPFA),* pp. 1-4 Incheon, Korea, 2011.

http://dx.doi.org/10.1109/IPFA.2011.5992756

[16]   M. Zhang, Z. Xia, W. Zhou, R. Chen, and M. Wong, "Significant Reduction of Dynamic Negative Bias Stress-Induced Degradation in Bridged-Grain Poly-Si TFTs", *IEEE Electron Device Lett.,* vol. 36, no. 2, pp. 141-143, 2015.

http://dx.doi.org/10.1109/LED.2014.2377040

[17]   H. Wang, M. Wang, Z. Yang, H. Hao, and M. Wong, "Stress Power Dependent Self-Heating Degradation of Metal-Induced Laterally Crystallized n-Type Polycrystalline Silicon Thin-Film Transistors", *IEEE Trans. Electron Dev.,* vol. 54, no. 12, pp. 3276-3284, 2007.

http://dx.doi.org/10.1109/TED.2007.908907

[18]   T.Y. Chan, J. Chen, P.K. Ko, and C. Hu, "The impact of gate-induced drain leakage current on MOSFET scaling", *1987 International Electron Devices Meeting,* pp. 718-721, 1987.

http://dx.doi.org/10.1109/IEDM.1987.191531

[19]   M. Zhang, M. Wang, H. Wang, and J. Zhou, "Degradation of Metal-Induced Laterally Crystallized n-Type Polycrystalline Silicon Thin-Film Transistors Under Synchronized Voltage Stress", *IEEE Trans. Electron Dev.,* vol. 56, no. 11, pp. 2726-2732, 2009.

http://dx.doi.org/10.1109/TED.2009.2030720

[20]   H. Wang, M. Wang, and Z. Yang, "Finite element analysis of temperature distribution of polycrystalline silicon thin film transistors under self-heating stress", *Front. Electr. Electron. Eng. China,* vol. 4, no. 2, pp. 227-233, 2009.

http://dx.doi.org/10.1007/s11460-009-0023-0

[21]   K. Takechi, M. Nakata, H. Kanoh, S. Otsuki, and S. Kaneko, "Dependence of self-heating effects on operation conditions and device structures for polycrystalline silicon TFTs", *IEEE Trans. Electron Dev.,* vol. 53, no. 2, pp. 251-257, 2006.
        http://dx.doi.org/10.1109/TED.2005.861729

[22]   S. Hashimoto, K. Kitajima, Y. Uraoka, T. Fuyuki, and Y. Morita, "Thermal Degradation Under Pulse Operation in Low-Temperature p-Channel Poly-Si Thin-Film Transistors", *IEEE Trans. Electron Dev.,* vol. 54, no. 2, pp. 297-300, 2007.
        http://dx.doi.org/10.1109/TED.2006.888724

[23]   A.G. Lewis, I-W. Wu, M. Hack, A. Chiang, and R.H. Bruce, "Degradation of polysilicon TFTs during dynamic stress", *International Electron Devices Meeting 1991 [Technical Digest],* pp. 575-578, 1991.
        http://dx.doi.org/10.1109/IEDM.1991.235404

[24]   Y. Uraoka, N. Hirai, H. Yano, T. Hatayama, and T. Fuyuki, "Hot carrier analysis in low-temperature poly-Si TFTs using picosecond emission microscope", *IEEE Trans. Electron Dev.,* vol. 51, no. 1, pp. 28-35, 2004.
        http://dx.doi.org/10.1109/TED.2003.820937

[25]   Y. Uraoka, T. Hatayama, T. Fuyuki, T. Kawamura, and Y. Tsuchihashi, "Reliability of low temperature poly-silicon TFTs under inverter operation", *IEEE Trans. Electron Dev.,* vol. 48, no. 10, pp. 2370-2374, 2001.
        http://dx.doi.org/10.1109/16.954479

[26]   K. M. Chang, Y. H. Chung, and G. M. Lin, "Hot Carrier Induced Degradation in the Low Temperature Processed Polycrystalline Silicon Thin Film Transistors Using the Dynamic Stress", *Jpn. J. Appl. Phys.,* vol. 41, no. Part 1, No. 4, pp. 1941-1946.
        http://dx.doi.org/10.1143/JJAP.41.1941

[27]   Y. Toyota, T. Shiba, and M. Ohkura, "A new model for device degradation in low-temperature N-channel polycrystalline silicon TFTs under AC stress", *IEEE Trans. Electron Dev.,* vol. 51, no. 6, pp. 927-933, 2004.
        http://dx.doi.org/10.1109/TED.2004.828163

[28]   Y. Toyota, T. Shiba, and M. Ohkura, "Effects of the timing of AC stress on device degradation produced by trap states in low-temperature polycrystalline-silicon TFTs", *IEEE Trans. Electron Dev.,* vol. 52, no. 8, pp. 1766-1771, 2005.
        http://dx.doi.org/10.1109/TED.2005.852726

[29]   Y.H. Tai, S.C. Huang, and C.K. Chen, "Analysis of Poly-Si TFT Degradation Under Gate Pulse Stress Using the Slicing Model", *IEEE Electron Device Lett.,* vol. 27, no. 12, pp. 981-983, 2006.
        http://dx.doi.org/10.1109/LED.2006.886416

[30]   B. Zhang, D. Zhang, M. Wang, H. Wang, and R. Wang, "Roles of Trap States in the Dynamic Degradation of Polycrystalline Silicon Thin-Film Transistors Under AC Gate Bias Stress", *IEEE Transactions on Electron Devices*, pp. 1-7, 2023.
        http://dx.doi.org/10.1109/ted.2023.3333289

[31]   M. Kimura, S. Inoue, T. Shimoda, and T. Eguchi, "Dependence of polycrystalline silicon thin-film transistor characteristics on the grain-boundary location", *J. Appl. Phys.,* vol. 89, no. 1, pp. 596-600, 2001.
       http://dx.doi.org/10.1063/1.1329141

[32]   L. Chen, Y. Wu, D. Zhang, H. Wang, and M. Wang, "Optimized Design of Carrier Injection Terminal for Reliable Low-Temperature Poly-Si TFTs Under Dynamic Hot-Carrier Stress", *IEEE Transactions on Electron Devices*, vol. 67, no. 5, pp. 1987-1992, 2020.
       http://dx.doi.org/10.1109/ted.2020.2977854

[33]   Han Hao, Mingxiang Wang, and Man Wong, "An Analytical Expression for Drain Saturation Voltage of Polycrystalline Silicon Thin-Film Transistors", *IEEE Electron Device Lett.,* vol. 29, no. 4, pp. 357-359, 2008.
       http://dx.doi.org/10.1109/LED.2008.917810

[34]   M.D. Jacunski, M.S. Shur, and M. Hack, "Threshold voltage, field effect mobility, and gate-to-channel capacitance in polysilicon TFTs", *IEEE Trans. Electron Dev.,* vol. 43, no. 9, pp. 1433-1440, 1996.
       http://dx.doi.org/10.1109/16.535329

[35]   M. Zhang, X. Ma, S. Deng, W. Zhou, Y. Yan, M. Wong, and H-S. Kwok, "Degradation Induced by Forward Synchronized Stress in Poly-Si TFTs and Its Reduction by a Bridged-Grain Structure", *IEEE Electron Device Lett.,* vol. 40, no. 9, pp. 1467-1470, 2019.
       http://dx.doi.org/10.1109/LED.2019.2931007

[36]   G. Zhu, M. Zhang, L. Lu, M. Wong, and H.-S. Kwok, "Hot Carrier Degradation Reduction in Metal Oxide Thin-Film Transistors by Implementing a Lightly Doped Drain-Like Structure", *IEEE Electron Device Letters*, vol. 45, no. 9, pp. 1602-1605, 2024.
       http://dx.doi.org/10.1109/led.2024.3424473

[37]   S. Li, H. Liu, Y. Jiang, J. He, L. Wang, and Y. Ji, "Annealing effects on the optical and structural properties of Si thin films deposited by ion beam sputtering technique", *Optik*, vol. 181, pp. 695-702, 2019.
       http://dx.doi.org/10.1016/j.ijleo.2018.12.100

[38]   L. Chen, M. Wang, D. Zhang, and H. Wang, "Gate Voltage Pulse Rising Edge Dependent Dynamic Hot Carrier Degradation in Poly-Si Thin-Film Transistors", *IEEE Electron Device Letters,* vol. 42, no. 11, pp. 1615-1618, 2021.
       http://dx.doi.org/10.1109/led.2021.3110916

[39]   F. Liu *et al.,* "Roles of Hot Carriers in Dynamic Self-Heating Degradation of a-InGaZnO Thin-Film Transistors", *IEEE Electron Device Letters*, vol. 43, no. 1, pp. 40-43, 2022.
       http://dx.doi.org/10.1109/led.2021.3133011

[40]   M. Zhang, and M. Wang, "An investigation of drain pulse induced hot carrier degradation in n-type low temperature polycrystalline silicon thin film transistors", *Microelectron. Reliab.,* vol. 50, no. 5, pp. 713-716, 2010.
       http://dx.doi.org/10.1016/j.microrel.2010.01.024

[41]   Y. Wang, Z. Jiang, L. Zou, and M. Zhang, "Analysis of degradation mechanism in polycrystalline silicon thin-film transistors under dynamic off-state stress with fast transition time", *Physica Scripta*, vol. 99, no. 12, 2024.

http://dx.doi.org/10.1088/1402-4896/ad896a

[42] M. Zhang, S. Deng, W. Zhou, Y. Yan, M. Wong, and H.S. Kwok, "Reversely-Synchronized-Stress-Induced Degradation in Polycrystalline Silicon Thin-Film Transistors and Its Suppression by a Bridged-Grain Structure", *IEEE Electron Device Lett.,* vol. 41, no. 8, pp. 1213-1216, 2020.
http://dx.doi.org/10.1109/LED.2020.3005046

[43] S. Zhao, Z. Meng, W. Zhou, J. Ho, M. Wong, and H.-S. Kwok, "Bridged-Grain Polycrystalline Silicon Thin-Film Transistors", *IEEE Transactions on Electron Devices*, vol. 60, no. 6, pp. 1965-1970, 2013.
http://dx.doi.org/10.1109/ted.2013.2258925

[44] H. Wang, M. Wang, D. Zhang, and Q. Shan, "Degradation of a-InGaZnO TFTs Under Synchronized Gate and Drain Voltage Pulses", *IEEE Trans. Electron Dev.,* vol. 65, no. 3, pp. 995-1001, 2018.
http://dx.doi.org/10.1109/TED.2018.2794416

[45] Y.-F. Tu *et al.*, "Analysis of self-heating-related instability in n-channel low-temperature polysilicon TFTs with different S/D contact hole densities", *Applied Physics Express*, vol. 15, no. 3, 2022.
http://dx.doi.org/10.35848/1882-0786/ac4e25

[46] N. Liang, D. Zhang, M. Wang, H. Wang, Y. Yu, and D. Qi, "Investigations on the Negative Shift of the Threshold Voltage of Polycrystalline Silicon Thin-Film Transistors Under Positive Gate Bias Stress", *IEEE Transactions on Electron Devices*, vol. 68, no. 2, pp. 550-555, 2021.
http://dx.doi.org/10.1109/ted.2020.3041568

[47] D. Ho, H. Jeong, H.-B. Park, S. K. Park, M.-G. Kim, and C. Kim, "Solution-processed amorphous zinc indium tin oxide thin-film transistors with high stability under AC stress", *Journal of Materials Chemistry C*, vol. 11, no. 39, pp. 13395-13402, 2023.
http://dx.doi.org/10.1039/d3tc02439c

[48] M. Zhang, Z. Jiang, L. Lu, M. Wong, and H.S. Kwok, "Analysis of Degradation Mechanism in Poly-Si TFTs Under Dynamic Gate Voltage Stress With Short Pulse Width Duration", *IEEE Electron Device Lett.,* vol. 45, no. 2, pp. 204-207, 2024.
http://dx.doi.org/10.1109/LED.2023.3345282

[49] Y. Yang, M. Zhang, L. Lu, M. Wong, and H.-S. Kwok, "Low-Frequency Noise in Bridged-Grain Polycrystalline Silicon Thin-Film Transistors", *IEEE Transactions on Electron Devices*, vol. 69, no. 4, pp. 1984-1988, 2022.
http://dx.doi.org/10.1109/ted.2022.3148697

[50] A.F. Tasch, and C.T. Sah, "Recombination-Generation and Optical Properties of Gold Acceptor in Silicon", *Phys. Rev., B, Solid State,* vol. 1, no. 2, pp. 800-809, 1970.
http://dx.doi.org/10.1103/PhysRevB.1.800

[51] Y. Uraoka, N. Hirai, H. Yano, T. Hatayama, and T. Fuyuki, "Hot Carrier Analysis in Low-TemperaturePoly-Si TFTs Using Picosecond Emission Microscope", *IEEE Trans. Electron Dev.,* vol. 51, no. 1, pp. 28-35, 2004.
http://dx.doi.org/10.1109/TED.2003.820937

[52]   Y.-Z. Zheng *et al.*, "Impact of AC Stress in Low Temperature Polycrystalline Silicon Thin Film Transistors Produced With Different Excimer Laser Annealing Energies", *IEEE Electron Device Letters*, vol. 42, no. 6, pp. 847-850, 2021.

http://dx.doi.org/10.1109/led.2021.3073200

[53]   Y. Oodate, Y. Tanimoto, H. Tanoue, H. Kikuchihara, H.J. Mattausch, and M. Miura-Mattausch, "Compact Modeling of the Transient Carrier Trap/Detrap Characteristics in Polysilicon TFTs", *IEEE Trans. Electron Dev.,* vol. 62, no. 3, pp. 862-868, 2015.

http://dx.doi.org/10.1109/TED.2015.2388799

[54]   Y. Zeng, Y. Yan, Z. Jiang, Z. Xia, M. Zhang, M. Wong, J.J. Liou, and H-S. Kwok, "Reliability of Poly-Si TFTs Under Voltage Pulse With Fast Transition Time", *IEEE Electron Device Lett.,* vol. 42, no. 12, pp. 1782-1785, 2021.

http://dx.doi.org/10.1109/LED.2021.3124755

CHAPTER 7

Environmental Stress-Induced Degradation in Thin-Film Transistors

Abstract. This chapter explores the impact of environmental factors on the reliability of thin-film transistors (TFTs), discussing the effects of temperature, illumination, and moisture on the performance and degradation of TFTs. It provides a comprehensive analysis of how these environmental conditions can influence the overall reliability and stability of electronic devices incorporating TFTs.

Keywords: Degradation, Environmental reliability, Illumination, Moisture, Temperature, TFT performance.

7.1. INTRODUCTION

In addition to electrical stress reliability, thin-film transistors (TFT) are often subject to various environmental factors such as temperature, humidity, and light exposure. Under the combined influence of electrical stress and environmental conditions, the degradation of devices may be suppressed or enhanced to varying degrees, and even new types of degradation mechanisms may emerge. Research into this type of degradation is typically more complex, requiring consideration of both the changes within the device caused by electrical stress, such as internal electric fields and current, as well as the impact of environmental factors on carrier movement, generation, and recombination. Moreover, the effects of certain environments, like humidity, may lead to additional and distinct degradation phenomena that necessitate separate investigation.

To study the impact of these environmental factors, researchers often examine the degradation of devices under the simultaneous influence of electrical stress and environmental conditions, probing the underlying degradation mechanisms. In this work, we focus on the degradation of silicon-based (Si-based) TFTs and metal oxide (MO) TFTs under various stress conditions induced by temperature, light exposure, and humidity.

7.2. EFFECTS OF TEMPERATURE ON THIN-FILM TRANSISTORS RELIABILITY

7.2.1. Effect of Temperature on the Reliability of Polycrystalline Silicon Thin-Film Transistors

As one of the most mainstream types of TFT devices, polycrystalline silicon (poly-Si) TFTs boast advantages such as high mobility and reliability. Since TFTs primarily operate in environments with complex electrical signals and are subject to temperature interference affecting reliability, research on TFT reliability has mainly focused on the electrical aspects, with temperature's impact on device performance studied across various electrical reliability scenarios.

Common degradation mechanisms affecting poly-Si TFT reliability include self-heating (SH), hot carrier (HC) effects, and positive/negative bias temperature instability (PBTI/NBTI). These are summarized for n-type poly-Si TFTs in Table **1**. We will now delve into these degradation mechanisms in more detail.

Table 1. Relationship between common degradation phenomena and temperature in poly-si TFT.

Degradation model	Gate voltage (V_g)	Drain voltage (V_d)	Temperature (T)
Self-heating	Large	Large	Worse as the temperature rises
Hot carrier	Slightly larger than V_{th}	Large	Worse as the temperature decreases
NBTI	Large	Ground	Worse as the temperature rises

(a) The Impact of Temperature on Poly-Si TFT Self-Heating Effects

SH degradation is generally believed to be caused by Joule heat from the on-state current (I_{on}), which raises the device temperature. Since the I_{on} is proportional to W/L, the heat power generated $P = I_{on} \cdot V_{ds}$ is also proportional to W/L. For devices with equal W/L ratios, small-sized devices generate more heat per unit area and have less heat dissipation area. Consequently, SH is more pronounced in smaller devices with larger W/L ratios. Satoshi Inoue and colleagues used an infrared thermometer to measure the surface temperature of devices with different dimensions under the same V_g and V_d electrical stress. They found that for a W/L of 100 μm/100 μm, the temperature was 70.6 °C, for 10 μm/10 μm, it was 104.6

°C, and for 100 μm/10 μm, it was 215.5 °C. For devices with the same channel length, as the channel width increases under the same V_g and V_d stress, the degradation of threshold voltage (V_{th}) becomes more severe. In experiments with devices in parallel, they also found that degradation was not uniformly distributed along the channel, with the center degrading more than the edges, similar to the non-uniform temperature distribution within the channel [1]. This indicates that device degradation is related to its temperature.

It is commonly accepted that Si-H bonds break down above 350 °C, so when the device temperature is high, many dangling bonds are formed at the Si/SiO$_2$ interface, increasing the interface state density and causing device performance degradation. Satoshi Inoue and colleagues confirmed this through two-dimensional device simulations using Atlas software, where increasing the state density in the channel region yielded results consistent with experimental data [2]. Mutsumi Kimura and colleagues' research on the interface and channel defect density in poly-Si TFTs also showed that SH degradation is due to the increase in interface state density caused by Si-H bond breakage at the Si/SiO$_2$ interface, with a slight increase in defect density at the grain boundaries in the channel [3]. This confirms that the generation of SH is indeed related to the Joule heat produced by the current during device operation and becomes more severe with increased environmental temperature.

(b) The Impact of Temperature on Poly-Si TFT Hot Carrier Effects

HC degradation typically occurs when devices, during operation, generate HCs due to the large electric field near the drain, leading to impact ionization and damage to the interface and channel lattice near the drain, increasing the trap state and causing device performance degradation [2, 4-13]. This is mainly manifested in the transfer curve as a decrease in I_{on} and an increase in off-state current (I_{off}), while V_{th} and subthreshold swing (SS) usually do not change significantly. This is because HC damage typically occurs near the drain, so it has a minimal impact on V_{th} and SS, which are related to the overall channel properties. This was confirmed by Toshiyuki Yoshida and colleagues [7]. They performed charge pump (CP) experiments on devices and found minimal changes in CP current, indicating little change in interface traps within the channel.

HC effects are most pronounced when V_d is large and the gate voltage is slightly above V_{th}. A large V_d significantly shields the longitudinal electric field near the drain, reducing the carriers near the drain and causing the device to "pinch off." At this point, the electric field generated by the large V_d mainly falls on the pinch-off region, creating an electric field greater than 10^5 V/cm. Carriers near the drain gain significant energy and are accelerated into HCs. Herein, the increase in temperature will intensify lattice vibrations, leading to augmented lattice scattering, which reduces the mean free path of carriers within the active layer. This results in a greater challenge for carriers to be accelerated into HCs under a high electric field [14, 15].

(c) The Impact of Temperature on poly-Si TFT Positive/Negative Bias Temperature Instability

positive bias stress (PBS) and negative gate bias (NBS) are applied by imposing a large positive or negative voltage on the gate. It is important to note that due to the high reliability of poly-Si TFTs, PBS and NBS studies usually require extremely high voltages (> 30 V) to cause significant degradation in the devices, which is not typical in operating conditions. Therefore, PBS/NBS studies on poly-Si TFTs often involve the combination of external environmental factors, such as positive/negative bias temperature reliability. This manifests as changes in parameters like V_{th} drift, which can ultimately lead to device failure.

The primary degradation principle involves the accumulation of a large number of carriers at the interface due to the application of gate voltage, with some carriers being captured by interface states and others being injected into the gate insulator (GI) under high electric fields. Charged defect states and fixed charges shield part of the gate electric field, leading to V_{th} shift. Environmental factors further exacerbate TFT degradation, with higher temperatures providing carriers with more energy, making them more likely to be injected into the GI, resulting in more severe V_{th} shifts [16-20]. Additionally, some poly-Si TFT devices experience a degradation of I_{on} and mobility under negative/positive bias temperature instability (PBTI/NBTI) stress, which can be attributed to high temperatures damaging poly-Si grain boundaries, further binding more carriers [16-20]. This not only reduces the device's I_{on} but also worsens its V_{th}. Therefore, PBTI/NBTI effects intensify with increasing environmental temperature.

7.2.2. The Impact of Temperature on the Reliability of Metal-Oxide Thin-Film Transistors

a) The Impact of Temperature on the Self-Heating Effect of InGaZnO TFTs

It is well-known that SH effects occur in metal-oxide-semiconductor field-effect transistors (MOSFETs) and poly-Si TFTs due to the difficulty of dissipating Joule heat generated in the channel when large currents pass through it, a situation that is very similar for IGZO TFTs. High drive currents can heat the IGZO channel layer and lead to SH effects [21]. Additionally, the thermal conductivity of IGZO is significantly lower than that of Si and is comparable to that of SiO_2, making heat dissipation in IGZO TFTs more challenging than in Si-based TFTs [22, 23].

SH stress is typically applied by simultaneously imposing large voltages on the source and gate electrodes, leading to substantial working currents and uneven carrier concentration distribution across the channel. Carriers near the drain are fewer in number, resulting in more Joule heat generation at the drain [24-28]. For a-IGZO TFTs, when SH occurs, some carriers are captured at the interface and injected into the oxide layer, causing a reduction in I_{on} and a positive shift in V_{th}. High temperatures can also lead to more defects and dangling bonds at the active layer and GI interface, worsening SS, although SS degradation may not occur if the power density is not high enough. All of these degradations are exacerbated by increased environmental temperatures. At sufficiently high temperatures, devices operating in the linear region may experience increased conductivity in the channel center due to the transformation of deep donor-like oxygen vacancies (V_oS) into shallow donor-like states, releasing electrons and potentially leading to a hump or even a stuck-on state in the transfer characteristics.

(b) The Impact of Temperature on the Hot Carrier Effects of InGaZnO TFTs

HC-induced degradation, common in MOSFETs and poly-Si TFTs, leads to a decrease in I_{on} and a shift in V_{th}. This degradation is caused by local depletion in the channel near the drain, equivalent to the formation of a barrier due to hot electron injection into the GI and the creation of interface traps between the channel and GI. The degradation due to HC effects includes a positive shift in V_{th} and a reduction in I_{on}. As the environmental temperature rises, the degree of this degradation decreases. This is primarily because increased temperatures intensify lattice scattering in the channel, reducing the acceleration time of HCs and their destructive impact on the lattice and grain boundaries.

Additionally, the transfer characteristics measured in both forward and reverse directions may exhibit asymmetric behavior during HC effects. This is mainly because the distribution of HCs is uneven, typically forming near large electric fields. When a large electric field is applied to the drain, more severe degradation occurs near the drain, leading to asymmetric transfer characteristics. If the measurement increases V_d, the degradation of reduced I_{on} may not be observed due to the large electric field's cancellation of the barrier.

(c) The Impact of Temperature on the Positive/Negative Bias Temperature Instability of InGaZnO TFTs

The thermal energy provided by increased temperatures can excite holes or electrons from traps or the valence band, causing curve shifts, such as the negative shift in curves under NBTI, or the accumulation of thermo-generated holes at the source under the combined effect of gate and drain voltages, reducing the source barrier and increasing I_{off}s and I_{on}s. In wide-channel devices, thermo-generated holes may also form parasitic channels on the back channel side, leading to a hump effect in the transfer characteristics after PBTI stress. Furthermore, when the external environment reaches certain temperatures, V_Os can more easily transition from deep to shallow donor-like states and release electrons.

7.3. EFFECTS OF ILLUMINATION ON THIN-FILM TRANSISTORS RELIABILITY

7.3.1. Photogenerated Carrier Excitation

When the active channel layer of a TFT is exposed to light, it generates a photocurrent. Unlike vertically structured photodiodes, photoelectric TFTs have a relatively longer channel length, and the photocurrent is produced through the generation of photocharges and the transport process of carriers. Initially, when the light energy exceeds the bandgap of the material, the material absorbs the light energy and generates electron-hole pairs. Since most amorphous oxide semiconductors have a bandgap energy of 3.0 eV, interband electron-hole pairs are generated through photoabsorption when exposed to ultraviolet radiation. In contrast, when long-wavelength visible light is incident, band-to-band electron-hole pair generation does not occur as the light is not absorbed.

However, if there are various sub-bandgap states due to defects in the bandgap, electrons can be excited into the conduction band through these sub-bandgap states, or defect states can be ionized to produce free electrons.

This means that even under long-wavelength illumination, there will be some electron-hole pairs in the light-absorbing layer material [29]. Secondly, since the basic structure of phototransistors is the same as that of TFTs, phototransistors operate in two driving modes: depletion and accumulation [30]. When the transistor operates in depletion mode under a negative gate bias, electrons are induced to move away from the channel/GI interface to suppress channel current. Conversely, when the transistor operates in accumulation mode under a positive gate bias, the accumulated electrons at the channel/GI interface increase the channel current. At this point, when a charge is generated by light exposure, depending on the accumulation operating mode of the transistor, a photoconductive effect can be observed [30]. In depletion mode, photoinduced charge generation occurs, and photogenerated holes accumulate near the source electrode under the gate and drain biases, reducing the electron injection barrier between the source and the semiconductor channel. This reduction in the electron injection barrier under light exposure causes a negative shift in V_{th}, leading to an increase in channel current, and the photocurrent of the phototransistor increases linearly with the intensity of the incident light.

7.3.2. Photodegradation of Metal-Oxide Thin-Film Transistors Under Stress

As shown in Fig. (**7.1**), the transfer curve degradation and long-term recovery of an ITZO TFT under 546.1 nm light stress at 14.92 mW/cm² are depicted. It can be observed that the transfer curve exhibits a negative shift. Concurrently, the SS also degrades [31]. After 7000 seconds of light stress, the V_{th} shifts negatively by 2.96 V, and SS degrades from 0.158 V/dec to 0.967 V/dec. Upon removal of the light stress, SS quickly recovers to its initial value (blue line). Then, the transfer curve remains almost stable for 10 hours during recovery [31]. As the recovery period extends, the transfer curve slowly recovers. After 36.88 days of recovery, V_{th} recovers from −3.65 V to −2.35 V. However, SS degrades from 0.158 V/dec to 0.428 V/dec.

Fig. (**7.2**) illustrates the experimental and fitted curves for ITZO at initial (black), after 7000 seconds of light exposure (red), immediately after stress removal (blue), and after 36.88 days of recovery (green). It can be observed that the simulated data fits the experimental data well. As shown in Fig. (**8**), the density of states distribution used for fitting the transfer curve is depicted. After 7000 seconds of light stress, both the tail states and the deep-level defect density increase. The generation of positive fixed charges and metastable defect states may be responsible for the positive V_{th} shift and SS degradation under light stress [32]. The immediate recovery of SS can be attributed to the immediate reduction of donor-like states in the tail distribution after the light stress is removed. The long recovery period may be due to the process of captured holes being de-trapped [31].

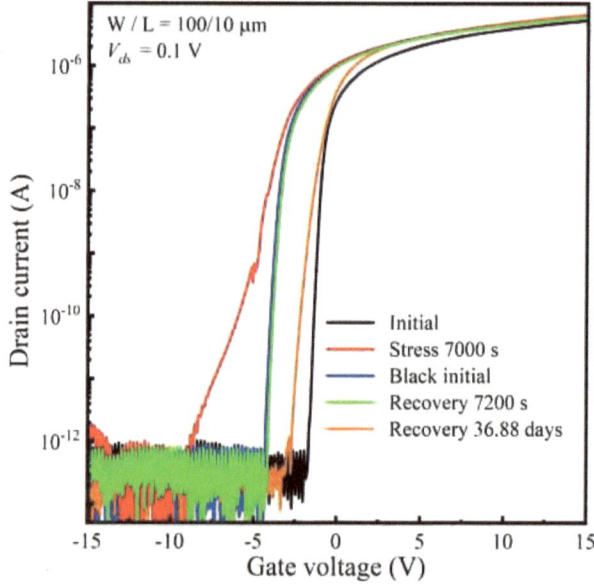

Fig. (7.1). IZO TFT for dynamic continuous detection of blue light with an intensity of 100 μW/cm^2 [31].

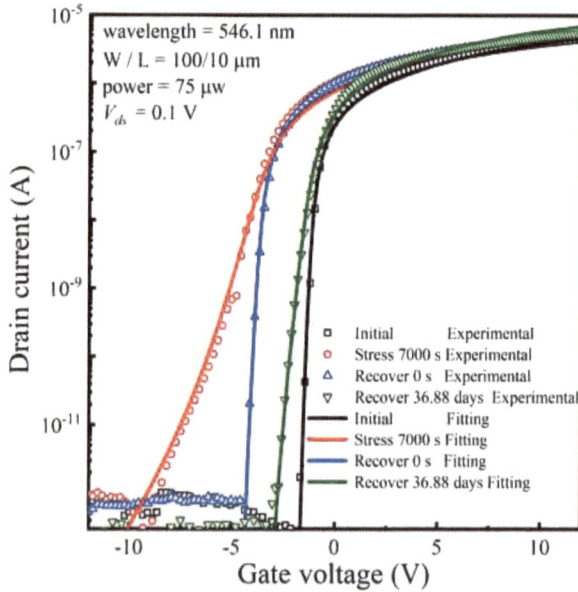

Fig. (7.2). Experimental curves and fitting curves of ITZO TFTs for the initial stage(black),7000 s light illumination (red), 0 s recovery (blue) and 36.88 days recovery(green) [31].

7.3.3. Photodegradation of Amorphous Silicon Thin-Film Transistors Under Stress

Fig. (7.3) displays the degradation states of the device in both light and dark conditions in amorphous silicon (a-Si) TFTs. In a dark environment, after the application of PBS (Positive Bias Stress), V_{th} shifts slightly in the positive direction due to the generation of defect states or electron capture [34]. Following PBIS (Positive Bias Temperature Stress), V_{th} moves negatively, which is more pronounced under higher gate voltages, indicating that this degradation is related to the vertical electric field. After NBS, V_{th} remains almost unchanged, but it shifts positively after negative bias illumination stress (NBIS). This instability of V_{th} during NBIS may be attributed to the deterioration of the SS [34]. To further analyze the characteristics after applying gate bias stress in both dark and light conditions, the variation of transconductance (g_m) over time is extracted, as shown in segment 1(b) of the figure. After PBS, g_m decreases with time, and this trend is more evident at $V_g = 50$ V. Comparing PBS with PBIS, the initial decrease in g_m appears similar. However, as time progresses, the trend for g_m during PBIS intensifies, especially under higher gate voltages. For NBS and NBIS, g_m decreases over time, with a more significant trend observed in NBIS. The substantial reduction in g_m after NBIS may be due to photogenerated carriers disrupting weak bonds, increasing the density of defect states [5].

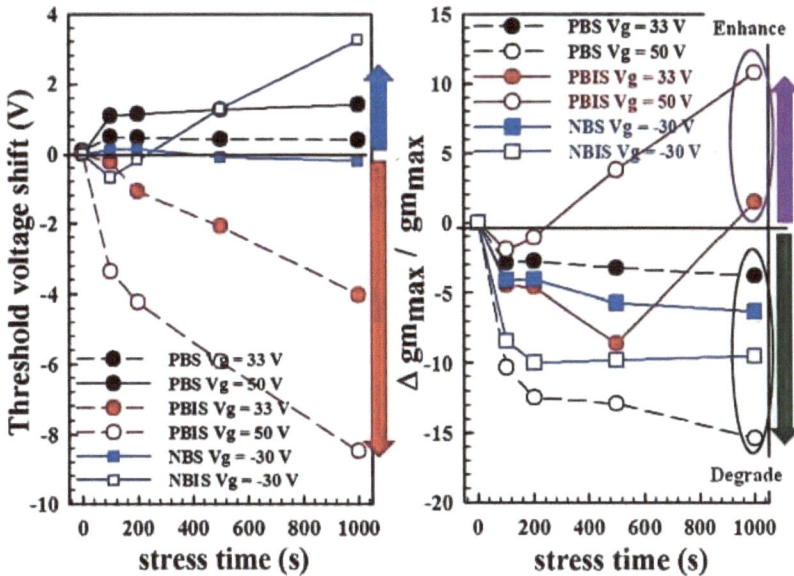

Fig. (7.3). Comparison of (a) V_{th} shift and (b) g_{mMAX} degradation after positive/negative gate bias stress conditions in the dark and under light illumination [33].

When a positive voltage is applied to the gate, the bands bend downward, and holes are generated through trap-assisted photoexcitation and drift toward the passivation layer. The lower process temperature in the passivation layer (SiN_x) leads to many defect states that capture holes, so holes are captured at the hydrogenated amorphous silicon (a-Si:H)/passivation layer interface or within the passivation layer itself. Hole capture in the passivation layer causes a negative shift in the V_{th}. During NBIS, the bands bend upward, inducing photogenerated electrons to drift towards the passivation layer [33]. Consequently, electron capture in the passivation layer results in a positive shift in V_{th}.

7.3.4. Photodegradation of Polycrystalline Silicon Thin-Film Transistors Under Stress

The distinct degradation states of poly-Si and MO TFTs are related to the crystal structure of the channel. Their activation energy values are nearly similar, indicating a comparable density of defects in both LTPS and oxide TFTs. The difference lies in the location of carrier generation due to structural differences, leading to different degradation states under light exposure. In a-IGZO TFTs, higher activation energy results in donor-like defects in a double-ionized state, creating V_os and shifting the Fermi level above the middle of the bandgap near the conduction band edge, as shown in Fig. (**7.4**) [35]. In contrast, no change in field-effect mobility is observed in LTPS TFTs after illumination, suggesting that defect generation is primarily due to SH effects within the TFT.

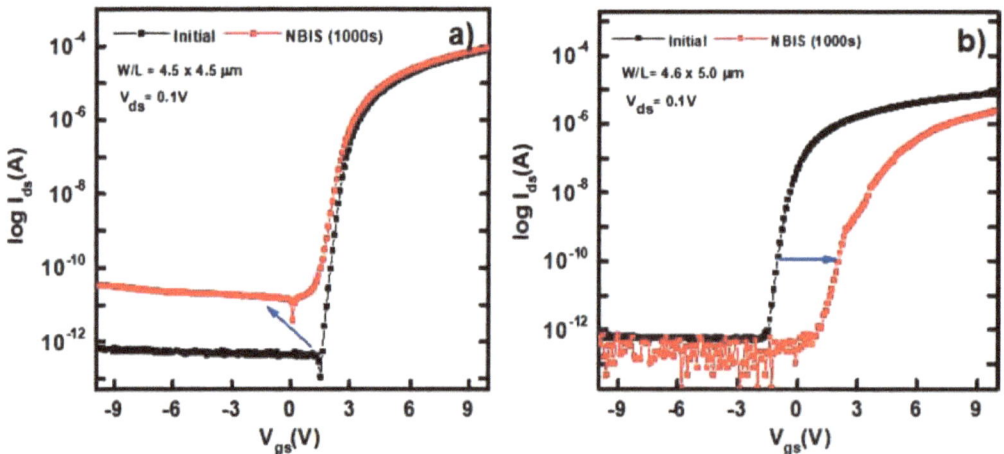

Fig. (7.4). I_{ds}-V_{gs} characteristic curve of (A) LTPS TFT (B) metal oxide before and after illumination under 20V negative bias voltage [35].

Negative bias and 400 Lux illumination lead to asymmetric degradation in both oxide and LTPS TFTs. Photogenerated carriers exhibit different photo sensitivities and locations. In oxide TFTs, carriers generate double donor levels under light, moving the Fermi level toward the conduction band, while in LTPS TFTs, photogenerated carriers increase the trap density in the poly-Si, leading to IOFF current degradation due to SH effects [35].

7.3.5. Persistent Photoconductivity Effect

After TFTs are exposed to light, the I_{ds} rapidly increases by four orders of magnitude and stabilizes. When the light is removed, I_{ds} quickly decreases but does not return to its initial state. After multiple cycles of photodetection, the dark current of the IZO TFT photodetector gradually rises, continually approaching its photocurrent. This phenomenon, known as the persistent photoconductivity effect, severely limits the real-time photodetection capability of TFTs [36].

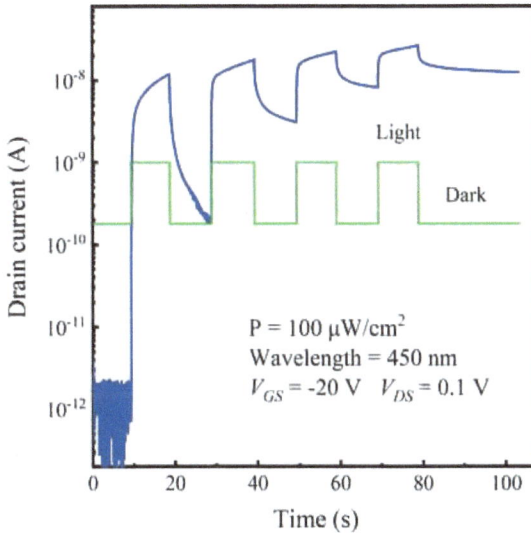

Fig. (7.5). Dynamic continuous detection of 100 μW/cm² blue light of IZO TFTs.

When dynamic real-time photodetection is performed on TFTs, the investigation to eliminate the persistent photoconductivity effect of the TFT photodetector by applying gate pulse voltage is shown in Fig. (7.5). A blue light with a wavelength of 450 nm and a photo power density of 100 μW/cm² is used as the detected light source, with a real-time detection cycle of 20 s, including 10 s of light exposure and 10 s without light. Voltages of −20 V and 0.1 V are applied to the gate and drain of the IZO TFT, respectively. After each cycle, when the light is removed, a 5 V gate pulse voltage with a duration of 10 μs is applied.

Upon light exposure, I_{ds} rapidly increases and stabilizes, allowing the IZO TFT photodetector to accurately detect blue light in real time. After 10 s, when the light is removed and a gate pulse voltage is applied, I_{ds} experiences a sharp increase and then quickly returns to its initial state. During this phase, the persistent photoconductivity effect of the IZO TFT photodetector is eliminated, allowing it to quickly return to its initial state and await the next photodetection, thus achieving dynamic continuous real-time photodetection.

As shown in Fig. (**7.7a**), the negative gate bias ($V_{gs} = -20$ V) applied after light removal causes the bands to bend upwards, with positively charged V_O^{2+} located in the front channel and free electrons in the rear channel. Under the influence of the drain voltage ($V_{ds} = 0.1$ V), unpaired ionized V_O^{2+} and electrons accumulate near the source side. The spatial separation of V_O^{2+} and electrons by the drain voltage and negative gate bias prevents the recombination into the initial neutral V_O. Consequently, the IZO TFT photodetector exhibits a strong, persistent photoconductivity effect. Fig. (**7.7c**) illustrates the process of eliminating the persistent photoconductivity effect in the IZO TFT photodetector by applying a gate pulse voltage. After light removal, a 5 V voltage pulse is applied to the gate, raising the Fermi level and causing the bands near the gate to bend downwards, inducing electron accumulation in the front channel and accelerating the recombination of electrons and V_O^{2+} into the initial neutral V_O. As shown in Fig. (**7.6**), the dark current quickly returns to its initial state after the gate pulse voltage is applied, enabling real-time continuous dynamic light detection with the IZO TFT by eliminating the persistent photoconductivity effect.

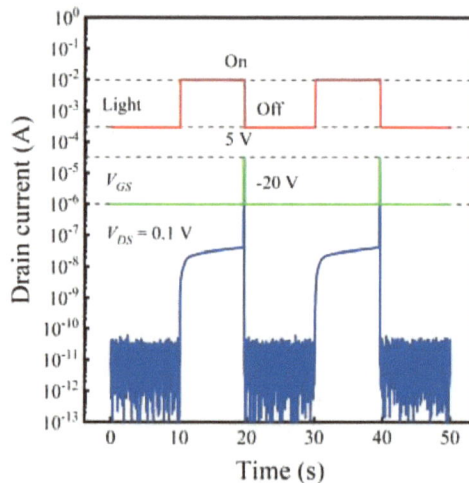

Fig. (7.6). Applying a 5 V pulse voltage to eliminate the sustained photoconductivity effect of IZO TFT during dynamic continuous detection.

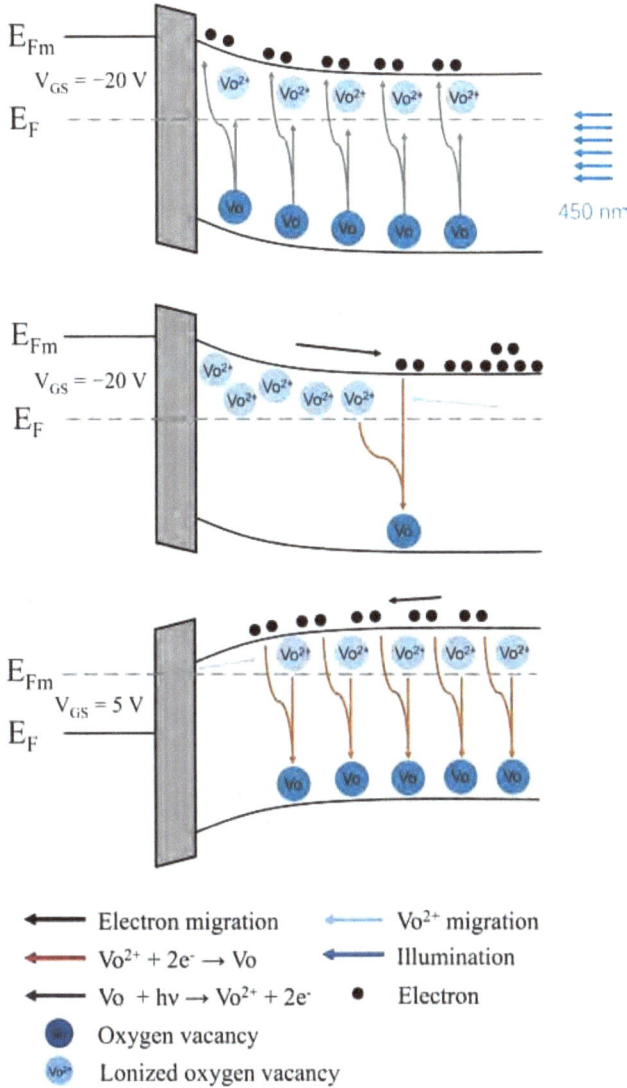

Fig. (7.7). The generation and recovery process of PPC effect.

7.4. EFFECTS OF MOISTURE ON THIN-FILM TRANSISTORS RELIABILITY

The impact of water vapor can be categorized into its effects on Si-based TFTs and its effects on Metal Oxide (MO) TFTs. In Si-based TFTs, water vapor typically reacts with oxysilicon to form hydroxyl groups. Under high fields, excited H^+ ions can then combine with these hydroxyl groups to produce H_2O, leaving behind Si^+ hanging bonds, which form positive fixed charges, ultimately leading to device degradation. Understanding the influence of water vapor on Si-based TFTs is facilitated by a comparative experiment we have conducted.

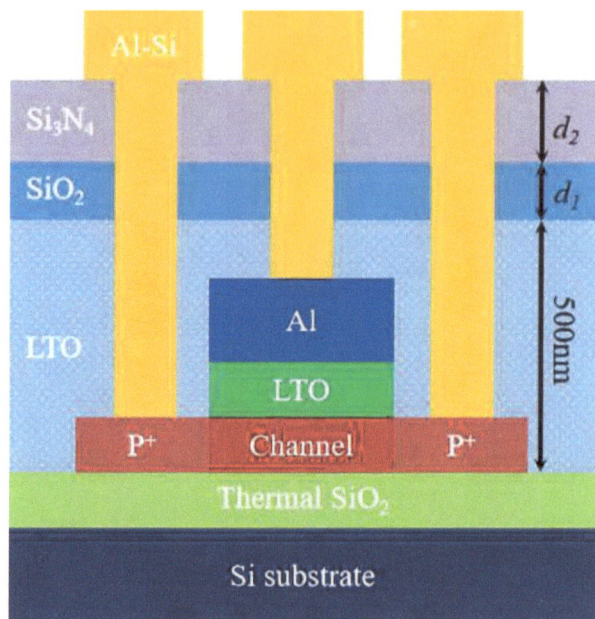

Fig. (7.8). The cross-sectional schematic diagram of a TFT used in the work. The d_1 and d_2 standsfor the thickness of PECVD SiO_2 and PECVD Si_3N_4, respectively [37].

Fig. (**7.8**) illustrates the schematic structure of the polysilicon device featuring a top-gate self-aligned structure. We conducted a series of tests on these devices, including Negative Bias Stress (NBS) evaluations in various conditions: initial state, after 15 days of storage in a dry box, after hard bake (H-B) treatments, after thermal-humidity (T-H) treatments, and after a second round of H-B treatment following T-H treatment. In Fig. (**7.9a**), the transfer curve degradation of a fresh metal-induced lateral crystallization (MILC) poly-Si TFT with $d_1 = d_2 = 0$ under NBTI stress is depicted, showcasing typical NBTI degradation with a continuous negative shift in V_{th} while the subthreshold slope remains nearly constant. The degradation slope (n) of 0.26, shown in Fig. (**7.9g**), aligns with typical NBTI degradation trends. Following a stress duration of 30,000 s, the V_{th} shifts approximately 0.32 V, as evidenced in Fig. (**7.9f**). Subsequent to the stress test, the wafer is stored in a dry box with 35% relative humidity and a temperature of 25 °C. After 15 days of storage, the same NBTI stress is applied to another device with identical W/L = 10/10 μm parameters. The degradation curve is presented in Fig. (**7.9b**). Compared to the fresh device in Fig. (**7.8a**), the device after 15 days' storage exhibits more severe NBTI degradation, with a V_{th} shift exceeding 2.0 V after

30,000 s stress, as shown in Fig. (**7.9f**). Moreover, the n value increases to 0.33, as illustrated in Fig. (**7.9g**). This notable disparity between the two test results suggests potential alterations to the wafer during storage.

Given that the test wafer lacks additional passivation layers ($d_1 = d_2 = 0$), atmospheric H_2O may infiltrate the porous low-temperature oxide (LTO) structures, potentially exacerbating NBTI. To ascertain whether the intensified NBTI degradation correlates with H_2O presence in the device structure, a series of experiments were devised and executed. Initially, an H-B treatment was administered to the wafer. Subsequently, following cool down to room temperature post-H-B treatment, an identical NBTI stress test was conducted on a TFT with matching W/L parameters, with resulting transfer curve degradation depicted in Fig. (**7.9c**). Remarkably, NBTI was observed to ameliorate following the 15-hour H-B treatment. The shift after 30,000 s stress exhibited recovery to approximately 1.0 V, as displayed in Fig. (**7.9f**), while the degradation slope (n) decreased to 0.31, as depicted in Fig. (**7.9g**). These findings suggest that H_2O within the device structure can evaporate at elevated temperatures, thereby improving NBTI.

After the preceding stress test, a T-H treatment was promptly administered to the wafer. During this treatment, the diffusion of H_2O into the gate oxide is expected to occur more readily compared to normal storage conditions due to the elevated air relative humidity and temperature. Subsequent to the T-H treatment, the wafer was swiftly dried using an N_2 gun. Following this, an identical NBTI stress was applied to a TFT with matching *W/L* parameters, with the resulting test outcomes depicted in Fig. (**7.9d**). Notably, NBTI deterioration was substantial following the T-H treatment, with the V_{th} shift reaching up to 4.1 V after 30000 s stress, as illustrated in Fig. (**7.9f**), and the degradation slope (n) dramatically increasing to 0.54, as shown in Fig. (**7.9g**). It is hypothesized that H_2O within the device structure exacerbates this NBTI. Subsequently, another H-B treatment was applied to the wafer after the aforementioned stress test. As consistent with (Fig. **7.9c**), NBTI of a TFT post-H-B treatment was significantly improved. The V_{th} shift recovered from 4.1 V to 1.9 V, as shown in Fig. (**7.9f**), and the degradation slope (n) decreased from 0.54 to 0.33, as depicted in Fig. (**7.9g**), attributed to H_2O evaporation.

In LTPS TFTs, the NBTI degradation is widely believed to be attributed to the generation of fixed charges and interface/GB trap states   [38-40] without the consideration of the H_2O effect. The water diffusion length in oxide for 15 days' storage and for thermal-humidity treatment is estimated as 4.8 nm and 10.7 nm, respectively, which are both much smaller than the field oxide thickness (500 nm). However, the results of the experiment indicate that H_2O must diffuse into the device structure. Thus it can be inferred that the LPCVD field oxide must be porous and even have pinholes, which will shorten the water diffusion path. Only with this porous layer of field oxide as a passivation layer the H_2O will diffuse into the gate oxide network by a reaction [41]:

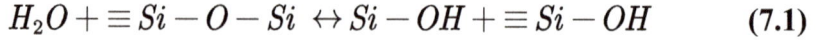

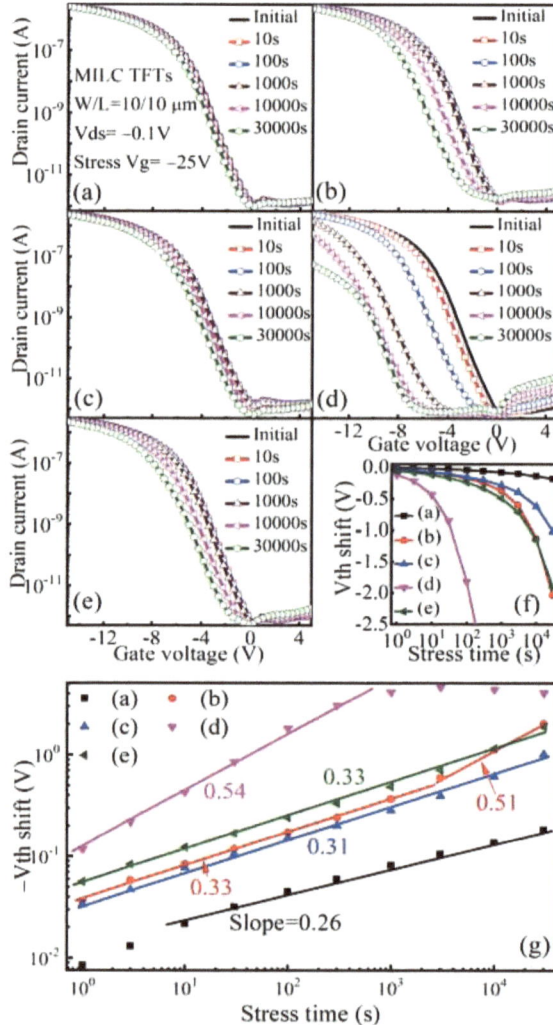

$$H_2O + \equiv Si - O - Si \leftrightarrow Si - OH + \equiv Si - OH \qquad (7.1)$$

Fig. (7.9). (a–e): Time evolution of transfer characteristics under NBTI stress for p-type MILC TFTs after a train of treatments. (a) A fresh TFT without any additional treatment. (b) A TFT after 15 days' normal storage. (c) A TFT after H-B treatment. (d) A TFT after T-H treatment. (e) A TFT after the second round of H-B treatment following T-H treatment. (f): V_{th} shift dependent on stress time extracted from (a–e) plotted in semi-logarithmic scale. (g): V_{th} shift dependent on stress time extracted from (a–e) plotted in logarithmic scale [37].

As shown in Fig. (**7.10**), if '≡Si–H' at interface/GBs is activated by the high field, H⁺ will be released from the '≡Si–H' [39], resulting in dangling bonds at interface/GBs. Then, the released H⁺ may react with the '≡Si–OH' nearby at the interface by:

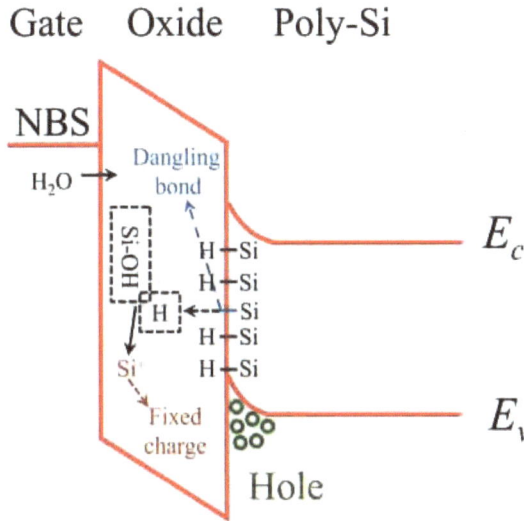

Fig. (7.10). Energy band diagram of p-type TFTs under NBTI stress.

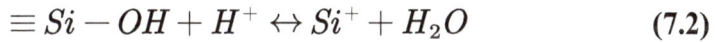

$$\equiv Si - OH + H^+ \leftrightarrow Si^+ + H_2O \tag{7.2}$$

leading to positive fixed charge (≡Si⁺) generation in the oxide at the interface, enhancing NBTI degradation. Based on the above experiment and discussion, it is thought that the factor of H₂O during the NBTI stress should be taken into consideration.

For the effect of water vapor on MO TFTs,   the impact of water vapor on the stability of MO TFTs results in degradation due to various factors. In the following sections, we will primarily utilize experimental observations to elucidate the internal mechanisms underlying this effect.

Fig. (**7.11c**) displays the transfer curves of ITZO TFTs from split wafers after low-temperature (LT) annealing at different temperatures, illustrating LT annealing's impact on device electrical characteristics. To provide clearer insights, we've extracted and plotted V_{on} and I_{on} dependence on annealing temperature in Fig. (**7.11d**). Notably, V_{on} initially shifts negatively before transitioning to a positive shift, with the most significant negative shift occurring at approximately 60°C and a turning point around 100°C. I_{on} follows a similar trend to V_{on}. These shifts may be attributed to variations in water content within the device. Varied moisture exchanges resulting from different LT annealing conditions between ITZO TFTs and ambient air could explain these phenomena.

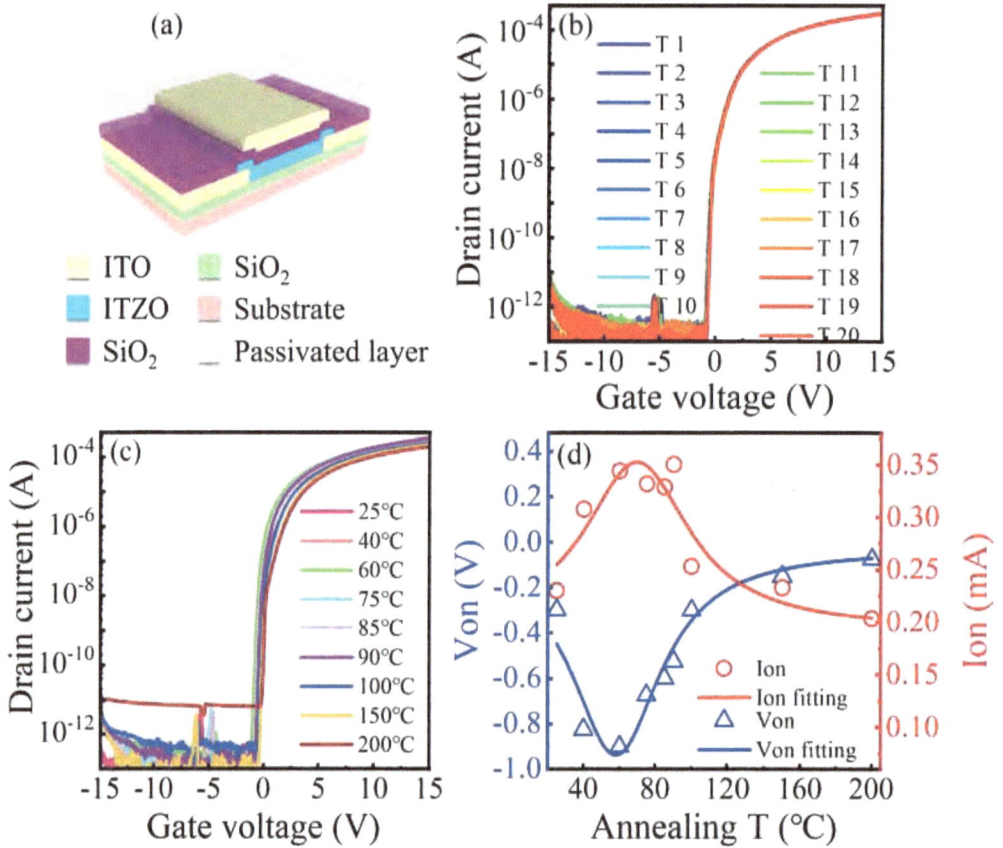

Fig. (7.11). (a) Schematic diagram of ITZO TFTs. (b) Transfer curves of 20 different devices, measured at $V_{ds} = 5$ V. (c) Device transfer curves after LT annealing with different annealing temperatures, measured at $V_{ds} = 5$ V. 9 devices were tested. (d) V_{on} and I_{on} dependent on the annealing temperature [42].

To validate this hypothesis, the stability of PBS was tested at different annealing temperatures. Fig. (**7.12**) illustrates the degradation of transfer curves under PBS for ITZO TFTs following LT annealing at temperatures of 25°C (untreated), 60°C, and 150°C. In contrast to the PBS-induced positive shift observed in MO TFTs, the transfer curves of ITZO TFTs consistently exhibit a negative shift under all LT annealing conditions. In our study, the passivation layer and gate oxide of ITZO

TFTs are composed of plasma-enhanced chemical vapor deposition (PECVD) SiO_2, which typically possesses porosity and may contain pinholes. Over time, moisture can diffuse into the device structure, potentially leading to water ionization during PBS application. This ionization negatively shifts V_{on} and increases I_{on}. The negative shifts vary significantly under different LT annealing conditions, showing a non-monotonous relationship with annealing temperature. Fig. (**7.13**) displays V_{on} shift and ΔI_{on} under PBS after LT annealing at various temperatures. Below ≈60°C, PBS-induced degradation intensifies with increasing annealing temperature, peaking around ≈ 60°C. Beyond ≈ 100°C, degradation improves compared to untreated devices (25°C). This suggests that moisture exchange, influenced by LT annealing, likely drives these effects on device characteristics. Additionally, relative humidity (RH) values decrease with temperature near the hot plate surface (Fig. **7.12a–c**), potentially affecting air-device moisture exchange.

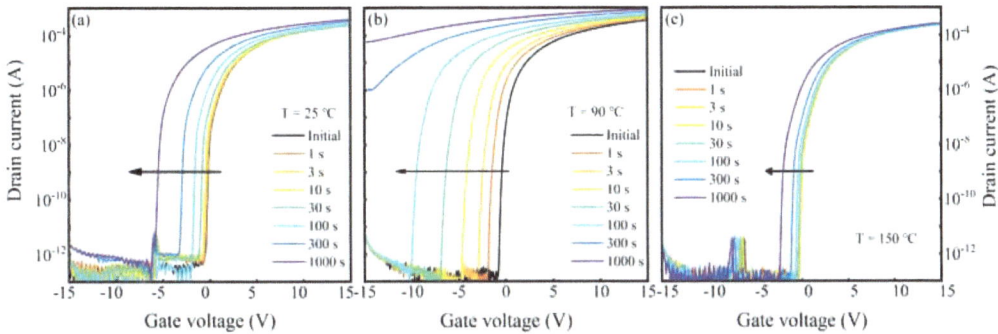

Fig. (7.12). Time evolution of transfer curves of ITZO TFTs under PBS after LT annealing of (a) 25 °C, (b) 60 °C, (c) 150 °C. The RH values were measured at 3 cm from the hot plate surface. V_{on} shift dependent on stress time under PBS after LT annealing with temperature [42].

To verify our hypothesis, X-ray photoelectron spectroscopy (XPS) was used to analyze the elemental compositions of ITZO films under various treatments. We characterized both the as-deposited ITZO thin film and those subjected to processes similar to ITZO TFTs, excluding ITO depositions (Fig. **7.13**). Generally, O 1 s spectra can be divided into three subpeaks representing metal-oxygen (M-O) bonds [43], metal-oxygen vacancies (M-V_o) [44], and metal-hydroxyl bonds (M-OH) [45]. After LT annealing at 60°C, the ratio of these peaks shifted, indicating an increase in M-OH concentration from 24.24% to 30.58% compared to the untreated ITZO thin film. Conversely, after LT annealing at 150°C, M-OH concentration decreased, suggesting water molecule diffusion from the ITZO thin film to air. To confirm moisture exchange during LT annealing, XPS Si 2p spectra were analyzed to estimate OH concentration in the LPCVD SiO_2 passivation layer. Following LT annealing at 60°C, relative OH concentration (peak 2) increased from 25.14% to

42.91%. After LT annealing at 150°C, it decreased to 7.23%. This trend aligned with M-OH in the ITZO thin film, indicating water diffusion from air to SiO_2 after annealing at 60°C and the reverse after annealing at 150°C. These XPS results highlight LT annealing's impact on moisture exchange between the device and air, affecting device characteristics and stability.

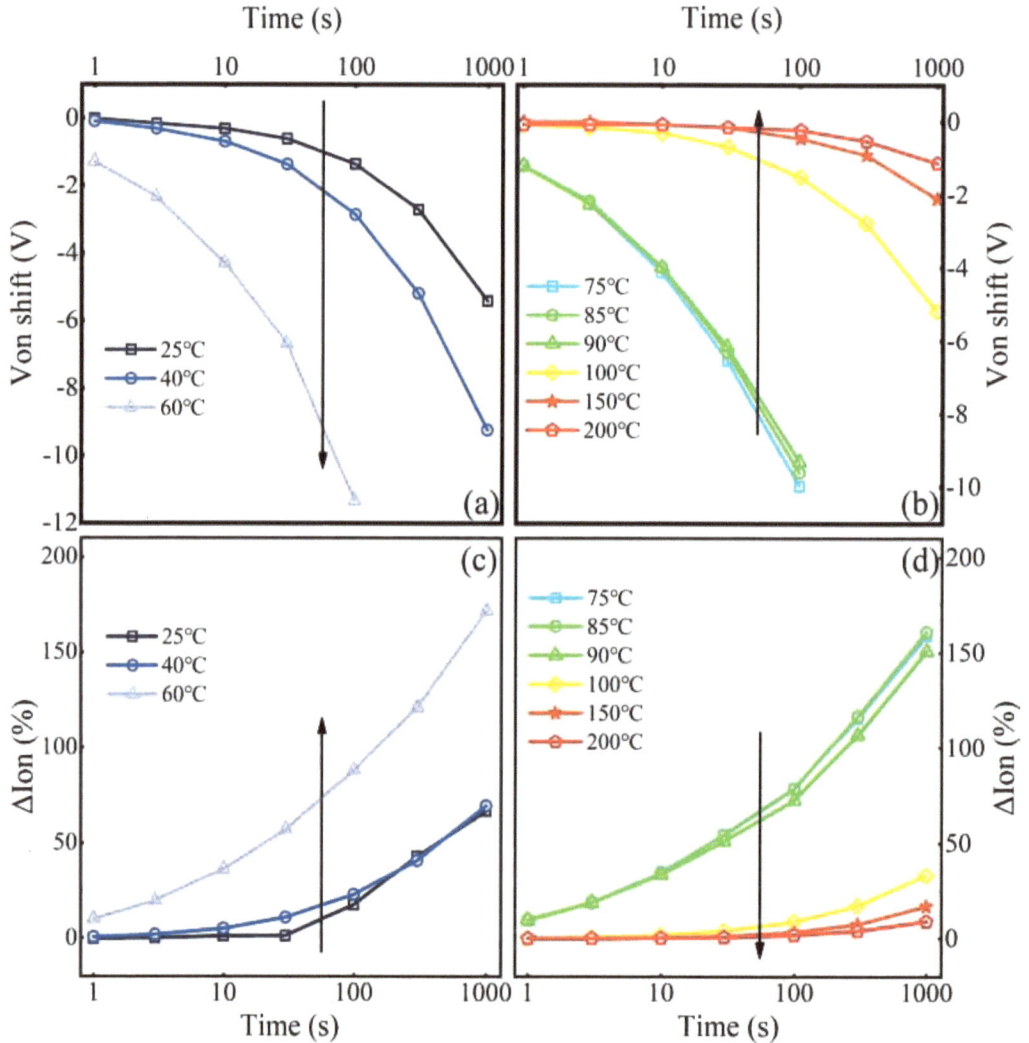

Fig. (7.13). (a) from 25 °C to 60 °C, and (b) from 75 °C to 200 °C. I_{on} degradation dependent on stress time under PBS after LT annealing with temperature (c) from 25 °C to 60 °C, and (d) from 75 °C to 200 °C.

To determine trap types, TCAD simulation is utilized. Fig. (**7.14e**) illustrates state density distributions in ITZO TFTs after 100 s PBS following different LT annealing. Donor-like states at the valence band edge remain constant across treatments. However, the density of acceptor-like states at the conduction band edge varies slightly with LT annealing. Initially increasing with annealing temperature, it then decreases, consistent with changes in device electrical characteristics and instability. The densities of donor-like states in a Gaussian distribution (N_{gd}) also fluctuate with treatments, likely corresponding to interstitial hydrogen (H_i) and substitutional hydrogen (H_O) [46, 47]. N_{gd} initially increases with annealing temperature, then decreases further. This trend aligns with changes in device electrical characteristics and instability.

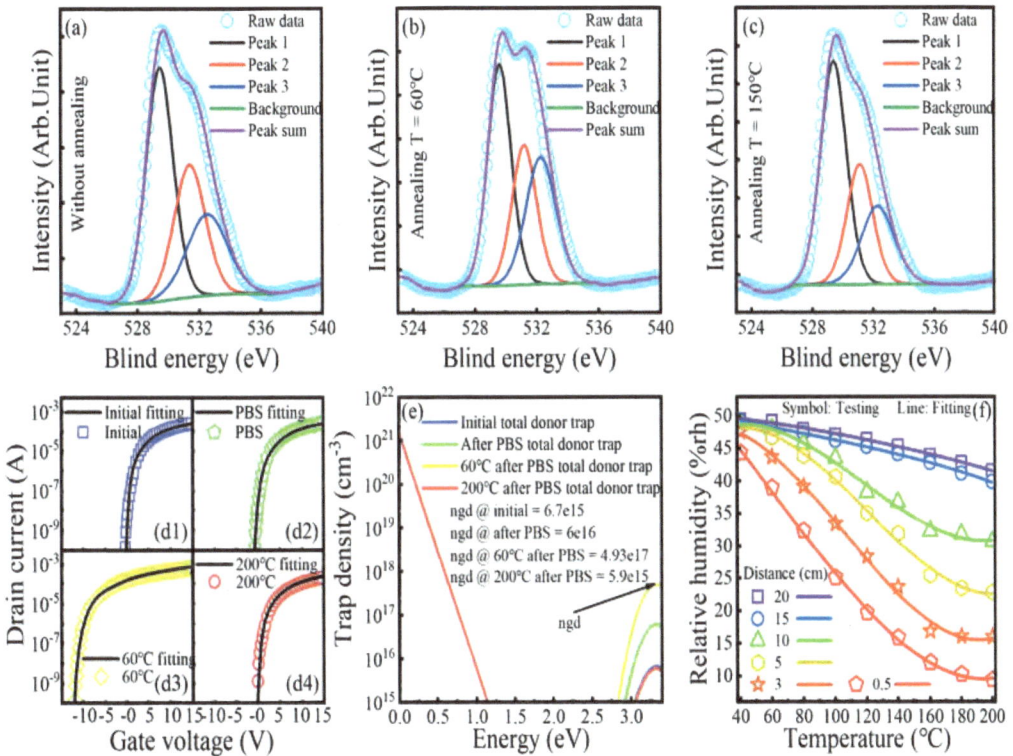

Fig. (7.14). XPS results of ITZO films (a) without annealing, (b) after LT annealing of 60 °C, and (c) after LT annealing of 150 °C. (d) TCAD simulation of ITZO TFTs under 100 s PBS after different annealing temperatures (symbols: experimental data, lines: fitting data). (e) Distributions of state density in ITZO TFTs under 100 s PBS after different annealing temperatures. (f) RH dependent on LT measured at different distances from the hot plate surface [42].

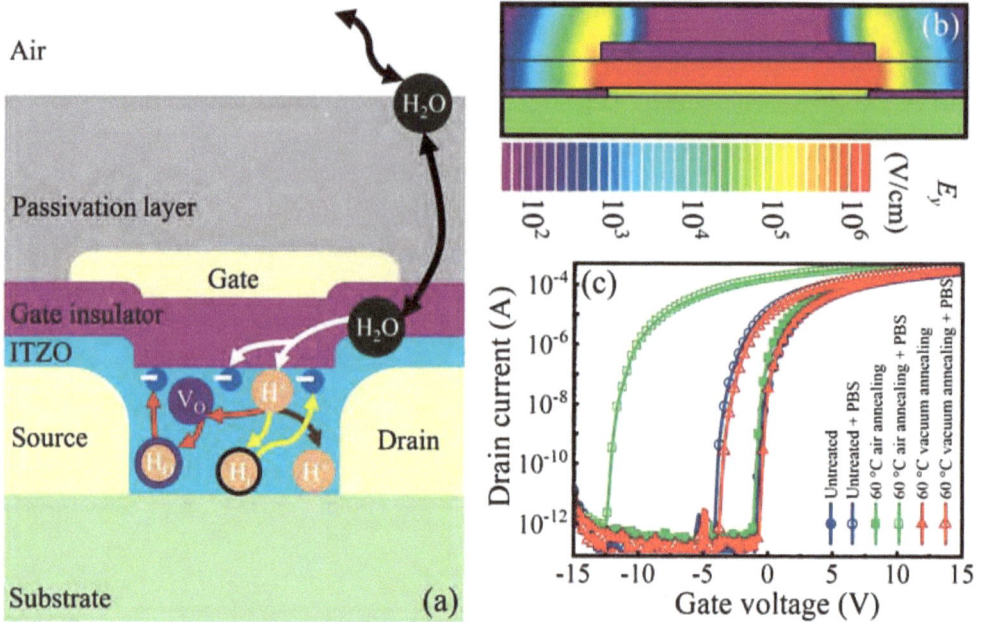

Fig. (7.15). (a) Schematic diagram of degradation model. (b) E_y distribution under Vg = 30 V in ITZO TFTs. (c) Comparison of transfer curve degradation under PBS of 100 s among the untreated ITZO TFT, the ITZO TFT treated by 12 h air annealing, and the ITZO TFT treated by 12 h vacuum annealing [42].

To elucidate the impact of moisture exchange on the characteristics and stability of ITZO TFTs, a tentative model was proposed. Moisture permeates the device structure through the porous passivation layer [37, 48], leading to increased water absorption and triggering the subsequent reaction (Fig. **7.15**) [49, 50]:

$$H_2O \leftrightarrow H_2O^+ + e^-\qquad(7.3)$$

The surplus electrons augment the free carrier concentration in the channel, facilitating device activation and resulting in a negative shift of the transfer curve. Introducing a PBS to the device complicates the scenario. As depicted in Chapter 5, the PBS degradation model introduces a substantial vertical electric field E_y within the device structure. Facilitated by E_y, water ionization occurs readily, as per the equation [51]:

$$H_2O \leftrightarrow 2H^+ + O^- + e^-\qquad(7.4)$$

The excess ionized electrons further elevate the free carrier concentration in the channel, exacerbating the negative shift in device transfer characteristics.

Additionally, H^+ ions resulting from water ionization diffuse into the channel under the influence of E_y. Analogous to H^+ behavior in zinc oxide (ZnO) thin films, H^+ in ITZO thin films exhibits varying characteristics at interstitial and substitutional sites [46]. At interstitial sites, H^+ combines with oxygen bonds in the channel, forming H_i at shallow energy levels. Conversely, at substitutional sites, H^+ can bond with V_o to form H_o, also at shallow energy levels [46]. Both H_i and H_o serve as shallow donors [52], injecting electrons into the channel and further shifting device transfer characteristics. Furthermore, a portion of H^+ ions continues to diffuse into the backchannel of ITZO TFTs under E_y. Accumulation of H^+ ions at the backchannel reduces source/drain barriers [52], facilitating easier device activation. Our experiments for ITZO TFTs applying different sizes of PBS also confirmed this. This negative V_{th} shift behavior contrasts with previously reported PBS-induced degradation in ITZO TFTs, where the transfer curve exhibits a positive shift over time. This discrepancy can be attributed to electron trapping at the interface for the positive shift [53], while the negative shift primarily arises from water ionization [51, 54, 55]. Additionally, a minor hump current emerges in the subthreshold region after 100 s of stress, suggesting the presence of two parallel transistors within the ITZO TFT with distinct V_{th} s. Both the increase in I_{on} and the negative shift of V_{on} are amplified with higher values of gate-source voltage (V_{gs}). These observations align with the influence of E_y and water ionization on the negative shift of the transfer curve.

The passivation layer utilized for ITZO TFTs is PECVD SiO_2, which typically exhibits porosity and may contain pinholes [37, 56]. Over time, atmospheric moisture can permeate into the device structure. Under E_y, water molecules can undergo ionization according to the equation (7.4).

The excess electrons subsequently augment the free carrier concentration within the channel, inducing a negative shift in the transfer curve. Concurrently, H^+ diffusion under E_y contributes to a further negative shift in the transfer curve. Increased stress applied to the gate results in larger E_y, exacerbating the negative shift in the transfer curve [57].

Furthermore, prolonged exposure to PBS leads to the accumulation of H^+ ions and V_o^{2+} species at the back channel of ITZO TFTs, thus fostering the formation of a parasitic transistor [58]. This parasitic transistor can activate earlier than the primary transistor, manifesting as a hump current in the subthreshold region.

For negative bias stress (NBS) of ITZO TFTs, in addition to the E_y, the electrons decomposed from the moisture are also affected by E_x to produce the SH effect in the source-drain region [59]. After simulating NBS stress in ALTAS, the negative V_{gs} induces a large E_x in the channel near the drain and the source. The width of the E_x distribution at each side is ~0.5 μm. As depicted in chapter 5.2, the $|E_x|$ becomes larger as it approaches the drain or the source, and a larger $|V_{gs}|$ brings a larger $|E_x|$. Under the influence of such high E_x, carriers can be accelerated into HC, creating defects and degenerating device performance [60, 61]. Under the influence of a heightened electric field induced by NBS, water molecules undergo the equation (7.4), thereby generating carriers in the channel and fostering an increase in ion concentration. This NBS-induced effect prompts a significant upward shift in the energy band of ITZO, consequently amplifying the barrier height between indium tin oxide (ITO) and ITZO. Consequently, the electric field (E_x) in the active layer, particularly near the source and drain regions, intensifies. These additional carriers, engendered by energy band bending and H_2O ionization, act as triggers for HC degradation. Exposed to the elevated E_x, these carriers gain energy and transform into HCs. Subsequently, the HCs disrupt the fragile bonds within the ITZO thin film [62], leading to the emergence of numerous acceptor-tail defects near the source and drain regions, thereby deteriorating ion concentration.

Greater absolute values of gate-source voltage ($|V_{gs}|$) correspond to larger $|E_x|$, resulting in a more pronounced decrease in ion concentration. Notably, the direct current (DC) HC degradation observed in ITZO TFTs under NBS differs from that observed in poly-Si TFTs, where source carriers for HC degradation are consistently supplied by an inverted channel [61].

The impact of moisture on TFTs varies depending on the device type and pressurization method. Therefore, specific discussions tailored to each situation are necessary. Different types of TFTs possess distinct material compositions and structural characteristics, leading to varying sensitivities to moisture. Additionally, the pressurization method (such as positive and negative pressure) influences the ingress and distribution of moisture within TFTs, thereby affecting their stability. Hence, when assessing and optimizing the stability of TFTs, it is essential to consider the specific properties of the devices and their operational environments. Adopting this tailored approach facilitates a better understanding and resolution of the challenges posed by moisture on TFT stability, thus providing guidance for TFT design and fabrication.

CONCLUSION

This chapter provides a comprehensive analysis of how environmental factors, including temperature, illumination, and moisture, significantly influence the performance and reliability of TFTs. It begins with an exploration of temperature effects, detailing how varying thermal conditions impact the reliability of both poly-Si and MO TFTs, leading to degradation mechanisms that can compromise device functionality. The discussion then shifts to illumination effects, examining the role of photogenerated carriers in the degradation processes of different types of TFTs, including photodegradation phenomena and persistent photoconductivity effects. Additionally, the chapter addresses the implications of moisture exposure on TFT reliability, emphasizing how humidity can lead to structural and electrical failures. By elucidating the interplay between these environmental stressors and TFT degradation, the chapter provides critical insights that inform the design and optimization of TFTs for diverse applications. These findings are essential for developing reliable TFT technologies that can withstand the challenges presented by various environmental conditions, ultimately contributing to enhanced performance and longevity in real-world applications.

REFERENCES

[1]　M. Kimura, S. Inoue, T. Shimoda, S.W-B. Tam, O.K.B. Lui, P. Migliorato, and R. Nozawa, "Extraction of trap states in laser-crystallized polycrystalline-silicon thin-film transistors and analysis of degradation by self-heating", *J. Appl. Phys.,* vol. 91, no. 6, pp. 3855-3858, 2002.
http://dx.doi.org/10.1063/1.1446238

[2]　S. Inoue, H. Ohshima, and T. Shimoda, ""Analysis of Degradation Phenomenon Caused by Self-Heating in Low-Temperature-Processed Polycrystalline Silicon Thin Film Transistors", *Jpn. J. Appl. Phys.,* vol. 41, no. Part 1, No. 11A, pp. 6313-6319, 2002.
http://dx.doi.org/10.1143/JJAP.41.6313

[3]　S. Inoue, M. Kimura, and T. Shimoda, "Analysis and Classification of Degradation Phenomena in Polycrystalline-Silicon Thin Film Transistors Fabricated by a Low-Temperature Process Using Emission Light Microscopy", *pn. J. Appl. Phys.,* vol. 42, no. Part 1, No. 3, pp. 1168-1172, 2003.
http://dx.doi.org/10.1143/JJAP.42.1168

[4]　Y. Uraoka, H. Yano, T. Hatayama, and T. Fuyuki, "Comprehensive Study on Reliability of Low-Temperature Poly-Si Thin-Film Transistors under Dynamic Complimentary Metal-Oxide Semiconductor Operations", *Jpn. J. Appl. Phys.,* vol. 41, no. Part 1, No. 4B, pp. 2414-2418, 2002.
http://dx.doi.org/10.1143/JJAP.41.2414

[5]　T. Yoshida, K. Yoshino, M. Takei, A. Hara, N. Sasaki, and T. Tsuchiya, "Experimental evidence of grain-boundary related hot-carrier degradation mechanism in low-temperature

poly-Si thin-film-transistors", *in IEEE International Electron Devices Meeting 2003,* Washington, DC, USA: IEEE, pp. 8.8.1-8.8.4, 2003.

http://dx.doi.org/10.1109/IEDM.2003.1269250

[6]   Y. Uraoka, Y. Morita, H. Yano, T. Hatayama, and T. Fuyuki, "Gate Length Dependence of Hot Carrier Reliability in Low-Temperature Polycrystalline-Silicon P-Channel Thin Film Transistors", *Jpn. J. Appl. Phys.,* vol. 41, no. Part 1, No. 10, pp. 5894-5899, 2002.

http://dx.doi.org/10.1143/JJAP.41.5894

[7]   T. Yoshida, Y. Ebiko, M. Take, N. Sasaki, and T. Tsuchiya, "Grain-boundary related hot carrier degradation mechanism in low-temperature polycrystalline silicon thin-film transistors", *JAPANESE JOURNAL OF APPLIED PHYSICS PART 1-REGULAR PAPERS SHORT NOTES & REVIEW PAPERS,* vol. 42, no. 4B, pp. 1999-2003, 2003.

http://dx.doi.org/10.1143/JJAP.42.1999

[8]   Y. Uraoka, N. Hirai, H. Yano, T. Hatayama, and T. Fuyuki, "Hot Carrier Analysis in Low-TemperaturePoly-Si TFTs Using Picosecond Emission Microscope", *IEEE Trans. Electron Dev.,* vol. 51, no. 1, pp. 28-35, 2004.

http://dx.doi.org/10.1109/TED.2003.820937

[9]   L. Mariucci, A. Pecora, S. Giovannini, R. Carluccio, F. Massussi, and G. Fortunato, "Hot carrier effects in polycrystalline silicon thin-film transistors: analysis of electrical characteristics and noise performance modifications", *Microelectron. Reliab.,* vol. 39, no. 1, pp. 45-52, 1999.

http://dx.doi.org/10.1016/S0026-2714(98)00162-0

[10]   K. Chang, Y. Chung, and G. Lin, "Hot carrier induced degradation in the low temperature processed polycrystalline silicon thin film transistors using the dynamic stress", *Japanese Journal Of Applied Physics Part 1-Regular Papers Short Notes & Review Papers,* vol. 41, no. 4A, pp. 1941-1946, 2002.

http://dx.doi.org/10.1143/JJAP.41.1941

[11]   F.V. Farmakis, C.A. Dimitriadis, J. Brini, G. Kamarinos, V.K. Gueorguiev, and T.E. Ivanov, "Photon emission and related hot-carrier effects in polycrystalline silicon thin-film transistors", *J. Appl. Phys.,* vol. 85, no. 9, pp. 6917-6919, 1999.

http://dx.doi.org/10.1063/1.370105

[12]   Y. Uraoka, T. Hatayama, T. Fuyuki, T. Kawamura, and Y. Tsuchihashi, "Reliability of low temperature poly-silicon TFTs under inverter operation", *IEEE Trans. Electron Dev.,* vol. 48, no. 10, pp. 2370-2374, 2001.

http://dx.doi.org/10.1109/16.954479

[13]   B.T. Chen, C-H. Tseng, H-C. Cheng, C-W. Chao, T-K. Chang, J-H. Lu, and A. Chin, "Symmetric gate-overlapped LDD poly-Si TFTs with selective and isotropic deposited Ni sub-gate", *Electrochem. Solid-State Lett.,* vol. 7, no. 2, pp. G37-G39, 2004.

http://dx.doi.org/10.1149/1.1635672

[14]   S. Tyaginov, M. Jech, J. Franco, P. Sharma, B. Kaczer, and T. Grasser, "Understanding and Modeling the Temperature Behavior of Hot-Carrier Degradation in SiON nMOSFETs", *IEEE Electron Device Lett.,* vol. 37, no. 1, pp. 84-87, 2016.

http://dx.doi.org/10.1109/LED.2015.2503920

[15] g. Vandenbosch, R. Bellens, G. Groeseneken, and H. Maes, "Temperature-dependence of the channel hot-carrier degradation of N-channel mosfets", *IEEE Trans. Electron Dev.*, vol. 37, no. 4, pp. 980-993, 1990.

http://dx.doi.org/10.1109/16.52433

[16] Y.X. Wang, T-C. Chang, M-C. Tai, C-C. Wu, Y-Z. Zheng, Y-F. Tu, J-J. Chen, K-J. Zhou, Y-S. Shih, Y-A. Chen, J-W. Huang, and S. Sze, "Improvement of Strained Negative Bias Temperature Instability in Flexible LTPS TFTs by a Stress-Release Design", *IEEE Trans. Electron Dev.*, vol. 69, no. 3, pp. 1532-1537, 2022.

http://dx.doi.org/10.1109/TED.2022.3140691

[17] H.C. Chen, H-Y. Tu, H-C. Huang, W-C. Lai, T-C. Chang, S-P. Huang, Y-F. Tu, C-W. Kuo, K-J. Zhou, J-J. Chen, Y-S. Shih, G-F. Chen, and W-C. Su, "Inhibiting the Kink Effect and Hot-Carrier Stress Degradation Using Dual-Gate Low-Temperature Poly-Si TFTs", *IEEE Electron Device Lett.*, vol. 41, no. 1, pp. 54-57, 2020.

http://dx.doi.org/10.1109/LED.2019.2951935

[18] Y.X. Wang, T-C. Chang, M-C. Tai, C-C. Wu, Y-F. Tu, J-J. Chen, W-C. Huang, Y-S. Shih, Y-A. Chen, J-W. Huang, and S. Sze, "Investigation of Degradation Behavior During Illuminated Negative Bias Temperature Stress in P-Channel Low-Temperature Polycrystalline Silicon Thin-Film Transistors", *IEEE Electron Device Lett.*, vol. 42, no. 5, pp. 712-715, 2021.

http://dx.doi.org/10.1109/LED.2021.3068423

[19] Y.X. Wang, M-C. Tai, T-C. Chang, S-P. Huang, Y-Z. Zheng, C-C. Wu, Y-S. Shih, Y-A. Chen, P-J. Sun, I-N. Lu, H-C. Huang, and S.M. Sze, "Suppression of Edge Effect Induced by Positive Gate Bias Stress in Low-Temperature Polycrystalline Silicon TFTs With Channel Width Extension Over Source/Drain Regions", *IEEE Trans. Electron Dev.*, vol. 67, no. 12, pp. 5552-5556, 2020.

http://dx.doi.org/10.1109/TED.2020.3033516

[20] W.C.Y. Ma, H.S. Hsu, and H.C. Wang, "Various Reliability Investigations of Low Temperature Polycrystalline Silicon Tunnel Field-Effect Thin-Film Transistor", *IEEE Trans. Device Mater. Reliab.*, vol. 20, no. 4, pp. 775-780, 2020.

http://dx.doi.org/10.1109/TDMR.2020.3035336

[21] M. Fujii, H. Yano, T. Hatayama, Y. Uraoka, T. Fuyuki, J.S. Jung, and J.Y. Kwon, "Thermal Analysis of Degradation in Ga_2O_3 –In_2O_3 –ZnO Thin-Film Transistors", *Jpn. J. Appl. Phys.*, vol. 47, no. 8R, pp. 6236-6240, 2008.

http://dx.doi.org/10.1143/JJAP.47.6236

[22] S.C. Andrews, M.A. Fardy, M.C. Moore, S. Aloni, M. Zhang, V. Radmilovic, and P. Yang, "Atomic-level control of the thermoelectric properties in polytypoid nanowires", *Chem. Sci. (Camb.)*, vol. 2, no. 4, pp. 706-714, 2011.

http://dx.doi.org/10.1039/c0sc00537a

[23] D.K. Seo, S. Shin, H.H. Cho, B.H. Kong, D.M. Whang, and H.K. Cho, "Drastic improvement of oxide thermoelectric performance using thermal and plasma treatments of the InGaZnO thin films grown by sputtering", *Acta Mater.*, vol. 59, no. 17, pp. 6743-6750, 2011.

http://dx.doi.org/10.1016/j.actamat.2011.07.032

[24]   D. Kim, J. Kim, H. Kang, J. W. Shim, and J. W. Lee, "Influence of flexible substrate in low temperature polycrystalline silicon thin-film transistors: temperature dependent characteristics and low frequency noise analysis", *Nanotechnology*, vol. 31, no. 43, 2020.
http://dx.doi.org/10.1088/1361-6528/ab98ba

[25]   S. Inoue, M. Kimura, and T. Shimoda, "Analysis and Classification of Degradation Phenomena in Polycrystalline-Silicon Thin Film Transistors Fabricated by a Low-Temperature Process Using Emission Light Microscopy", *Jpn. J. Appl. Phys.,* vol. 42, no. Part 1, No. 3, pp. 1168-1172, 2003.
http://dx.doi.org/10.1143/JJAP.42.1168

[26]   M. Xue, M. Wang, Z. Zhu, D. Zhang, and M. Wong, "Degradation Behaviors of Metal-Induced Laterally Crystallized n-Type Polycrystalline Silicon Thin-Film Transistors Under DC Bias Stresses", *IEEE Trans. Electron Dev.,* vol. 54, no. 2, pp. 225-232, 2007.
http://dx.doi.org/10.1109/TED.2006.888723

[27]   L. Mariucci, P. Gaucci, A. Valletta, A. Pecora, L. Maiolo, M. Cuscuna, and G. Fortunato, "Edge Effects in Self-Heating-Related Instabilities in p-Channel Polycrystalline-Silicon Thin-Film Transistors", *IEEE Electron Device Lett.,* vol. 32, no. 12, pp. 1707-1709, 2011.
http://dx.doi.org/10.1109/LED.2011.2169040

[28]   A. Valletta, P. Gaucci, L. Mariucci, A. Pecora, L. Maiolo, and G. Fortunato, "Downscaling effects on self-heating related instabilities in p-channel polycrystalline silicon thin film transistors", *Appl. Phys. Lett.,* vol. 99, no. 5, 2011.053503
http://dx.doi.org/10.1063/1.3621874

[29]   T. Agostinelli, M. Caironi, D. Natali, M. Sampietro, P. Biagioni, M. Finazzi, and L. Duò, "Space charge effects on the active region of a planar organic photodetector", *J. Appl. Phys.,* vol. 101, no. 11, 2007.114504
http://dx.doi.org/10.1063/1.2738429

[30]   W. Cai, J. Wilson, and A. Song, "Present status of electric-double-layer thin-film transistors and their applications", *Flexible and Printed Electronics,* vol. 6, no. 4, 2021.043001
http://dx.doi.org/10.1088/2058-8585/ac039f

[31]   M. Zhang, "Light-Illumination-Induced Degradation and Its Long-Term Recovery in Indium-Tin-Zinc Oxide Thin-Film Transistors", *in 2019 IEEE 26th International Symposium on Physical and Failure Analysis of Integrated Circuits (IPFA),* pp. 1-3, 2019.*Hangzhou, China: IEEE,* pp. 1-3, 2019.
http://dx.doi.org/10.1109/IPFA47161.2019.8984822

[32]   M.D.H. Chowdhury, P. Migliorato, and J. Jang, "Light induced instabilities in amorphous indium–gallium–zinc–oxide thin-film transistors", *Appl. Phys. Lett.,* vol. 97, no. 17, 2010.173506
http://dx.doi.org/10.1063/1.3503971

[33]   M.Y. Tsai, T-C. Chang, A-K. Chu, T-Y. Hsieh, K-Y. Lin, Y-C. Wu, S-F. Huang, C-L. Chiang, P-L. Chen, T-C. Lai, C-C. Lo, and A. Lien, "Anomalous degradation behaviors under illuminated gate bias stress in a-Si:H thin film transistor", *Thin Solid Films,* vol. 572, pp. 79-84, 2014.
http://dx.doi.org/10.1016/j.tsf.2014.09.050

[34]   Y.H. Tai, J.W. Tsai, H.C. Cheng, and F.C. Su, "Instability mechanisms for the hydrogenated amorphous silicon thin-film transistors with negative and positive bias stresses on the gate electrodes", *Appl. Phys. Lett.*, vol. 67, no. 1, pp. 76-78, 1995.
http://dx.doi.org/10.1063/1.115512

[35]   K.S. Agrawal, V.S. Patil, E.C. Cho, and J. Yi, "Analysis of Negative Bias Illumination Stress Induced Effect on LTPS and a-IGZO TFT", *ECS J. Solid State Sci. Technol.*, vol. 9, no. 10, 2020.106005
http://dx.doi.org/10.1149/2162-8777/abc6f0

[36]   J. Choi, J. Park, K-H. Lim, N. Cho, J. Lee, S. Jeon, and Y.S. Kim, "Photosensitivity of InZnO thin-film transistors using a solution process", *Appl. Phys. Lett.*, vol. 109, no. 13, 2016.132105
http://dx.doi.org/10.1063/1.4963881

[37]   M. Zhang, W. Zhou, R. Chen, M. Wong, and H.S. Kwok, "Water-enhanced negative bias temperature instability in p-type low temperature polycrystalline silicon thin film transistors", *Microelectron. Reliab.*, vol. 54, no. 1, pp. 30-32, 2014.
http://dx.doi.org/10.1016/j.microrel.2013.07.082

[38]   C. Hu, M. Wang, B. Zhang, and M. Wong, "Negative Bias Temperature Instability Dominated Degradation of Metal-Induced Laterally Crystallized p-Type Polycrystalline Silicon Thin-Film Transistors", *IEEE Trans. Electron Dev.*, vol. 56, no. 4, pp. 587-594, 2009.
http://dx.doi.org/10.1109/TED.2009.2014428

[39]   C.Y. Chen, J-W. Lee, S-D. Wang, M-S. Shieh, P-H. Lee, W-C. Chen, H-Y. Lin, K-L. Yeh, and T-F. Lei, "Negative Bias Temperature Instability in Low-Temperature Polycrystalline Silicon Thin-Film Transistors", *IEEE Trans. Electron Dev.*, vol. 53, no. 12, pp. 2993-3000, 2006.
http://dx.doi.org/10.1109/TED.2006.885543

[40]   J. Zhou, M. Wang, and M. Wong, "Two-Stage Degradation of p-Channel Poly-Si Thin-Film Transistors Under Dynamic Negative Bias Temperature Stress", *IEEE Trans. Electron Dev.*, vol. 58, no. 9, pp. 3034-3041, 2011.
http://dx.doi.org/10.1109/TED.2011.2158582

[41]   K.M. Davis, and M. Tomozawa, "Water diffusion into silica glass: Structural changes in silica glass and their effect on water solubility and diffusivity", *J. Non-Cryst. Solids,* vol. 185, no. 3, pp. 203-220, 1995.
http://dx.doi.org/10.1016/0022-3093(95)00015-1

[42]   Z. Chen, M. Zhang, S. Deng, Z. Jiang, Y. Yan, S. Han, Y. Zhou, M. Wong, and H-S. Kwok, "Effect of Moisture Exchange Caused by Low-Temperature Annealing on Device Characteristics and Instability in InSnZnO Thin-Film Transistors", *Adv. Mater. Interfaces,* vol. 9, no. 14, 2022.2102584
http://dx.doi.org/10.1002/admi.202102584

[43]   H.S. Kim, J.S. Park, H.K. Jeong, K.S. Son, T.S. Kim, J.B. Seon, E. Lee, J.G. Chung, D.H. Kim, M. Ryu, and S.Y. Lee, "Density of states-based design of metal oxide thin-film transistors for high mobility and superior photostability", *ACS Appl. Mater. Interfaces,* vol. 4, no. 10, pp. 5416-5421, 2012.
http://dx.doi.org/10.1021/am301342x PMID: 22957907

[44]　K.L. Han, K.C. Ok, H.S. Cho, S. Oh, and J.S. Park, "Effect of hydrogen on the device performance and stability characteristics of amorphous InGaZnO thin-film transistors with a SiO2/SiNx/SiO2 buffer", *Appl. Phys. Lett.,* vol. 111, no. 6, 2017.063502

http://dx.doi.org/10.1063/1.4997926

[45]　H. Bong, W.H. Lee, D.Y. Lee, B.J. Kim, J.H. Cho, and K. Cho, "High-mobility low-temperature ZnO transistors with low-voltage operation", *Appl. Phys. Lett.,* vol. 96, no. 19, 2010.192115

http://dx.doi.org/10.1063/1.3428357

[46]　Y. Kang, B.D. Ahn, J.H. Song, Y.G. Mo, H-H. Nahm, S. Han, and J.K. Jeong, "Hydrogen Bistability as the Origin of Photo-Bias-Thermal Instabilities in Amorphous Oxide Semiconductors", *Adv. Electron. Mater.,* vol. 1, no. 7, 2015.1400006

http://dx.doi.org/10.1002/aelm.201400006

[47]　K. Ide, K. Nomura, H. Hosono, and T. Kamiya, "Electronic Defects in Amorphous Oxide Semiconductors: A Review", *Phys. Status Solidi., A Appl. Mater. Sci.,* vol. 216, no. 5, 2019.1800372

http://dx.doi.org/10.1002/pssa.201800372

[48]　Y.G. Tropsha, and N.G. Harvey, "Activated Rate Theory Treatment of Oxygen and Water Transport through Silicon Oxide/Poly(ethylene terephthalate) Composite Barrier Structures", *J. Phys. Chem. B,* vol. 101, no. 13, pp. 2259-2266, 1997.

http://dx.doi.org/10.1021/jp9629856

[49]　J.S. Park, J.K. Jeong, H.J. Chung, Y.G. Mo, and H.D. Kim, "Electronic transport properties of amorphous indium-gallium-zinc oxide semiconductor upon exposure to water", *Appl. Phys. Lett.,* vol. 92, no. 7, 2008.072104

http://dx.doi.org/10.1063/1.2838380

[50]　W.-F. Chung, "Influence of H2O Dipole on Subthreshold Swing of Amorphous Indium–Gallium–Zinc-Oxide Thin Film Transistors",

[51]　C.L. Fan, F-P. Tseng, B-J. Li, Y-Z. Lin, S-J. Wang, W-D. Lee, and B-R. Huang, "Improvement in reliability of amorphous indium–gallium–zinc oxide thin-film transistors with Teflon/SiO 2 bilayer passivation under gate bias stress", *Jpn. J. Appl. Phys.,* vol. 55, no. 2S, 2016.02BC17

http://dx.doi.org/10.7567/JJAP.55.02BC17

[52]　H.C. Chen, C.W. Kuo, T.C. Chang, W.C. Lai, P.H. Chen, G.F. Chen, S.P. Huang, J.J. Chen, K.J. Zhou, C.C. Shih, Y.C. Tsao, H.C. Huang, and S.M. Sze, "Investigation of the Capacitance–Voltage Electrical Characteristics of Thin-Film Transistors Caused by Hydrogen Diffusion under Negative Bias Stress in a Moist Environment", *ACS Appl. Mater. Interfaces,* vol. 11, no. 43, pp. 40196-40203, 2019.

http://dx.doi.org/10.1021/acsami.9b11637 PMID: 31573173

[53]　S. Maeng, H. Kim, G. Choi, Y. Choi, S. Oh, and J. Kim, "Investigation of electrical performance and operation stability of RF-sputtered InSnZnO thin film transistors by oxygen-ambient rapid thermal annealing", *Semicond. Sci. Technol.,* vol. 35, no. 12, 2020.125019

http://dx.doi.org/10.1088/1361-6641/abbc8f

[54] F.H. Chen, T.M. Pan, C.H. Chen, J.H. Liu, W.H. Lin, and P.H. Chen, "Two-step Electrical Degradation Behavior in α-InGaZnO Thin-film Transistor Under Gate-bias Stress", *IEEE Electron Device Lett.,* vol. 34, no. 5, pp. 635-637, 2013.
http://dx.doi.org/10.1109/LED.2013.2248115

[55] C-L. Fan, M-C. Shang, B-J. Li, S-J. Wang, and W-D. Lee, "Self-aligned amorphous indium-gallium-zinc-oxide thin-film transistor using a two-mask process without etching-stop layer", *in 2014 21st International Workshop on Active-Matrix Flatpanel Displays and Devices (AM-FPD),* Kyoto, Japan: IEEE, pp. 129-132, 2014.
http://dx.doi.org/10.1109/AM-FPD.2014.6867147

[56] G. Rochat, A. Delachaux, Y. Leterrier, J.A.E. Månson, and P. Fayet, "Influence of substrate morphology on the cohesion and adhesion of thin PECVD oxide films on semi-crystalline polymers", *Surf. Interface Anal.,* vol. 35, no. 12, pp. 948-952, 2003.
http://dx.doi.org/10.1002/sia.1621

[57] H.C. Chen, G-F. Chen, P-H. Chen, S-P. Huang, J-J. Chen, K-J. Zhou, C-W. Kuo, Y-C. Tsao, A-K. Chu, H-C. Huang, W-C. Lai, and T-C. Chang, "A Novel Heat Dissipation Structure for Inhibiting Hydrogen Diffusion in Top-Gate a-InGaZnO TFTs", *IEEE Electron Device Lett.,* vol. 40, no. 9, pp. 1447-1450, 2019.
http://dx.doi.org/10.1109/LED.2019.2927422

[58] Z. Jiang, M. Zhang, S. Deng, Y. Yang, M. Wong, and H.S. Kwok, "Evaluation of Positive-Bias-Stress-Induced Degradation in InSnZnO Thin-Film Transistors by Low Frequency Noise Measurement", *IEEE Electron Device Lett.,* vol. 43, no. 6, pp. 886-889, 2022.
http://dx.doi.org/10.1109/LED.2022.3165558

[59] Z. Jiang, M. Zhang, S. Deng, M. Wong, and H.S. Kwok, "Degradation of InSnZnO Thin-Film Transistors Under Negative Bias Stress", *IEEE Trans. Electron Dev.,* vol. 70, no. 12, pp. 6381-6386, 2023.
http://dx.doi.org/10.1109/TED.2023.3327975

[60] M. Zhang, M. Wang, X. Lu, M. Wong, and H.S. Kwok, "Analysis of Degradation Mechanisms in Low-Temperature Polycrystalline Silicon Thin-Film Transistors under Dynamic Drain Stress", *IEEE Trans. Electron Dev.,* vol. 59, no. 6, pp. 1730-1737, 2012.
http://dx.doi.org/10.1109/TED.2012.2189218

[61] Meng Zhang, Wei Zhou, Rongsheng Chen, Man Wong, and Hoi-Sing Kwok, "Characterization of DC-Stress-Induced Degradation in Bridged-Grain Polycrystalline Silicon Thin-Film Transistors", *IEEE Trans. Electron Dev.,* vol. 61, no. 9, pp. 3206-3212, 2014.
http://dx.doi.org/10.1109/TED.2014.2341676

[62] N. On, B.K. Kim, S. Lee, E.H. Kim, J.H. Lim, and J.K. Jeong, "Hot Carrier Effect in Self-Aligned In–Ga–Zn–O Thin-Film Transistors With Short Channel Length", *IEEE Trans. Electron Dev.,* vol. 67, no. 12, pp. 5544-5551, 2020.
http://dx.doi.org/10.1109/TED.2020.3032383

CHAPTER 8

Strategies for Improving Thin-Film Transistor Reliability

Abstract. This chapter presents strategies for enhancing the reliability of thin-film transistors (TFTs), discussing the implementation of special structures and other improvement methods. This chapter focuses on the reliability improvement of polysilicon TFT in lightly doped drain (LDD) and bridge-grain (BG) structures. The method to improve the reliability of metal oxide TFT is also described, such as the elevated-metal metal oxide (EMMO) structure.

Keywords: Bridge-grain (BG), Elevated-metal metal oxide (EMMO), Lightly doped drain (LDD), Reliability improvement strategies.

8.1. INTRODUCTION

The previous chapters of this book have detailed the reliability variations of thin-film transistors (TFT) under different operating conditions and the various degrees of damage caused by the application of multiple stresses. If such damage occurs directly within the display, it will lead to a degradation of the display's overall performance. Furthermore, as devices are typically integrated within systems, this decline in overall performance is challenging to remedy.

To address the different types of degradation encountered in devices, it is necessary to improve their reliability in a targeted manner. Based on the common causes of degradation, methods to enhance the reliability of devices can generally be categorized into adjusting the signal voltage levels or modifying the TFT devices themselves. However, adjusting signal sizes can impact the overall performance of the display to some extent, especially in circuit design aspects. Therefore, improving device reliability often requires adjustments to the devices themselves. This chapter focuses on two typical structures: the lightly doped drain (LDD) structure and the bridged grain (BG) structure. Both of these structures can significantly enhance the inherent reliability of TFTs. The chapter also introduces additional methods for improving device reliability.

8.2. LIGHTLY DOPED DRAIN STRUCTURE IN THIN-FILM TRANSISTORS

In the pursuit of continually improving device performance and increasing the number of devices per unit area, the dimensions of devices are consistently scaled down proportionally. However, this proportional scaling is not ideal, as not all parameters reduce proportionally. The lateral electric field (E_x) strength in the channel of a device increases as the dimensions are scaled down, with the strongest field occurring near the drain. When the characteristic dimensions of a device are reduced to a certain extent, a high electric field is generated between the source and drain, leading to carriers being accelerated into hot carriers (HCs) as they move. When these HCs have enough energy to overcome the Si-SiO$_2$ barrier height (3.5 eV), they can directly inject or tunnel into SiO$_2$, causing damage to the gate oxide and leading to device performance degradation or damage. This phenomenon is known as the HC effect.

The LDD structure utilizes two different doping concentrations in the source and drain regions. The doping concentration is lower near the channel and higher near the source and drain electrodes. This creates a gradient in doping concentration between the source/drain region and the channel, acting as a buffer for the source/drain electric field. It can reduce the peak electric field at the source and drain electrodes, decrease the generation and injection of hot electrons, and thereby enhance the stability of the device [1].

8.2.1. Fabrication Process of Lightly Doped Drain Structure Thin-Film Transistors

Initially, a 100 nm layer of SiO$_2$ is deposited on a glass substrate using plasma-enhanced chemical vapor deposition (PECVD) as a buffer layer. Subsequently, a 50 nm thick amorphous silicon (a-Si) film is deposited at 380°C using PECVD. The a-Si film is then subjected to dehydrogenation annealing at 450°C in a furnace. The a-Si is patterned through a sequence of processes, including spin-coating photoresist, pre-baking, exposure, development, etching, and cleaning. The a-Si film is then crystallized using a 308 nm XeCl excimer laser with a linear beam power of 350 mJ/cm². Another 100 nm thick SiO$_2$ layer is deposited via PECVD to serve as the gate insulator (GI).

A 300 nm thick layer of molybdenum is then deposited using magnetron sputtering and patterned as the gate electrode through photolithography and etching. Phosphorus ions are implanted using a mass-separated ion implanter to form n-regions in the LDD areas and n$^+$ regions at the source and drain, followed by a furnace anneal to activate the dopants. The final LDD structure TFT is shown in Fig. (**8.1**).

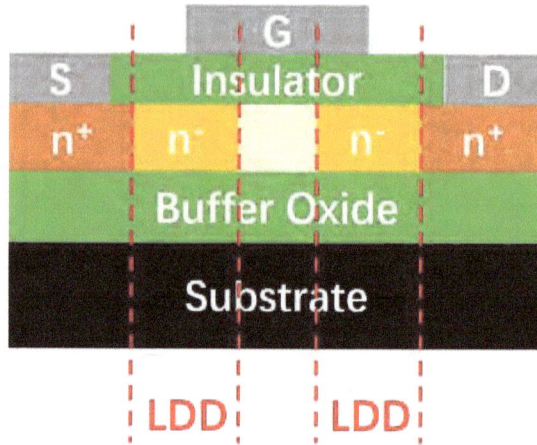

Fig. (8.1). LDD structure in TFT.

8.2.2. Effect of Lightly Doped Drain Structure on Thin-Film Transistor Reliability

When discussing the reliability of TFTs, two key parameters of the LDD structure play a crucial role: the length of the LDD and the doping concentration. These parameters significantly affect the electrical performance and reliability of TFTs.

The LDD structure introduces a lightly doped region that moderately increases resistance, leading to a more uniform electric field distribution near the drain and preventing the occurrence of high electric fields. The transfer curve of a TFT without an LDD structure is shown in Fig. (**8.2a**), where the curve tails up due to the high electric field at the drain, resulting in a pronounced drain-induced barrier lowering (DIBL) effect and a higher off-state current (I_{off}). The transfer characteristic curve of a TFT with an LDD structure is shown in Fig. (**8.2b**). The presence of the lightly doped source and drain regions disperses the electric field, making the distribution more uniform and reducing the electric field at the drain. Consequently, the I_{off} of the LDD structure TFT is lower. The transfer curves exhibit consistent behavior across V_{ds} of 0.1 V, 5 V, and 10 V [2]. Similarly, by incorporating the LDD structure, the shift in threshold voltage (V_{th}) can be reduced, but this also results in lower mobility and on-state current (I_{on}) [3].

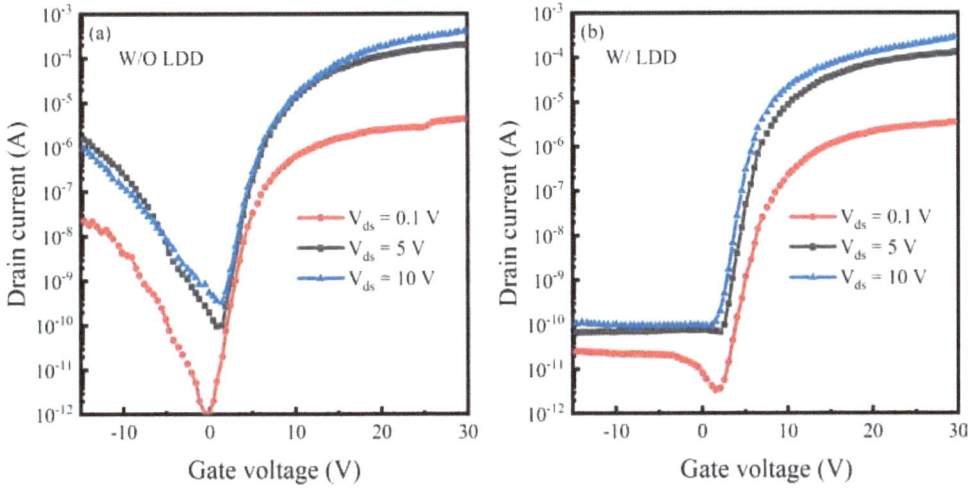

Fig. (8.2). Transfer characteristic curves of TFTs with different LDD widths [2].

8.2.3. Effect of Lightly Doped Drain Length on Thin-Film Transistor Reliability

The impact of LDD structure length on device performance is illustrated in Fig. (**8.3**), with LDD lengths ranging from 0 to 5 μm in increments of 0.5 μm. As shown in Fig. (**8.3a**), I_{on} decreases with increasing LDD length. This is because HCs are injected into the inter-layer dielectric (ILD) layer of the LDD near the drain, increasing the resistance of the LDD and consequently reducing I_{on}. As the LDD length increases, both V_{th} and mobility decrease, as depicted in Fig. (**8.3b**) and Fig. (**8.3c**). During stress application, LDD structures with appropriate lengths can reduce V_{th} shift due to their ability to disperse the E_x, lowering the kinetic energy of HCs and the incidence of impact ionization.

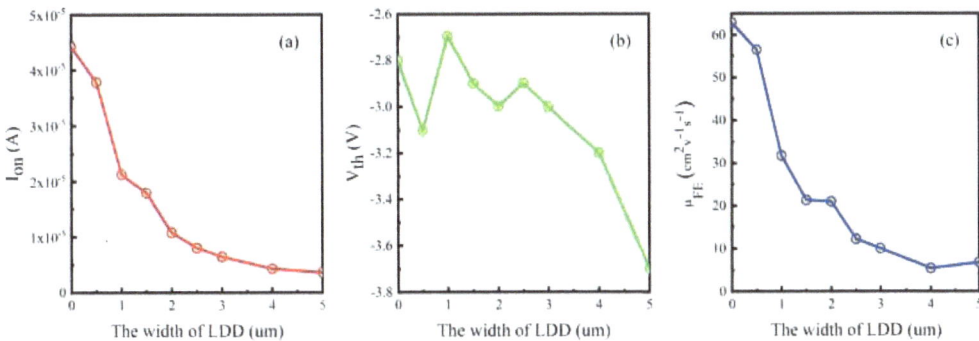

Fig. (8.3). Effect of LDD structures with different widths on (a) I_{on} (b) V_{th} (c) mobility of TFT [3].

8.2.4. Effect of Lightly Doped Drain Doping on Thin-Film Transistor Reliability

Fig. (**8.4**) presents the results for non-LDD structured TFTs with doping concentrations of 1×10^{13} cm^{-2}, 2×10^{13} cm^{-2}, and 3×10^{13} cm^{-2}. As indicated in Fig. (**8.4a**), the degree of I_{on} decay reduces with increasing doping concentration. When the doping concentration is 1×10^{13} cm^{-2} and 2×10^{13} cm^{-2}, the I_{on} of the device remains relatively stable, maintaining high performance. However, at a doping concentration of 3×10^{13} cm^{-2}, a significant decay in I_{on} is observed. This is due to excessively high doping concentrations leading to low resistance in the source and drain regions, which fails to reduce the electric field near these areas, resulting in greater degradation. When the doping concentration is 1×10^{13} cm^{-2}, the mobility of the device decays significantly, as shown in Fig. (**8.4b**), but when the doping concentration is 2×10^{13} cm^{-2}, the mobility is better maintained. It is evident that the doping concentration of the LDD has a significant impact on the reliability of TFTs [4].

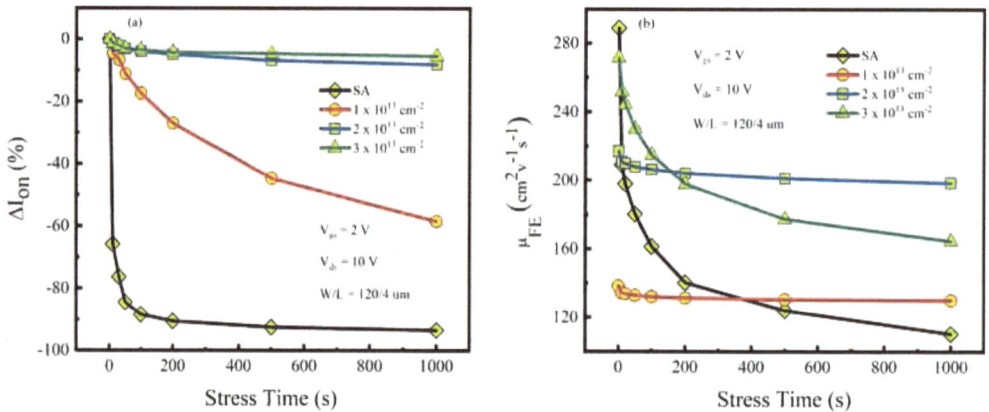

Fig. (8.4). Effect of LDD doping concentration on (a) I_{on} (b) mobility of TFT [4].

By introducing a lightly doped region, LDD structure TFTs, through appropriate LDD lengths and doping concentrations, increase the resistance near the source and drain regions, reduce the peak voltage at the source and drain electrodes, and achieve a more uniform electric field distribution. This reduces the HC injection effect and enhances the stability of the device.

8.2.5. Problems with Lightly Doped Drain Structures

LDD structure TFTs introduce an additional ion implantation step for light doping compared to traditional TFTs, increasing the complexity of the manufacturing process and raising costs. Furthermore, while the LDD structure improves reliability, it also increases parasitic capacitance and reduces the device's mobility and I_{on}. The processes and design of LDD structures differ significantly from new device structures, such as fin-shaped TFT, presenting new process challenges.

8.3. LIGHTLY DOPED DRAIN-LIKE STRUCTURE IN METAL-OXIDE THIN-FILM TRANSISTORS

In the previous section, the LDD structure under the polysilicon thin film transistor system was briefly introduced, and the degradation of the device was reduced by reducing the lateral electric field. However, metal oxides often make it difficult to perform similar doping behavior. However, a similar effect can be achieved by manipulating the oxygen vacancy defect inside the metal oxide. A lower conductivity structure can be set near the channel, thereby greatly increasing the reliability of the device.

8.3.1. Fabrication Process of Lightly Doped Drain-Like Structure Metal-Oxide Thin-Film Transistors

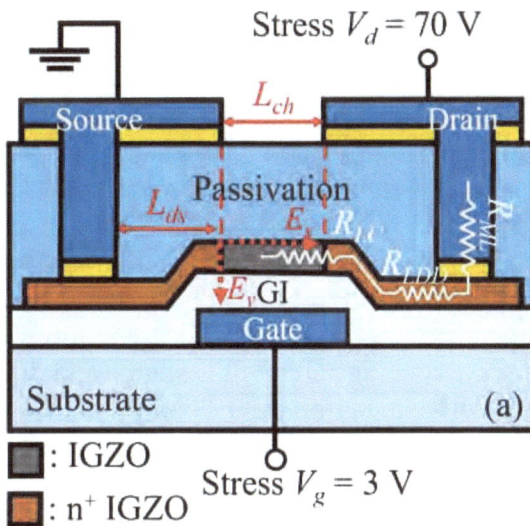

Fig. (8.5). LDD-like structure in MO TFT.

The LDD-like structure is shown in Fig. (**8.5**). Initially, 175 nm of aluminum (Al) was deposited onto a glass substrate that had a 500 nm layer of silicon dioxide (SiO$_2$), which was then patterned to function as the gate. Following this, a 125 nm SiO$_2$ layer was applied using low-pressure chemical vapor deposition (LPCVD) to serve as the gate insulator (GI). Subsequently, a 20 nm layer of amorphous indium gallium zinc oxide (a-IGZO) was sputtered at room temperature to create the active layer. A 300 nm SiO$_2$ passivation layer was then added through LPCVD. Contact holes were etched to allow access, and then 50 nm of molybdenum (Mo) and 300 nm of Al were sequentially deposited and patterned to form the source and drain electrodes. A subsequent annealing process at 400 degrees in an oxygen (O$_2$) environment for a duration of 2 hours was conducted. The oxygen vacancies (V_{oS}) in the IGZO layer, which is covered by the source and drain electrodes, will not be filled by O$_2$ during this high-temperature treatment due to the excellent barrier properties of the Al film. This results in the creation of a significant number of electrons, leading to the formation of n$^+$ IGZO regions that serve as the source and drain for the IGZO thin-film transistor (TFT). The resistivity of n$^+$ IGZO is remarkably lower, at approximately 1.0×10^{-2} $\Omega \cdot$cm, compared to that of standard IGZO, which is around 1.7×10^4 $\Omega \cdot$cm. This pronounced difference is attributed to the presence of V_{oS} rather than crystallization, as evidenced by X-ray photoelectron spectroscopy and X-ray diffraction analyses, as shown in Fig (**8.6** and **8.7**). In contrast, the oxygen vacancies in the IGZO areas not shielded by the Al film are repaired by O2 at high temperatures, which contributes to the formation of the active layer. Consequently, the concentration of oxygen vacancies in the n$^+$ IGZO is significantly higher than in the active layer. Due to the lateral diffusion of oxygen, a transitional region with a varying concentration of Vo exists between the n$^+$ IGZO region, resembling a lightly doped drain (LDD) and the active layer.

During the device manufacturing process, the metal electrodes of the source/drain act as a barrier to oxygen, preventing the oxygen vacancies (V_{oS}) in IGZO covered by the metal electrodes from being passivated by oxygen during the annealing process; meanwhile, under high T, a great number of carriers are ionized from oxygen vacancies, forming the LDD-like region. In contrast, the V_{oS} in the IGZO not covered by the metal are passivated by oxygen, forming the active layer. Therefore, by adjusting the coverage area of the source and drain metal electrodes, the size of the device's LDD-like region can be controlled.

For the diffusion process, the lateral diffusion distance is typically about 0.75 times the vertical diffusion distance [5], especially under conditions of high oxygen concentration. According to Ref [6], at 400 °C, the D of oxygen in IGZO is 4.7×10^{-17} cm^2/s. Additionally, Ref [6] indicates that at 400 °C, after 60 min of oxygen treatment, the diffusion depth of oxygen in IGZO is 20 nm. Using a similar methodology, we have determined the vertical diffusion profile of oxygen in IGZO after 120 min of oxygen treatment at 400 °C, as depicted in Fig. (**8.8**). The vertical diffusion depth is approximately 22 nm, which is similar to the thickness of the IGZO thin film used in our work.

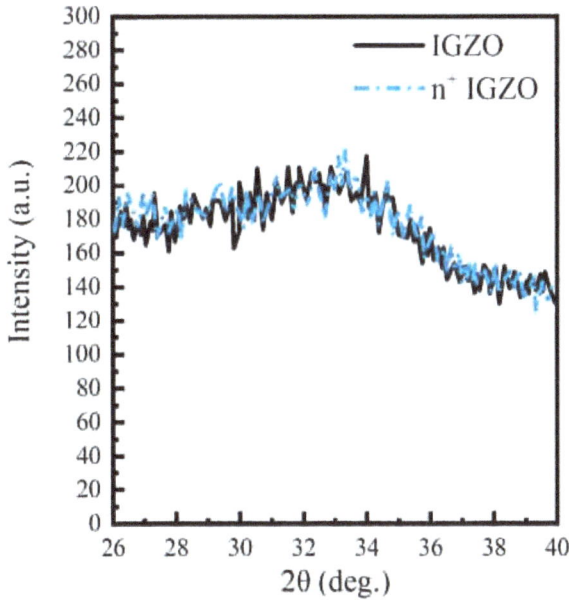

Fig. (8.6). XRD spectra of n$^+$ IGZO and IGZO.

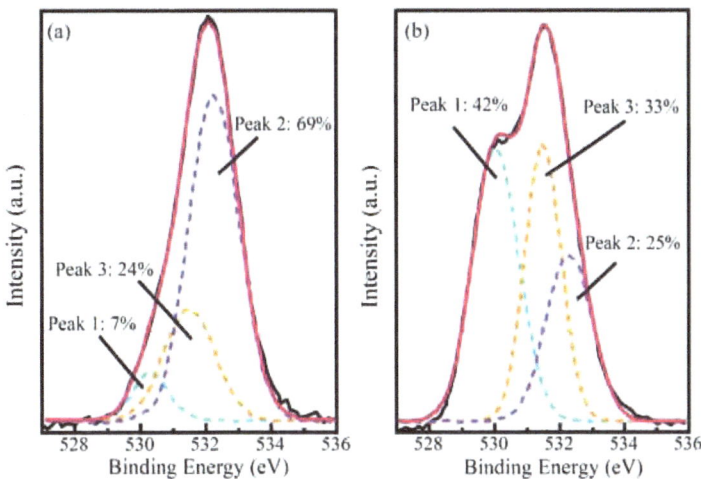

Fig. (8.7). The XPS spectra of O $1s$ in (a) n$^+$ IGZO and (b) IGZO.

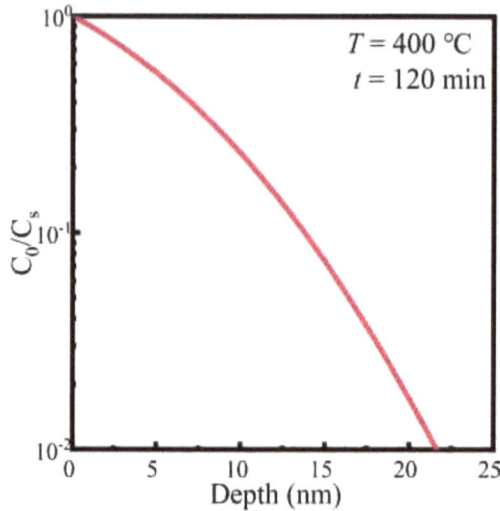

Fig. (8.8). O_2 vertical diffusion profile of IGZO annealed at 400 for 120 min.

Furthermore, since the lateral diffusion distance is generally less than the vertical diffusion distance [5], the lateral diffusion distance of O in IGZO is less than 22 nm. Compared to the μm scale of the LDD-like region, the lateral diffusion distance can be negligible. Therefore, it can be considered that the boundary between the areas with and without metal electrode coverage corresponds to the boundary between the active layer and the LDD-like region.

8.3.2. Effect of Lightly Doped Drain-Like Structure on Metal-Oxide Thin-Film Transistor Reliability

Similar to the LDD structure in poly TFTs, the LDD-like structure in MO TFTs can greatly improve the HC degradation of MO TFTs. As shown in Fig. (**8.9**), when extending the length of L_{ds}, the HC degradation of the device significantly decreases. When $L_{ds} = 100$ μm, the device has almost no degradation under 10000 s hot carrier stress.

After simulation by SILVACO, it is found that the LDD-like structure of the MO TFT is able to distribute the large voltage applied to the drain, as shown in Fig. (**8.10**). With the extension of L_{ds}, the E_x in the active layer decreases, and the reduction of the E_x effectively suppresses the generation of HCs in the active layer. Therefore, the LDD-like structure can effectively improve the hot carrier degradation of MO TFT.

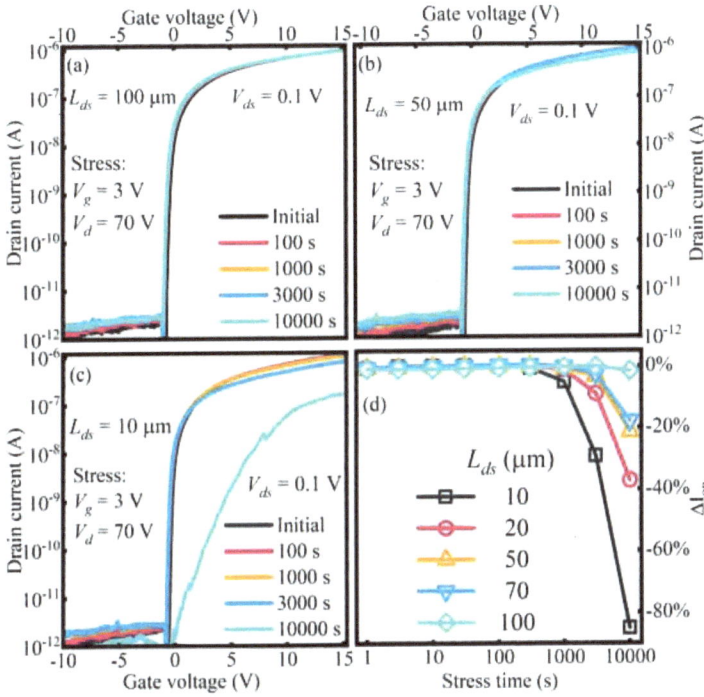

Fig. (8.9). Time evolution of transfer curves of IGZO TFT with (a) L_{ds} = 100 μm, (b) L_{ds} = 50 μm, (c) L_{ds} = 10 μm under HC stress of V_g = 3 V and V_d = 70 V. (d) ΔI_{on} dependent on stress time in IGZO TFTs with different L_{ds}.

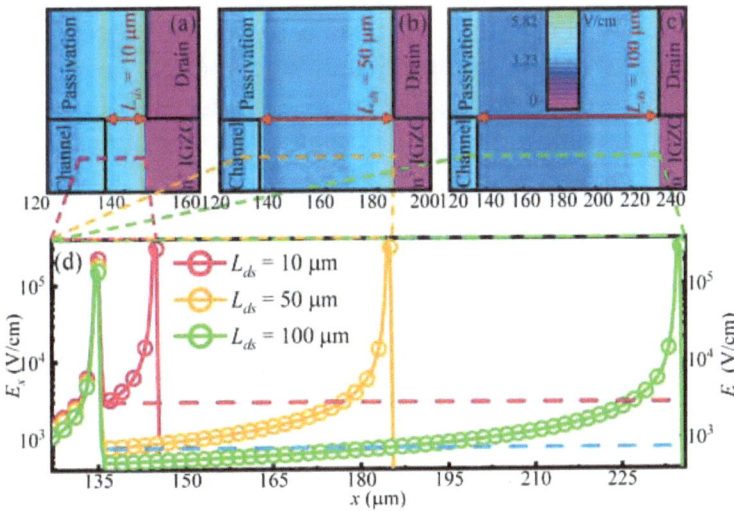

Fig. (8.10). Calculated Ex distribution under HC stress of Vg = 3 V and Vd = 70 V in IGZO TFT with (a) Lds = 10 μm, (b) Lds = 50 μm, (c) Lds = 100 μm (d) The extracted Ex in the middle of the channel from the middle of the device to the drain in IGZO TFTs with various Lds.

The LDD-like region functions as a resistor, connected between the active layer and the drain electrode, to distribute the voltage applied to the drain and thereby reduce the electric field in the active layer, improving the HC reliability in MO TFTs [7].

8.4. BRIDGED-GRAIN STRUCTURE IN THIN-FILM TRANSISTORS

High-performance polycrystalline silicon (poly-Si) TFTs with good reliability are thus required to accomplish the above purpose. In the past few decades, different kinds of treatments on active channels have been proposed and applied to fabricate high-performance poly-Si TFTs, such as plasma passivation [8], high-temperature annealing [9, 10], *etc*. However, these methods either bring process variation/reliability issues or are incompatible with the low-temperature process/standard process in the industry.

Recently, BG technology has been proposed and introduced to generate high-performance poly-Si TFTs [8]. By selectively doping BG lines inside the active channel, grain size effect, short channel effect (SCE), and multijunction effect are beneficially exploited, resulting in excellent device electrical characteristics in terms of field-effect carrier mobility (μ_{FE}), V_{th}, subthreshold swing (SS), and ON-current/OFF-current ratio (I_{ON}/I_{OFF}).

8.4.1 Fabrication Process

In the BG structure, the thin-film transistors (TFTs) use a conventional self-aligned top-gate configuration. Fig (**8.5**) illustrates a diagram of the device fabrication process. Initially, a 100 nm-thick amorphous silicon (a-Si) layer was grown on a silicon dioxide wafer using low-pressure chemical vapor deposition (LPCVD). This a-Si layer was then crystallized using solid-phase crystallization (SPC) at 600°C in a nitrogen atmosphere for 24 hours (Fig. **8.11**).

After crystallization, a photoresist (PR) layer was deposited and patterned to form a number of lines, as shown in Fig. (**8.5a**). The red rectangles represent the photoresist left after photolithography. The BG lines have a period of approximately 1 micron, with approximately 50% of each period covered by photoresist, as shown in Fig. (**8.5a**). Boron ions were then implanted at a dose of 10^{15} cm^{-2}. After implantation, the photoresist was removed to form doped BG polysilicon lines, as shown in Fig. (**8.5b**).

The active layer was then patterned using wet etching. A 70 nm thick aluminum oxide (Al_2O_3) layer was then grown at room temperature by DC magnetron sputtering at a deposition pressure of 3 mTorr and a power setting of 120 W. The surface roughness was measured to be 0.489 nm. For a 50 μm × 50 μm capacitor consisting of a Si/53 nm-thick Al_2O_3/Al stack, the measured capacitance density was approximately 130 nF/cm². The dielectric constant of Al_2O_3 film was measured to be 8.14, about twice that of LPCVD deposited SiO_2.

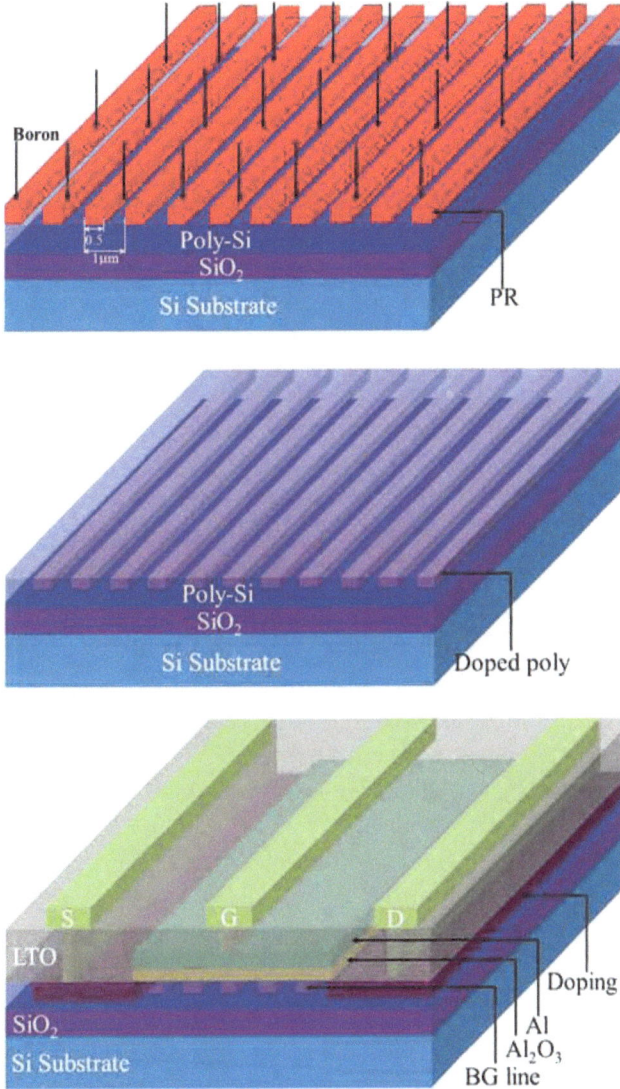

Fig. (8.11). Schematic diagram of the manufacturing process for BG SPC TFTs using Al_2O_3 as the gate dielectric. (a) Formation of BG line patterns, with red blocks indicating photoresist. (b) Active channel formation through boron doping of BG lines. (c) Design of BG SPC TFTs featuring Al_2O_3 as the gate oxide layer [11].

Following the deposition of the gate dielectric, a 300 nm thick layer of pure aluminum was sputtered to act as the gate electrode, which was then patterned. The source/drain regions were then formed by self-aligned boron ion deposition at a dose of 4×10^{15} cm^{-2}. A 500 nm thick oxide layer was then grown by LPCVD. After the contact holes were opened, a 700 nm thick aluminum layer containing 1% silicon was sputtered and patterned for the electrodes. Finally, the wafer was annealed in the forming gas at 420°C. The resulting TFT is referred to as a BG SPC TFT with an Al$_2$O$_3$ gate oxide.

Two other types of TFTs were also fabricated for comparison. One type is a conventional SPC polysilicon TFT with a 70 nm thick sputtered Al$_2$O$_3$ gate oxide, referred to as "SPC TFT with Al$_2$O$_3$ gate dielectric". The other type is a conventional SPC polysilicon TFT with a 70 nm thick LPCVD SiO$_2$ gate oxide, referred to as "SPC TFT with SiO$_2$ gate dielectric". After fabrication, these TFTs underwent no further processing [11].

8.4.2. Suppression of Direct-Current Degradation

Fig. (**8.12**) illustrates the transfer curve comparison between standard polysilicon TFTs and BG polysilicon TFTs. For a channel length of $L = 12$ μm, it is evident that the BG polysilicon TFT demonstrates significantly superior electrical performance, particularly in terms of I_{on}, SS, I_{off}, and V_{th}, compared to the conventional polysilicon TFT. When the effective channel length is reduced to $L_{eff} = 6$ μm, the BG polysilicon TFT still outperforms the standard polysilicon TFT across all device parameters. These substantial enhancements, brought about by the BG structure, are largely due to the influence of grain size, SCE, and the multijunction effect [12].

The influence of grain size in the doped BG segment of the channel is observed to reduce grain boundary (GB) trapping sites and lower the GB potential barrier in the path of electrical current. The analysis of the data presented in Fig. (**8.12**), employing the technique developed by Proano and Levinson, reveals that the grain boundary trap state density (N_t) for standard polysilicon TFTs is 3.48×10^{12} cm^{-2}, compared to a value of 2.98×10^{12} cm^{-2} for BG polysilicon TFTs. This supports the earlier point that the BG lines can reduce the GB N_t density. Essentially, the BG regions act as shortcuts for carriers, guiding them along more conductive paths, which enhances I_{on}, SS, I_{off}, and V_{th}. In terms of SCE, this structure is advantageous for boosting I_{on}, leading to higher I_{on} and lower V_{th}. The multijunction effect, inherent in the BG structure, is particularly effective at suppressing I_{off}.

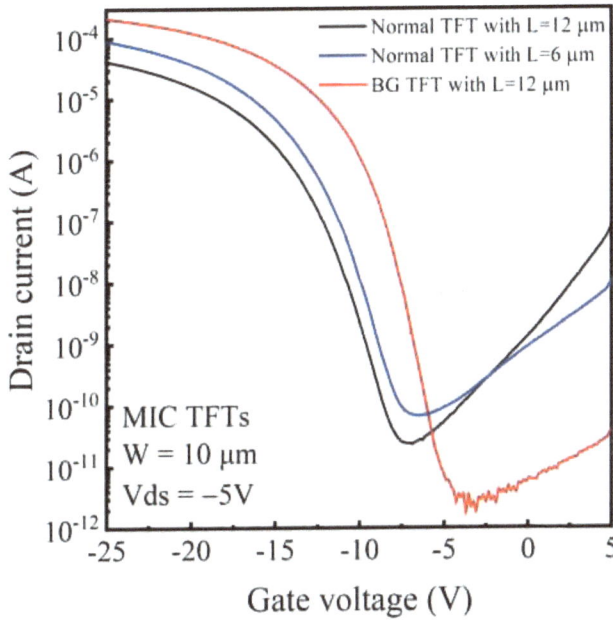

Fig. (8.12). Transfer curves of a normal poly-Si TFT and a BG poly-Si TFT.

8.4.2.1. Hot Carrier

Fig. (**8.13a**) illustrates the transfer curve degradation observed in both standard polysilicon TFTs and BG polysilicon TFTs under identical HC stress conditions. For the standard polysilicon TFTs, classic HC degradation is evident, characterized by a reduction in I_{on}, while SS remains unaffected [13, 14]. The high E_x induced by the applied stress provides carriers with the energy to transform into HCs, resulting in defect creation at grain boundaries and the interface on the drain side, which heightens the trap potential [13]. In Fig. (**8.13a**), with $L = 12$ μm, the BG polysilicon TFT demonstrates notably enhanced HC reliability when contrasted with the conventional polysilicon TFT. The inset of the figure presents the extracted ΔI_{ON} values. After 10^4 seconds of stress, the ΔI_{ON} for the standard polysilicon TFT approaches 100%, while for the BG polysilicon TFT, it remains below 12%. To maintain an equitable comparison, the HC degradation of both types of TFTs is also evaluated at an effective channel length of $L_{eff} = 6$ μm, as illustrated in Fig. (**8.13b**). In line with the findings at $L = 12$ μm, the BG polysilicon TFT again exhibits significantly greater resistance to HC stress compared to the standard polysilicon TFT. Additionally, a reduction in L for standard polysilicon TFTs correlates with an increase in HC degradation, which is linked to a higher E_x under the same stress conditions. The improved HC reliability observed in BG polysilicon TFTs is likely due to the reduced E_x at the drain side, a consequence of the BG lines.

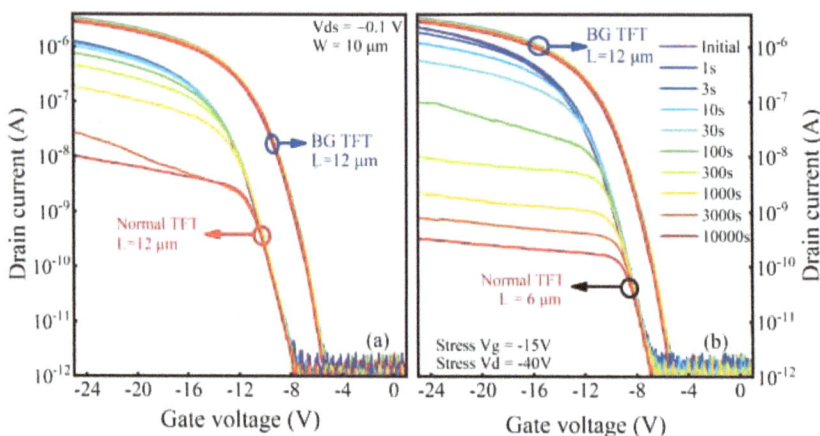

Fig. (8.13). Comparison of transfer curve degradation between normal polysilicon TFT and BG polysilicon TFT under the same HC stress for (a) L = 12 μm and (b) L_{eff} = 6 μm.

Using Silvaco Atlas, simulations were carried out to examine the E_x under HC stress in both conventional polysilicon TFTs and BG polysilicon TFTs, employing a continuous defect polysilicon model, as shown in Fig. (**8.14**). The inset of the figure provides a comparison of the E_x distributions for the two types of TFTs. It is clear that the intensity of E_x at the drain end is notably reduced with the incorporation of the BG architecture. To enhance understanding, E_x data was sampled 5 nm beneath the oxide/channel interface, ranging from the drain to the source region, for both the standard polysilicon TFT (indicated by the red line) and the BG polysilicon TFT (indicated by the blue line). Importantly, the peak value of E_x, which is a key contributor to HC generation, is decreased by the BG structure, thereby enhancing the HC reliability of BG polysilicon TFTs.

The I_{on} current degradation over stress time was investigated at different stress V_g levels while maintaining a constant stress V_d of −40 V for both conventional polysilicon TFTs (represented by red lines) and BG polysilicon TFTs (represented by blue lines), as shown in Fig. (**8.14**). During the ON-state stress condition (with stress $|V_g|$ exceeding stress $|V_{th}|$), the degradation observed in BG polysilicon TFTs is markedly less pronounced compared to that in standard polysilicon TFTs across all tested V_g stress levels, underscoring the enhanced reliability of BG polysilicon TFTs. Additionally, it is observed that for both TFT types, an increase in stress $|V_g|$ correlates with a more significant I_{on} degradation, which is in contrast to the behavior seen in n-type polysilicon TFTs. In n-type devices subjected to a constant stress V_d, HC degradation versus V_g typically exhibits a bell-shaped trend, with the most severe degradation occurring when the stress V_g is close to V_{th}, a phenomenon attributed to the decrease in Ex beyond V_{th} [15]. However, for p-type polysilicon TFTs, raising the stress $|V_g|$ in the negative direction induces negative bias temperature instability (NBTI) at the source, thus intensifying the overall degradation. Under OFF-state stress conditions (where stress $|V_g|$ is less than stress $|V_{th}|$), HC degradation is not detected; instead, I_{on} experiences a slight uptick rather than a decline. Since stress $|V_g|$ is below stress $|V_{th}|$, no holes are induced in the channel, precluding HC generation. The modest rise in I_{on} is primarily attributed to the trapping of negative charges in the gate oxide adjacent to the drain side [16].

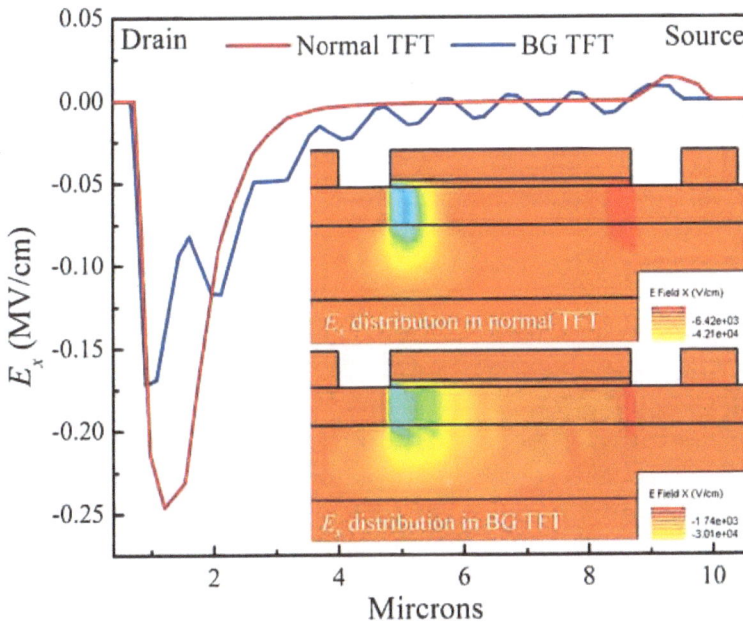

Fig. (8.14). E_x is extracted along the drain side to the source side at 5 nm below the oxide/channel interface under HC stress in normal polysilicon TFT and BG polysilicon TFT. The inset shows the E_x distribution of normal polysilicon TFT and BG polysilicon TFT [16].

8.4.2.2. Self-Heating

The degradation due to self-heating (SH) is affected by Joule heating, which is a consequence of the high-stress power, as well as by the E_{ox} induced by the stress V_g [17-20]. In order to conduct an equitable comparison of the SH degradation between conventional polysilicon TFTs and BG polysilicon TFTs, the stress power density (p), which is calculated by dividing the stress power by the channel area, in conjunction with E_{ox}, is employed for the SH analysis.

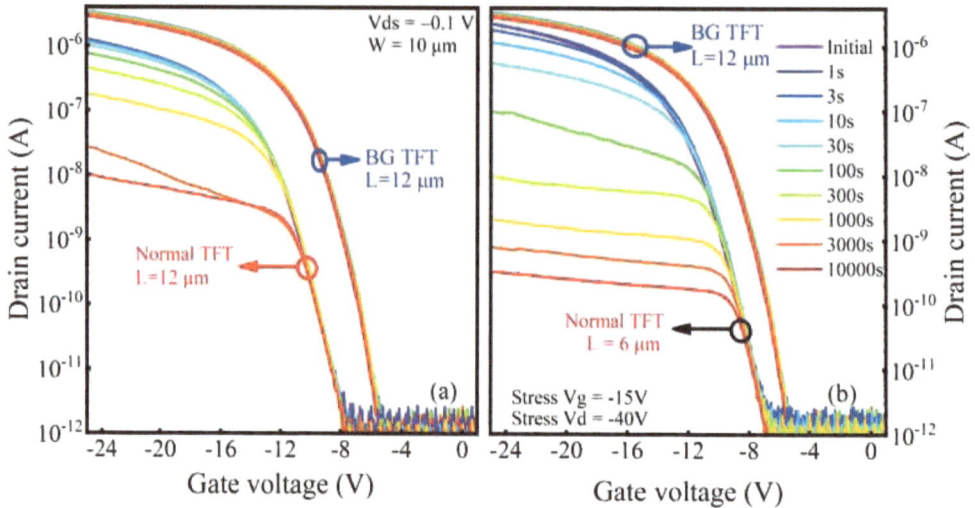

Fig. (8.15). Comparison of transfer curve degradation between normal polysilicon TFT and BG polysilicon TFT at the same p for (a) L = 12 μm and (b) L_{eff} = 6 μm. The inset shows the ion degradation versus stress time for normal polysilicon TFT and BG polysilicon TFT.

Fig. (**8.15**) provides a side-by-side comparison of the transfer curve degradation for both conventional polysilicon TFTs and BG polysilicon TFTs under identical power density conditions, p = 75.8 μW/μm². The analysis reveals that self-heating stress, unlike hot-carrier stress, leads to degradation in both the ON-state and subthreshold regions. The combination of high power and high $|E_{ox}|$ values can cause the distortion or fracturing of robust Si–Si bonds at grain boundaries across the channel [17, 21], as well as the breaking of Si–H bonds at interfaces and grain boundaries [21, 22]. This process renews dangling bonds within the channel layer and generates positive oxide charges, resulting in a decline in I_{on} and a shift in V_{th}. As illustrated in Fig. (**8.15a**) for a channel length of 12 μm, the BG polysilicon TFT exhibits enhanced self-heating reliability compared to the standard polysilicon TFT, given the same p and E_{ox}. A similar trend is observed in Fig. (**8.15b**) for an effective channel length of 6 μm, where the BG polysilicon TFT again demonstrates superior self-heating reliability, even in the face of higher $|E_{ox}|$ values. An effective strategy to combat self-heating effects involves the rapid dissipation of heat [23]. In BG polysilicon TFTs, the highly doped BG lines within the channel have low electrical resistance, which minimizes Joule heating and significantly improves heat diffusion along the L direction [24], thus mitigating self-heating degradation. Moreover, since some self-heating degradation occurs at grain boundaries, the reduced GB N_t in unstressed BG polysilicon TFTs may also play a role in lowering self-heating degradation. Furthermore, it is noted that in standard polysilicon TFTs, as clearly depicted in the inset, an increase in L leads to more pronounced self-heating degradation at the same p, which can be attributed in part to geometric effects [22] and in part to the higher $|E_{ox}|$ experienced by the standard polysilicon TFT with $L = 12$ µm.

The dependence of I_{on} degradation on p at various E_{ox} levels was also analyzed for both standard and BG polysilicon TFTs. As expected, a higher p combined with greater $|E_{ox}|$ leads to more pronounced SH degradation in both types of devices. When analyzing the impact of varying p while maintaining a constant E_{ox} of 4.6 MV/cm, it is observed that both standard and BG polysilicon TFTs experience enhanced SH degradation as p increases. However, the slope of the SH degradation curve for different p values at a constant $|E_{ox}|$ is less steep for both TFT types compared to the degradation slope seen with high p and high $|E_{ox}|$. This suggests that the combined effect of p and $|E_{ox}|$ is more detrimental to SH degradation than either factor alone. Consistent with the findings in Fig. (**8.15**), the BG polysilicon TFT shows better SH reliability under the same p and E_{ox} conditions, and, as expected, longer L results in greater SH degradation.

8.4.2.3. Negative Bias Temperature Instability

Fig. (**8.16**) displays the dependence of V_{mg} on L before NBTI stress and the V_{th} shift after NBTI stress in standard polysilicon TFTs. For V_{mg} (black line), $|V_{mg}|$ remains constant when L is greater than approximately 5 µm but begins to decrease as L drops below this threshold. Upon 10^4 seconds of stress with V_g set at -40 V, the V_{th} shift (indicated by the blue line) is constant for L greater than approximately 5 µm, yet it rises as L decreases when L is less than approximately 5 µm, mirroring the trend of L-dependent V_{mg}. The intensified NBTI observed for L less than approximately 5 µm can be linked to the rise in $|E_{ox}|$ due to the decrease in $|V_{mg}|$. Additionally, NBTI degradation is more pronounced at the channel edges [25]. Therefore, devices with shorter L are likely more vulnerable to damage near the source/drain regions, partially explaining the L-dependent NBTI degradation in short L TFTs. To counteract the degradation caused by variations in V_{mg} and channel edge effects in both standard and BG polysilicon TFTs, a consistent E_{ox} induced by the stress V_g is preserved, and TFTs with larger L ($L > \sim 5$ µm) are selected.

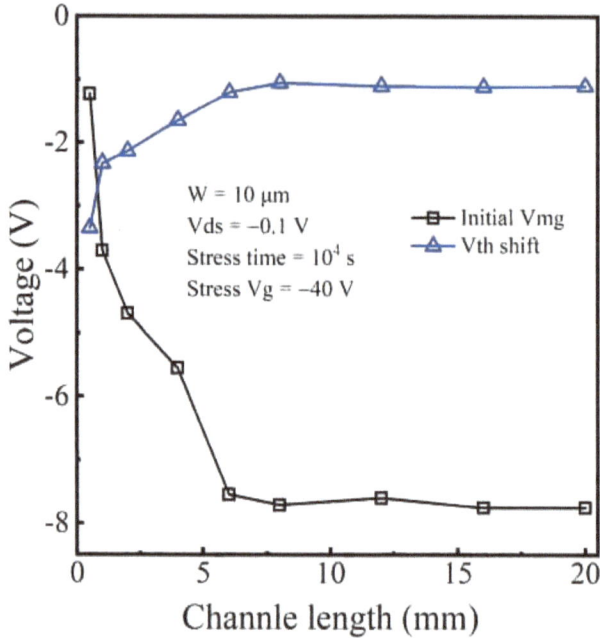

Fig. (8.16). In normal polysilicon TFTs, V_{mg} depends on L before NBTI stress and V_{th} offset depends on L after NBTI stress.

Fig. (**8.17**) illustrates the contrast in transfer curve degradation between conventional polysilicon TFTs and BG polysilicon TFTs under NBTI stress. The figure and its right inset clearly display typical NBTI degradation traits, such as a negative V_{th} shift and a degradation slope of 0.23, as reported in several studies [26, 27]. The main drivers of NBTI degradation are the emergence of trap states at interfaces and grain boundaries, along with the generation of fixed oxide charges, as detailed in some studies [26, 27]. Fig. (**8.17a**) depicts the transfer curve degradation for both standard and BG polysilicon TFTs with a channel length of L = 12 μm, where the V_{th} shift in the BG polysilicon TFT is observed to be less pronounced than in the standard TFT, indicating enhanced NBTI reliability. A similar trend is observed in Fig. (**8.17b**) for an effective channel length of L_{eff} = 6 μm, with the BG polysilicon TFT demonstrating superior NBTI reliability once more. The formation of trap states and fixed charges underlying NBTI degradation is attributed to the depassivation of weak Si-H bonds at interfaces and grain boundaries, as discussed in these studies [26, 27]. The selective doping of the active channel with boron leads to the localization of hydrogen near boron atoms in the polysilicon due to the creation of B-H bonds [28], which are more stable than Si-H bonds. As a result, a reduced fraction of hydrogen bonds dissociate under the same E_{ox}, contributing to improved NBTI reliability. Furthermore, the reduced GB N_t in non-stressed BG polysilicon TFTs may also contribute to better NBTI performance. Additionally, for standard polysilicon TFTs, NBTI degradation does not vary with L when L exceeds 5 µm, as depicted in the insets of Fig. (**8.17**), in alignment with the findings presented earlier in Fig. (**8.16**).

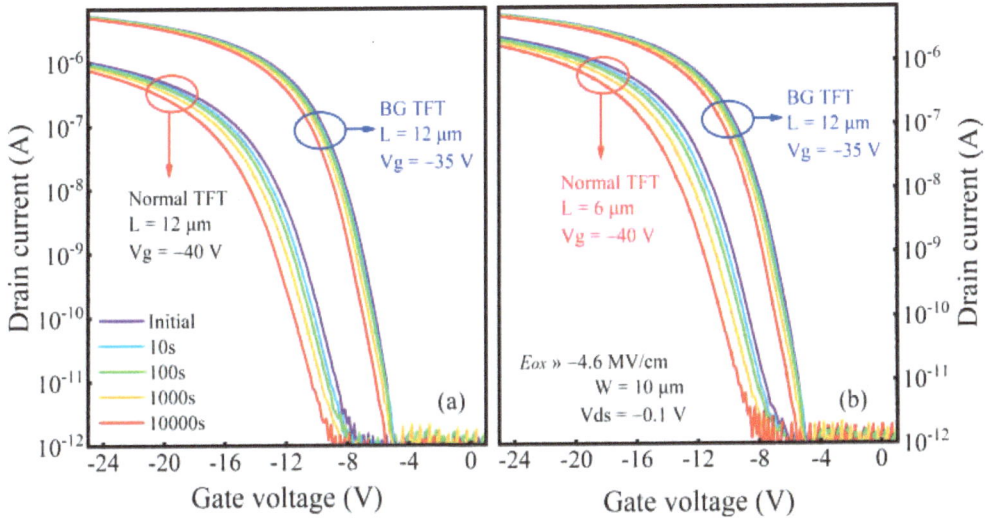

Fig. (8.17). Comparison of transfer curve degradation between normal polysilicon TFT and BG polysilicon TFT under NBTI stress for (a) L = 12 µm and (b) L_{eff} = 6 µm. The left and right insets show the stress time-dependent I_{on} degradation and V_{th} shift for normal\BG polysilicon TFTs, respectively.

NBTI degradation under various E_{ox} levels for both standard polysilicon TFTs and BG polysilicon TFTs is also investigated, as depicted in Fig. (**8.18**). The degradation slope is approximately 0.23, with higher $|E_{ox}|$ leading to increased NBTI degradation. Notably, BG TFTs demonstrate superior NBTI reliability compared to standard TFTs at the same $|E_{ox}|$. Additionally, it is important to mention that BG polysilicon TFTs show enhanced NBTI reliability even when subjected to the same stress V_g.

8.4.3. Suppression of Alternating Current Degradation

Recently, the alternating current (AC) degradation and AC stress [29] with the "driving" stress [30], OFF-state stress [31], forward synchronized stress (FSS) [32], reversely synchronized stress (RSS) [33] and AC stress with short pulse width duration were investigated to simulate the operating conditions of TFTs in pixel circuits in AM displays, which can almost simulate the working conditions of switching TFTs. The BG structure can not only suppress the degradation under direct current (DC) stress, but it can also improve the reliability of poly-Si TFT under the AC stress mentioned in the above contexts.

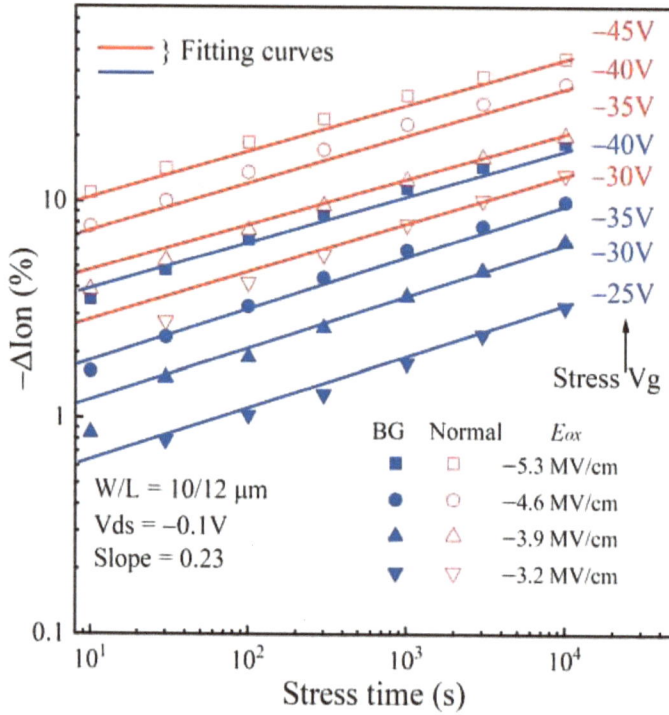

Fig. (8.18). For normal polysilicon TFTs and BG polysilicon TFTs, the I_{on} degradation depends on the stress time at different E_{ox}.

8.4.3.1. Normal Alternating-Current Stress

The degradation of transfer curves for both normal TFTs and BG TFTs under static and dynamic gate stresses is illustrated in Fig. (**8.19**). In the context of TFT technology, a discernible advantage of BG TFTs over their normal counterparts is evident from their superior electrical performance characteristics. Specifically, unstressed BG TFTs exhibit a lower V_{th}, steeper SS, and an enhanced I_{on}/OFF-current ratio, indicative of their superiority. This remarkable performance can be attributed to the strategic doping of BG lines within the active channel region, which effectively harnesses the grain size effect, mitigates short channel effects, and mitigates multijunction effects, thereby yielding exceptional electrical properties in BG TFTs [34, 35].

Fig. (**8.19b**) presents a comparative analysis of the temporal evolution for both normal TFTs and BG TFTs subjected to a static gate stress of -20 V, with the source and drain electrodes grounded. Notably, both types of TFTs display minimal degradation under these conditions, as evidenced by the Ion degradation of merely -0.42% and -0.44%, respectively, after 10^4 seconds of stress application (Fig. **8.19c**). This observation underscores the resilience of both TFT variants against static gate stress at -20 V.

However, when subjected to dynamic gate stress with Vgh = -20 V, as depicted in Fig. (**8.19a**), a stark contrast emerges. The I_{on} of normal TFTs undergoes a pronounced decline with increasing stress duration, indicating significant degradation. In stark contrast, the transfer curve of BG TFTs remains virtually unchanged, demonstrating their remarkable resistance to degradation under identical dynamic gate stress conditions. This finding underscores the pivotal role of the BG structure in enhancing the reliability and durability of poly-Si TFTs in dynamic operating environments.

Upon a closer inspection of device degradation, it becomes evident that normal TFTs and BG TFTs exhibit a distinct two-phase degradation pattern, as illustrated in Fig. (**8.19c**). This two-stage behavior is characterized by an initial increase in the I_{on} value, followed by a subsequent decrease. Such degradation trends have been previously documented in the literature [36], indicating a recurring phenomenon.

The first stage of degradation can be attributed to electron trapping or injection within the gate oxide layer, which temporarily enhances the channel conductivity, leading to an increase in Ion. Conversely, the second stage of degradation is likely governed by the dynamic HC effect, which progressively degrades the device's performance, resulting in a decrease in I_{on}.

When comparing the degradation profiles, BG TFTs display a more pronounced I_{on} increase during the first stage, suggesting enhanced electron trapping/injection dynamics. However, in the second stage, these devices exhibit a significantly smaller Ion decrease, indicating improved resilience against the dynamic HC effect. Furthermore, the inflection point marking the transition between the two stages occurs earlier in normal TFTs compared to BG TFTs, underscoring the different degradation mechanisms and timescales at play in these two types of devices.

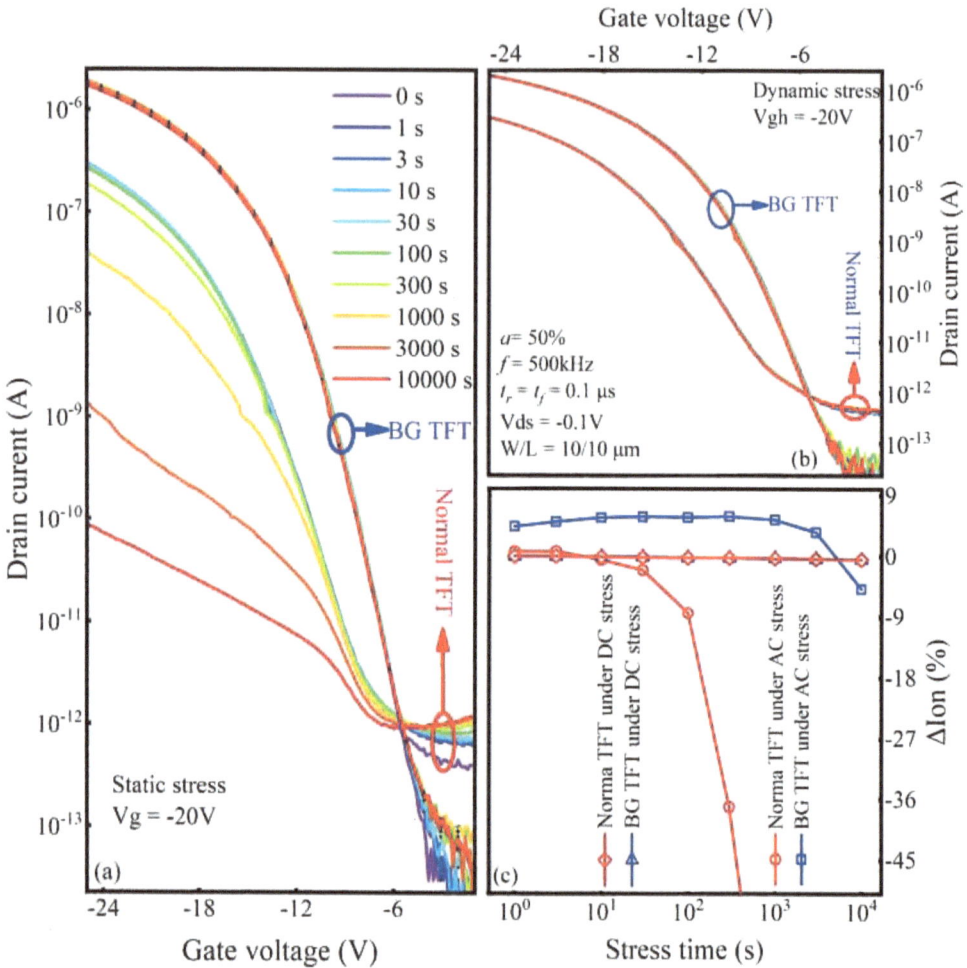

Fig. (8.19). Time evolution of transfer characteristics for normal & BG TFTs under (a) dynamic & (b) static gate stress. (c) I_{on} degradation vs. stress time for both types under dynamic & static gate stress.

To elucidate the paramount parameter within dynamic gate stress conditions that governs device degradation, a series of stress tests were conducted, each tailored with varying αs, frequencies (f_s), rise times (t_r), fall times (t_f), and gate-to-high-state voltages (V_{ghs}). Initially, the focus was on assessing device degradation with differing αs values, specifically in normal TFTs and BG TFTs, as depicted in Fig. (**8.20**). The test conditions were standardized at $V_{gh} = -20$ V, a frequency of 30 kHz, and rise/fall times of 0.1 microseconds.

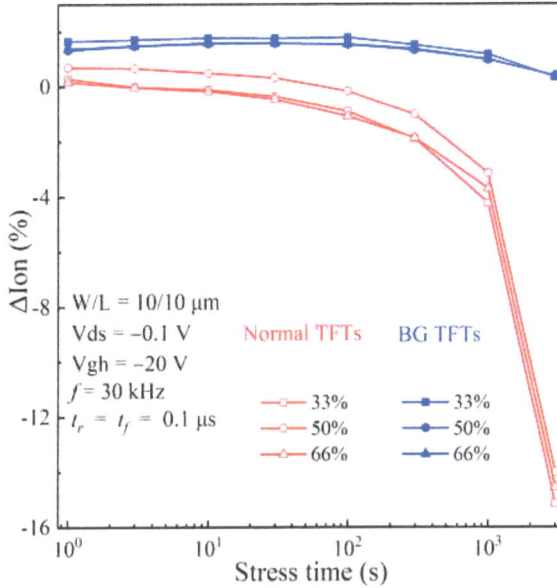

Fig. (8.20). ΔI_{on} versus stress time for various αs.

The results reveal a striking overlap of degradation curves for both normal and BG TFTs subjected to dynamic gate stresses with varying αs. This phenomenon indicates that the device degradation is independent of the duration time of the gate-to-high voltage (t_{gh}), suggesting the absence of any significant static effects under $V_{gh} = -20$ V. This observation aligns with the findings presented in Fig. (**8.19a**).

Further analysis under identical dynamic gate stress conditions underscores the superior reliability of BG TFTs. Specifically, these devices demonstrate a more pronounced Ion increase during the initial stage of degradation, coupled with a significantly smaller ΔI_{on} in the subsequent stage. This behavior highlights the resilience of BG TFTs against degradation mechanisms, particularly under dynamic stress conditions.

Subsequently, the investigation delves into degradation under dynamic stress conditions, with a particular emphasis on varying frequencies (f_s), as presented in Fig. (**8.21**). The frequency range spans from 50 Hz to 500 kHz while maintaining a constant gate-to-high voltage (V_{gh}) of -20 V, a duty cycle (α) of 50%, and rise/fall times (t_r, t_f) of 0.1 microseconds.

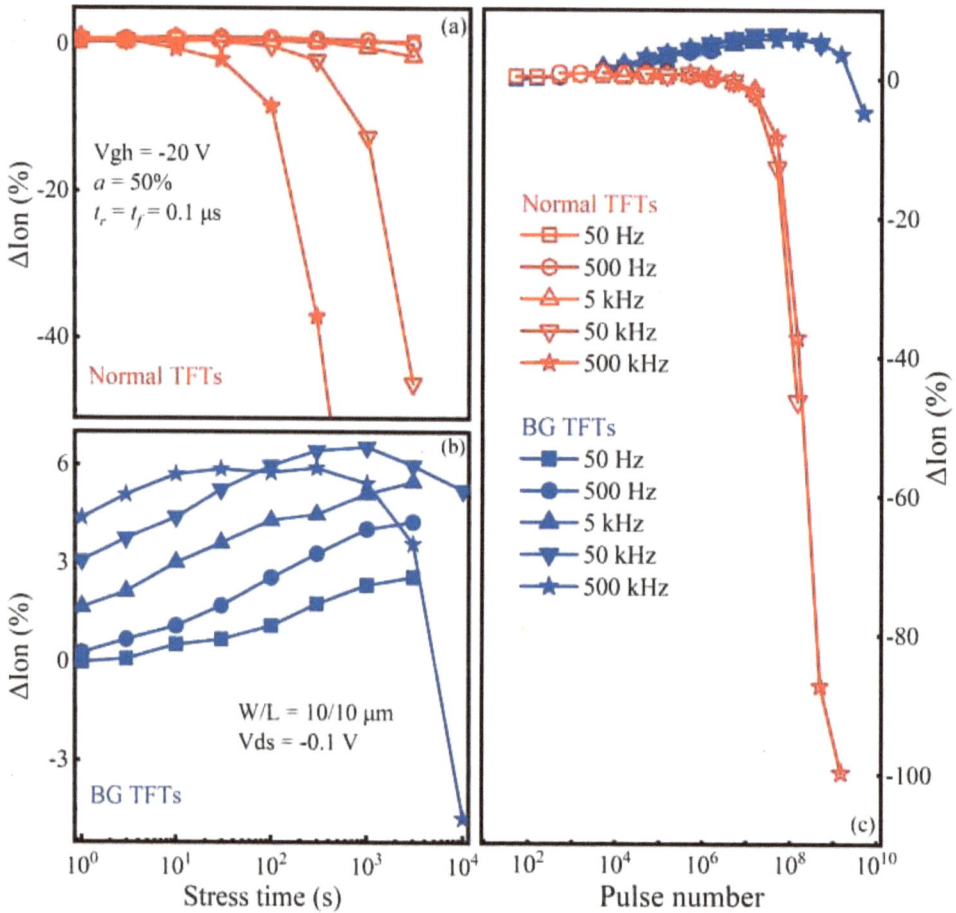

Fig. (8.21). ΔI_{on} with various *f*s in (a) normal and (b) BG TFTs. (c) ΔI_{on} dependent on pulse number with various *f*s.

Fig. (**8.21a**) illustrates the I_{on} degradation as a function of stress time for normal TFTs subjected to dynamic stresses with different f_s values. A notable trend emerges, where the inflection point marking the transition between degradation stages occurs earlier for higher frequencies. Additionally, the degradation in the second stage intensifies with increasing frequency.

Analogous degradation behaviors are observed in BG TFTs, as depicted in Fig. (**8.21b**). Here, dynamic gate stress with higher frequencies results in a more pronounced Ion increase during the initial stage and a correspondingly larger Ion decrease in the subsequent stage. The inflection point also shifts earlier with increasing frequency. Notably, under identical dynamic gate stress conditions, BG

TFTs exhibit a larger Ion increase in the first stage and a significantly smaller Ion decrease in the second stage compared to normal TFTs. Furthermore, the inflection point in BG TFTs appears later.

To gain further insights, the I_{on} degradation curves are plotted against the pulse number, as shown in Fig. (**8.21c**). This analysis reveals that, regardless of the frequency (f_s), the degradation curves for both normal and BG TFTs follow a consistent trend in both degradation stages. This finding underscores the correlation between device degradation and the pulse transition edges, suggesting that the degradation mechanisms are intimately linked to the dynamic nature of the gate stress pulses.

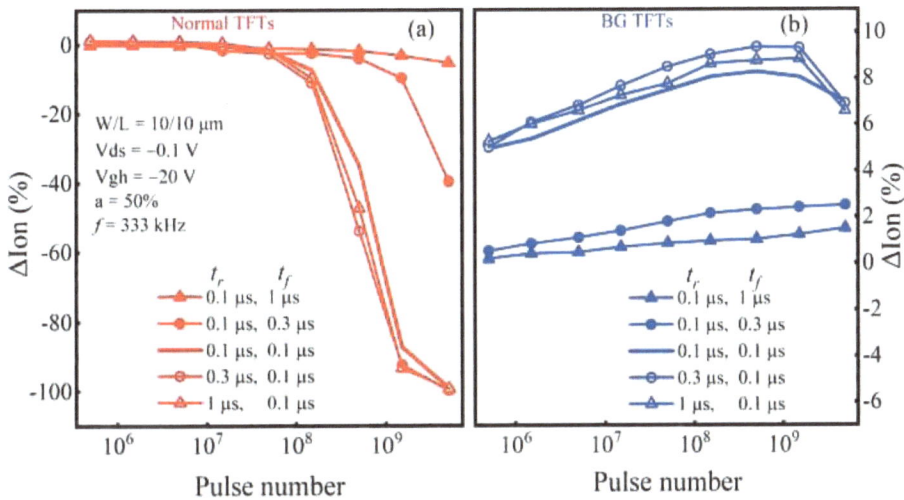

Fig. (8.22). ΔI_{on} with various t_r and t_f in (a) normal and (b) BG TFTs.

Subsequently, the investigation shifts focus to the impact of pulse transition edges on device degradation. Fig. (**8.22**) presents a comprehensive analysis of I_{on} degradation as a function of pulse number, with rise time (t_r) and falling time (t_f), for both normal TFTs (Fig. **8.22a**) and BG TFTs (Fig. **8.22b**). Throughout this analysis, the gate-to-high voltage (V_{gh}) is maintained at −20 V, the duty cycle (α) is set to 50%, and the frequency (f) is fixed at 333 kHz.

For both normal and BG TFTs, the I_{on} degradation curves exhibit a remarkable overlap when subjected to dynamic gate stresses with varying tr values but a fixed tf. This observation underscores the independence of degradation from the t_r, suggesting that the t_f plays a more pivotal role in the degradation process.

In the case of normal TFTs (Fig. **8.22a**), a shorter t_f leads to a more pronounced ΔI_{on} in the second stage of degradation, with the inflection point occurring earlier. Conversely, in BG TFTs (Fig. **8.22b**), a shorter t_f results in an enhanced I_{on} increase during the initial stage, while the second-stage degradation is notably absent within the 10^4-second stress duration.

These conclusively demonstrate that the dynamic-gate-stress-induced degradation in both stages is primarily associated with the t_f rather than the t_r. Furthermore, BG TFTs exhibit significantly superior reliability under identical conditions, reinforcing the findings presented earlier.

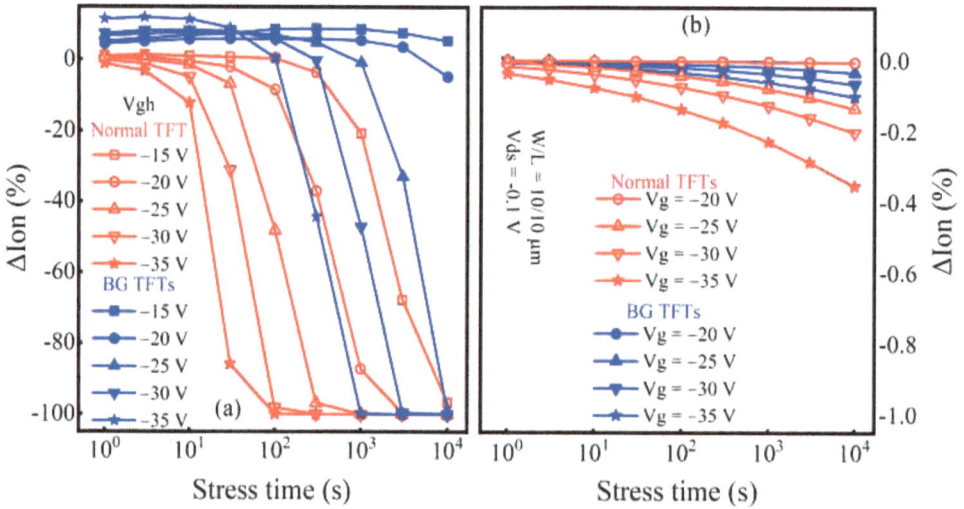

Fig. (8.23). ΔI_{on} degradation (a) with various V_{gh}s and (b) under static gate stresses.

In conclusion, Fig. (**8.23a**) showcases the degradation characteristics of normal TFTs (depicted in red) and BG TFTs (depicted in blue) under dynamic stresses featuring varied V_{gh}s. The conditions are set at $\alpha = 50\%$, $f = 500$ kHz, and $t_r = t_f = 0.1$ μs. For normal TFTs, a higher absolute value of V_{gh} results in an earlier occurrence of the turnaround point and an intensified I_{on} degradation during the second stage. Conversely, BG TFTs exhibit similar degradation trends but with a delayed turnaround point and significantly mitigated degradation in the second stage. Notably, when $|V_{gh}|$ exceeds 20 V, the second-stage degradation intensifies due to the involvement of the static NBTI effect.

Furthermore, Fig. (**8.23b**) illustrates the relationship between Ion degradation and stress time under various static gate stresses for both TFT types. It is evident that for $|V_g|$ values greater than 20 V, both TFTs undergo static NBTI degradation, with

BG TFTs displaying superior resistance to this phenomenon, attributed primarily to the formation of boron–hydrogen bonds at the interface/grain boundaries [13]. This indicates that BG TFTs possess more reliable NBTI compared to their normal counterparts.

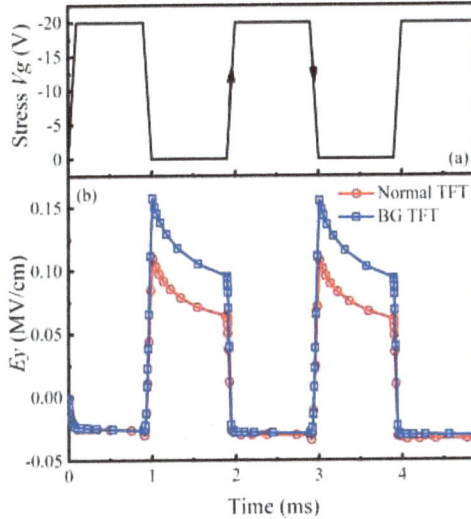

Fig. (8.24). (a) Dynamic gate pulse applied to the gate. (b) Time-dependent E_y at channel/drain edge near gate oxide.

To delve into the fundamental mechanisms behind dynamic gate-stress-induced degradation in normal TFTs and BG TFTs, a simulation was conducted using Silvaco ATLAS. The initial ΔI_{on} observed in the first stage is attributed to electron trapping/injection, facilitated by the presence of an electric field. To visualize this process, a simulation was first performed to analyze the transient electric field E_y dependent on pulse time at the channel/drain interface in both TFT types, as depicted in Fig. (**8.24**). Here, the positive direction is from the gate electrode towards the substrate. Notably, when the gate pulse transitions from -20 to 0 V (during t_f), a positive E_y is induced in both TFT types. This positive E_y aids in electron trapping/injection into the gate oxide, thereby contributing to the I_{on} increase observed in the first stage. Additionally, the E_y in BG TFTs is observed to be stronger than that in normal TFTs, providing an explanation for the more pronounced ΔI_{on} increase in BG TFTs during this initial stage.

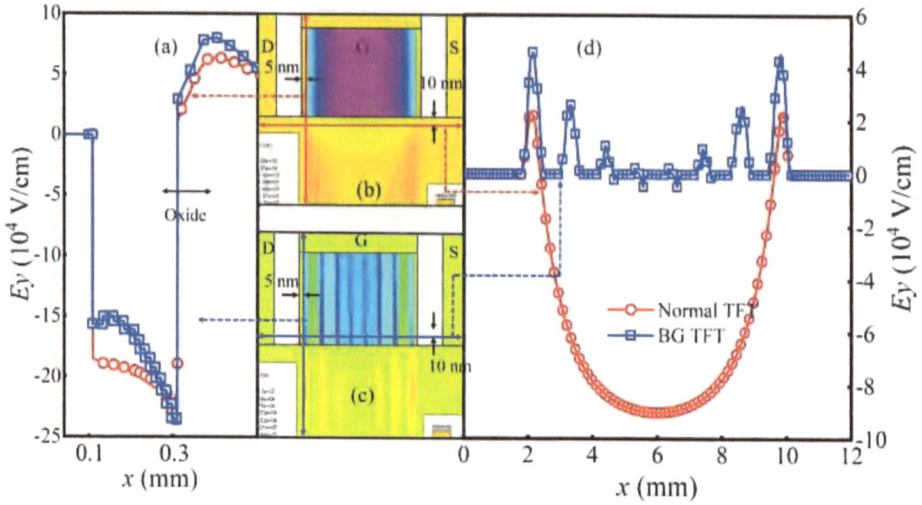

Fig. (8.25). (a) E_y near gate oxide edge at t_f end in normal and BG TFTs. (b-c) Ey dist. in normal and BG TFTs. (d) E_y below oxide/channel int. along drain-to-source at t_f end in both.

To gain a more comprehensive and precise understanding of electric field (E_y) distributions in transistor architectures, simulations were conducted for both normal TFTs and BG TFTs at the end of t_f. The results are presented in Figs. (**8.25b**) and 8.25c, respectively. The analysis reveals an enhancement in E_y intensity within the channel region proximal to the gate oxide in the BG configuration. For clarity, a detailed comparison of E_y profiles at 5 nm from the gate oxide edge, spanning from the gate electrode to the channel, is extracted and displayed in Fig. (**8.25a**) for both types of TFTs. Specifically, the red line represents the normal TFT, while the blue line corresponds to the BG TFT. Notably, a positive E_y is induced in the channel near the source/drain regions in both cases. However, the magnitude of E_y in the BG TFT channel exceeds that of the normal TFT at identical positions. Furthermore, E_y values at a depth of 10 nm below the oxide/channel interface, along the direction from the drain to the source, are extracted and plotted in Fig. (**8.25d**). In the normal TFT, E_y exhibits a positive peak near the source/drain edges and gradually diminishes towards the channel center. Conversely, in the BG TFT, E_y behavior diverges significantly. Within the heavily doped BG regions, E_y values approach zero due to the low resistance of the BG lines. In the undoped channel regions, E_y peaks, with the maximum positive E_y also observed near the source/drain sides. Notably, the positive E_y value at equivalent positions within the BG TFT channel is higher than that in the normal TFT.

Based on these simulation outcomes, it can be deduced that in normal TFTs, electron trapping/injection into the gate oxide is predominantly confined to the vicinity of the source/drain regions. In contrast, for BG TFTs, this phenomenon may extend to the undoped regions across the entire channel, albeit with a stronger prevalence near the source/drain sides. This unique behavior explains the observed larger initial increase in I_{on} in BG TFTs, underscoring the influence of the back-gate architecture on electric field distribution and charge transport dynamics.

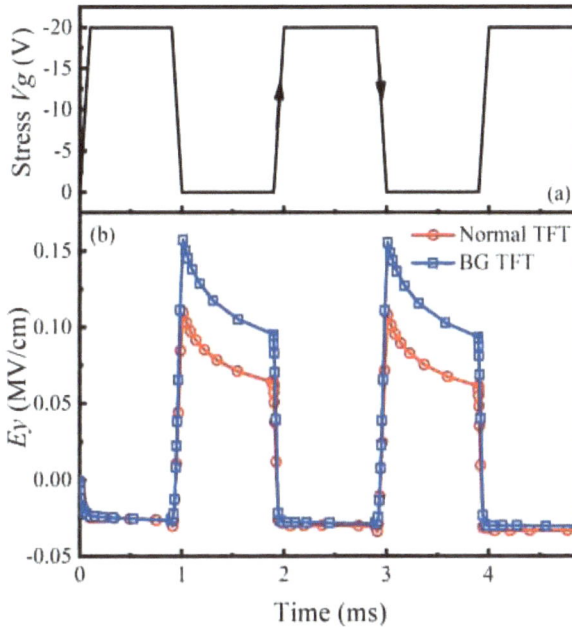

Fig. (8.26). Dependence of E_y on (a) t_r transient and (b) t_f transient at the channel/drain edge.

Fig. (8.26) illustrates the relationship between Ey and the normalized t_r as well as the normalized t_f transient, specifically at the channel/drain interface proximal to the gate oxide in a BG thin-film transistor TFT. For the t_r transient (Fig. 8.26a), E_y exhibits an increase towards the negative direction, precluding electron trapping/injection into the gate oxide. Consequently, the initial stage increase in I_{on} is independent of t_r. Conversely, for the t_f transient, a shorter t_f duration corresponds to a higher positive E_y, leading to a more pronounced I_{on} increase during the initial stage.

During dynamic gate voltage application, transient voltage differentials arise within the channel, at the channel/source edge, and at the channel/drain edge [37, 38]. When the dynamic stress transitions from 0 to −20 V during t_r, the induced transient voltage difference is negative (depicted in Fig. 8.26), inhibiting electron trapping/injection into the gate oxide. Hence, the initial I_{on} increase is not influenced by t_r. Alternatively, during t_f, the transient voltage difference shifts towards positivity. Facilitated by the positive E_y present towards the latter stages of t_f, electrons are trapped/injected into the gate oxide, enhancing I_{on}. A faster t_f translates to a steeper voltage slope, resulting in a higher E_y (Fig. **8.26b**), which in turn augments the I_{on} increase during the initial stage. In normal TFTs, the electric field tends to be diverted towards these regions, giving rise to peak E_y values at the channel/source and channel/drain edges, as evidenced by the red line in Fig. (**8.25d**). Conversely, in BG TFTs, the presence of heavily doped BG lines within the active channel allows E_y to be shunted towards both the source/drain regions and these BG lines, enhancing E_y throughout the entire channel (blue line in Fig. (**8.25d**). Consequently, the initial stage I_{on} increase in BG TFTs surpasses that observed in normal TFTs.

In essence, as the channel length (L) diminishes, the gate's influence over the channel weakens, allowing the source/drain regions to exert a more pronounced control. In the context of this study, the source/drain terminals are grounded during dynamic gate stress experiments, leading to an augmentation in the potential difference along the y-axis proximate to the source/drain edges as L decreases. Consequently, the electric field in the E_y within the channel intensifies with decreasing L. Notably, the BG TFT can be conceptually regarded as an array of short-channel TFTs connected in series, owing to the strategically positioned heavily doped BG lines within the channel. This configuration fosters the induction of a heightened E_y within the undoped regions sandwiched between the doped BG segments, thereby contributing to an enhanced I_{on} increase during the initial stage of operation in BG TFTs. Regarding the I_{on} decrease observed in the second stage, the dynamic HC effect emerges as the primary culprit. The generation of HCs is fundamentally governed by two key factors: the electric field in the E_x and carrier dynamics. To elucidate the mechanisms underlying this second-stage degradation, transient simulations of E_x and J_x were performed, as depicted in Fig. (**8.27**). During the t_r transition, a marked decrease in $|E_x|$ is discernible, with J_x flowing in opposition to E_x, precluding the generation of HCs [13]. Hence, the dynamic HC degradation in the second stage is independent of t_r dynamics. Conversely, during the t_f transition, a substantial E_x is induced, with J_x aligning with E_x. In this scenario, carriers immersed in such intense E_x acquire sufficient energy to transform into HCs. A reduction in t_f further amplifies $|E_x|$ [29], exacerbating the dynamic HC degradation in the second stage. Notably, the $|E_x|$ magnitudes within BG TFTs are significantly lower than those in normal TFTs, accounting for the mitigated second-stage degradation observed in BG TFTs.

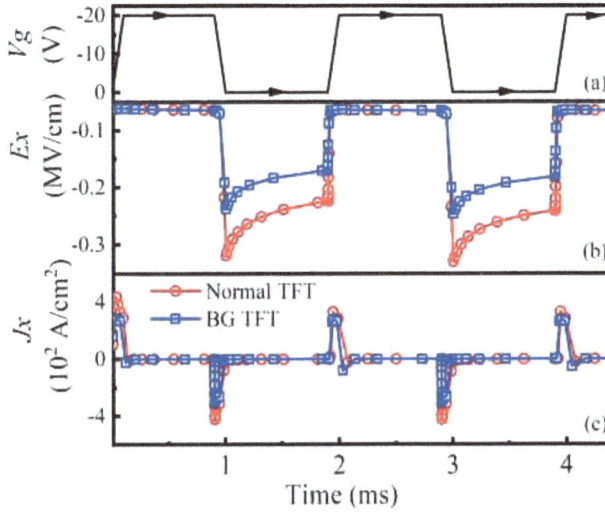

Fig. (8.27). (a) Dynamic gate pulse ($t_r = t_f = 0.1$ μs). The dependence of (b) E_x and (c) J_x on time.

To provide further insight, simulations of E_x distributions at the conclusion of the t_f phase were conducted for normal and BG TFTs, as presented in Fig. (**8.28**). These simulations reveal a pronounced attenuation of E_x intensity in the vicinity of the source/drain edges in BG TFTs, attributable to the mitigating effect of the BG structure. For a clearer comparison, E_x profiles were extracted 10 nm below the oxide/channel interface, traversing from the drain to the source for both TFT types. The peak E_x value in BG TFTs is markedly reduced, underscoring their superior dynamic HC reliability during the second stage of operation.

To elucidate the intricacies of the second-stage degradation mechanism pertaining to HC effects in poly-Si TFTs, we adopt and extend the framework of the nonequilibrium drain junction degradation model [13]. This approach reveals that dynamic stress imparted upon the gate electrode triggers a transient voltage disparity between the channel and p+ source/drain regions [37], effectively positing the existence of two nonequilibrium junctions at the channel's extremities. Specifically, when the dynamic gate stress transitions from 0 to −20 V over a time interval t_r, the source/drain junctions assume a forward bias configuration. Consequently, the depletion zones adjacent to these junctions in the channel undergo contraction through carrier recombination processes, initiating from the depletion region's outer fringes. Within this scenario, the electric field remains subdued, precluding HC generation. Hence, the degradation induced by dynamic stress in the second stage of poly-Si TFTs exhibits independence from t_r.

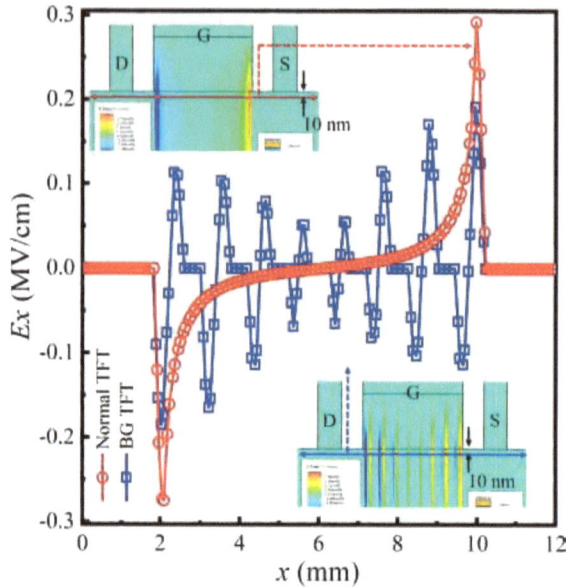

Fig. (8.28). Extracted E_x of t_f in TFTs. Insets: E_x distribution in normal (top) and BG TFT (bottom).

Conversely, during the transition from −20 to 0 V over t_f, the source/drain junctions reverse their bias, necessitating the expansion of the depletion zones within the channel by the emission of trap-associated carriers. Notably, carriers emanating from deep trap states later in the process encounter an E_x across the depleted regions, acquiring sufficient energy to transform into HCs. A shortened t_f period leads to an augmented $|E_x|$ in the channel's vicinity to the source/drain sides, while only traps with relatively shallow energy levels can emit within this abbreviated timeframe. Consequently, a significant portion of deep trap emissions occurs subsequent to the establishment of higher $|E_x|$, intensifying degradation for shorter t_f intervals.

Distinguishingly, normal TFTs possess only two reverse-biased junctions (channel/source and channel/drain) to distribute the voltage drop, whereas BG TFTs exhibit a series of such junctions, as depicted in Fig. **(8.28)**. This configuration in BG devices mitigates dynamic HC degradation in the second stage by reducing E_x at both the channel/source and channel/drain edges. With the degradation mechanisms of both stages now clarified for both normal and BG TFTs, a natural question arises: how does first-stage degradation transition to the second stage? In the initial phase, electron trapping/injection prevails, diminishing local E_x via electric field line dispersion. This initial suppression of dynamic HC degradation stems from the reduced local E_x. However, as stress duration increases, electron trapping/injection and HC generation at the source/drain interfaces and/or grain boundaries near the channel foster the creation of additional traps within the depletion region. This narrowing of source/drain junctions enhances E_x, enabling greater carrier emission from newly formed deep traps during t_f. Consequently, the dynamic HC effect intensifies, culminating in a tipping point where HCs assume primacy in device degradation. For BG TFTs, the combined effects of heightened E_y and diminished E_x, attributable to the heavily doped BG lines within the channel, collaboratively delay the emergence of this transition point.

8.4.3.2. *"Driving" Stress*

The phenomenon of dynamic HC degradation under "driving" stress conditions is predominantly attributed to the E_x at the channel/source and channel/drain interfaces. This E_x is identified as a pivotal factor contributing to the degradation process. A reduction in the magnitude of E_x is postulated to mitigate the severity of dynamic HC degradation. The introduction of a BG architecture in poly-Si TFTs offers a viable solution by virtue of its ability to distribute the E_x across the multiple p-n junctions inherent within the channel region. This sharing mechanism effectively reduces the localized intensity of E_x at the aforementioned junctions. Consequently, the BG TFTs exhibit potential for improved resilience against "driving"-stress-induced HC degradation.

To substantiate this hypothesis, simulations were conducted to analyze the distribution of E_x in both normal poly-Si TFTs and their BG counterparts at the conclusion of the transistor t_r during an alternating current (AC) gate pulse excitation. As depicted in the insets of Fig. (**8.29a**), these simulations were performed under conditions of $t_r = 0.1$ microseconds and a direct current (DC) source-drain voltage (V_s) of 10 volts. The results reveal a notable attenuation in the intensity of E_x at the channel/source and channel/drain interfaces in BG poly-Si TFTs, attributable to the redistribution of the electric field across the multiple reverse-biased junctions within the channel. This finding, as reported in a study [30], underscores the efficacy of the BG structure in mitigating dynamic HC degradation in poly-Si TFTs.

To elucidate further, the extraction of the E_x profile within the region spanning 10 nanometers below the oxide/channel interface, traversing from the source to the drain side, is conducted for both normal poly-Si TFTs (represented by the red line) and BG TFTs (depicted by the blue line). The adoption of a BG architecture within the active channel region leads to a notable reduction in the peak E_x values at both the channel/source and channel/drain interfaces. This observation underscores the enhanced dynamic HC reliability exhibited by BG TFTs.

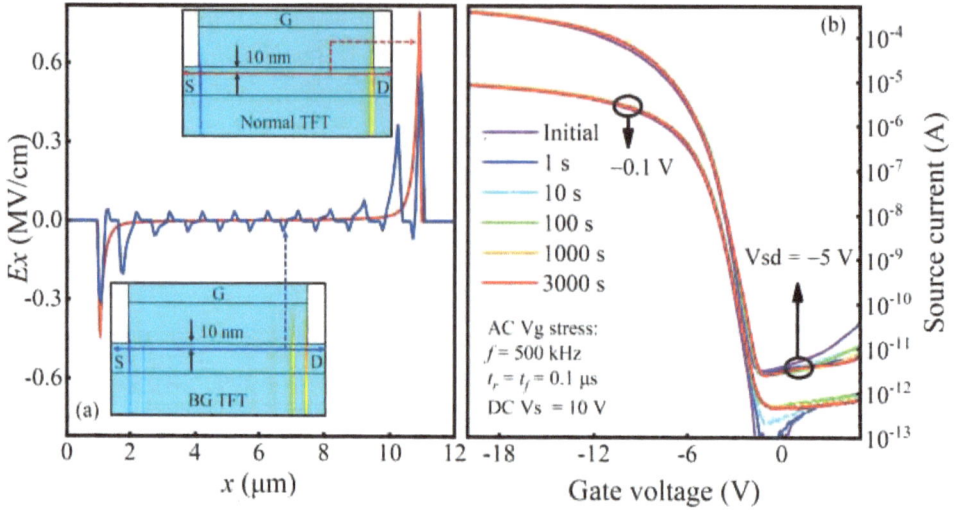

Fig. (8.29). (a) E_x of t_r of an AC gate pulse in TFTs. The insets are the E_x distribution in normal (upper) and BG TFT (lower). (c) Time evolution under AC gate stress [30].

Subsequently, a reliability assessment under identical "driving" stress conditions is performed on BG TFTs, with the results presented in Fig. (**8.29b**). In contrast to normal TFTs, the transfer characteristic curve of BG TFTs remains virtually unchanged post-"driving" stress, signifying superior stability. An analysis of the transfer curve degradation, following 3000 seconds of "driving" stress, reveals that the Ion degradation for normal poly-Si TFTs at $V_{sd} = -0.1$ V and -5 V amounts to -76.7% and -41.0%, respectively. In contrast, BG TFTs exhibit significantly lower I_{on} degradation rates of merely -3.6% and -1.60% under the same conditions. This remarkable resilience stems primarily from the presence of a series of reversed junctions in BG TFTs, which effectively distribute the voltage drop, as illustrated in Fig. (**8.29a**). The extent of HC degradation is intimately tied to the magnitude of E_x [39]. Notably, under AC V_g stress cycling between -14.7 V and 14.7 V (with a direct current source-drain voltage $V_s = 10$ V), BG TFTs exhibit comparable levels of substantial degradation to normal TFTs, where the peak E_x at the end of the t_r in BG devices aligns with that observed in standard devices. This underscores the complex interplay between device architecture, stress conditions, and the resulting electric field distribution, which ultimately governs the reliability performance of these TFTs.

8.4.3.3. OFF-State Stress

Drawing insights from the aforementioned chapter on dynamic HC degradation mechanisms, it becomes evident that the E_x within the channel region, particularly in proximity to the junctions, assumes a pivotal role in exacerbating dynamic HC degradation under off-state stress conditions. A reduction in the magnitude of this electric field is expected to positively impact the mitigation of dynamic HC degradation. The incorporation of a BG architecture in poly-Si TFTs presents a solution by virtue of its capability to distribute the E_x across the multiple reverse-biased junctions residing within the channel [11]. This feature of the BG structure potentially enhances the resilience of BG poly-Si TFTs against off-state stress-induced dynamic HC degradation. To further elucidate this point, simulations were conducted to compare the distribution of E_x at the t_f under a positive DC V_g of 15 volts for normal TFTs and BG counterparts. The results, presented in Figs (**8.30b** and **8.30c**), offer a clear visualization of the differences in E_x distribution. In the case of normal poly-Si TFTs (Fig. **8.30b**), the E_x is predominantly concentrated near the drain and source junctions, in accordance with previous findings [31]. Conversely, in BG TFTs, the distribution of E_x is more evenly spread due to the sharing effect across the multiple reverse-biased junctions within the channel, suggesting a potential improvement in the resistance to off-state stress-induced dynamic HC degradation.

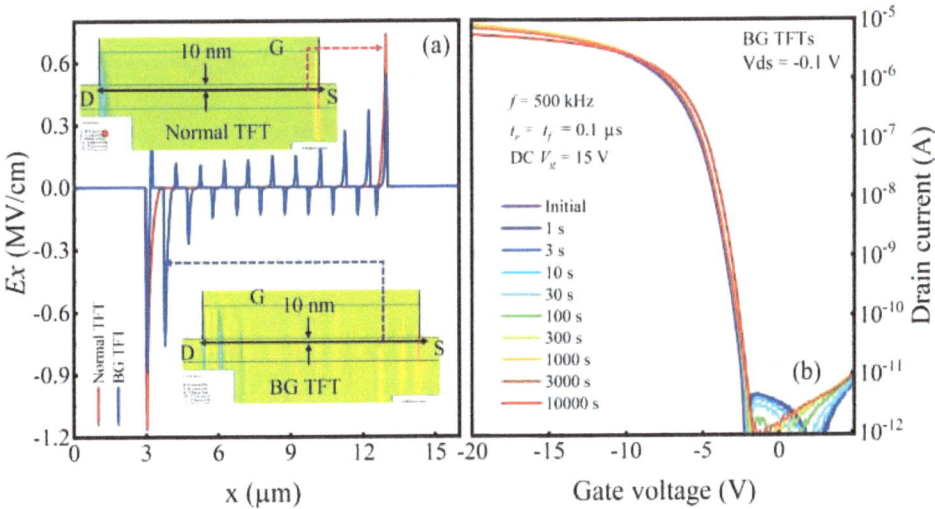

Fig. (8.30). (a) Extracted E_x below the oxide/channel interface along the drain side to the source side in TFTs. (b) Time evolution of BG poly-Si TFTs [31].

In contrast to the behavior observed in normal poly-Si TFTs, the E_x distribution in BG poly-Si TFTs exhibits a periodic pattern spanning the entire channel length, with a notable decrease in E_x intensity towards the channel center (as depicted in Fig. (**8.30c**). Furthermore, a comparative analysis reveals that the intensity of E_x proximate to the junctions is attenuated in BG TFTs compared to their normal counterparts. To clarify, Fig. (**8.30a**) presents the extracted and plotted E_x profiles, measured 10 nanometers below the oxide/channel interface, extending from the drain to the source side, for normal and BG TFTs. The introduction of the BG structure leads to a reduction in the peak E_x values near the junctions, signifying improved dynamic HC reliability under OFF-state stress conditions in BG TFTs.

A reliability assessment under OFF-state stress was subsequently conducted, with the results presented in Fig. (**8.30d**). Unsurprisingly, BG TFTs exhibit significantly more reliable characteristics under identical OFF-state stress conditions compared to normal poly-Si TFTs. An analysis of the transfer curves, following 10,000 seconds of OFF-state stress, reveals that the Ion degradation at a V_{ds} of −0.1 V and a V_{gs} of −20 V is substantially mitigated, improving from −75.5% in normal poly-Si TFTs to −28.8% in BG poly-Si TFTs. This enhanced reliability is primarily attributed to the sharing of the electric field across a series of reversed junctions, as illustrated in Figs. (**8.30a-8.30c**) [31].

8.4.3.4. Forward Synchronized Stress

Upon thorough analysis, it is established that the E_x within the channel region, particularly at the source and drain junctions, assumes a paramount significance in the context of FSS-induced HC degradation. A reduction in the magnitude of this electric field is anticipated to mitigate the severity of FSS-induced HC degradation. Notably, the implementation of a BG structure can effectively diminish the electric field at these critical junctions by virtue of its ability to distribute the voltage drop across multiple reverse-biased junctions within the channel. This feature is hypothesized to contribute favorably towards the reduction of FSS-induced HC degradation.

To substantiate this hypothesis, simulations were performed to compare the electric field distribution at the conclusion of the t_r and t_f of FSS in both normal TFTs and their BG counterparts. The results, depicted in Figs (**8.31a** and **8.31b**), respectively, provide a quantitative assessment of the electric field distribution. Specifically, at the end of the rise time, a notable weakening of the electric field intensity at the source and drain junctions is observed in BG TFTs, as highlighted in the insets of Fig. (**8.31a**). For a more comprehensive understanding, the electric field profiles, 5 nanometers below the oxide/channel interface, were extracted along the path from the drain to the source side for both normal TFTs (represented by the red line) and BG TFTs (represented by the blue line) [32]. This analysis underscores the potential advantages offered by the BG structure in mitigating FSS-induced HC degradation through the reduction of electric field strength at critical junctions.

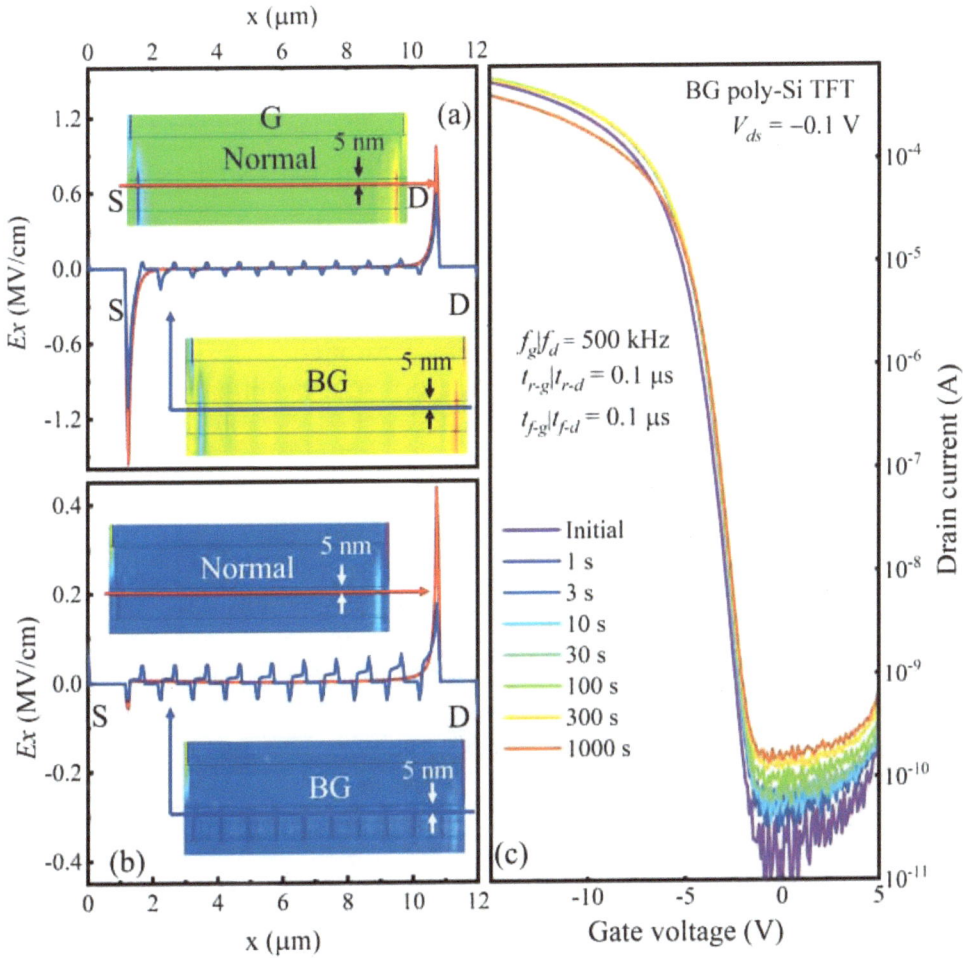

Fig. (8.31). E_x in TFTs (a) at the end of t_r and (b) at the end of t_f. The upper inset is E_x distribution in the normal TFT and the lower inset is in the BG TFT. (c) Time evolution of BG TFTs under FSS [32].

Evidently, the implementation of the BG structure in poly-Si TFTs results in a suppression of the peak E_x within the channel, particularly pronounced at the source/drain regions, thereby indicating an enhanced HC reliability. A comparable phenomenon is discernible at t_f in FSS configurations for both standard and BG TFTs, as evidenced in Fig. (**8.31b**). Notably, the substantial peak of E_x, originally confined to the drain side, is effectively mitigated by the incorporation of the BG structure, further underlining its positive impact on HC reliability.

To substantiate the aforementioned deductions, an identical FSS was applied to BG TFTs, and the subsequent degradation of their transfer characteristics is documented in Fig. (**8.31c**). Consistent with the predictions derived from simulations, the BG TFTs exhibit significantly superior reliability characteristics under the same FSS conditions, in stark contrast to their normal counterparts. Specifically, under an equivalent FSS, the I_{on} of BG TFTs degrades by approximately 25% after 1000 seconds of stress, a marked improvement over the degradation rate of approximately 100% observed in standard poly-Si TFTs.

The remarkable amelioration of FSS-induced HC degradation in BG TFTs is primarily credited to the reduction of E_x achieved through the BG structure's unique design, as elaborated in [32]. This reduction effectively mitigates the deleterious effects of hot carriers, thereby enhancing the overall reliability and longevity of the transistor devices.

8.4.3.5. Reversely Synchronized Stress

Reducing the E_x at the source and drain regions of TFTs is anticipated to significantly enhance HC degradation mitigation. In the case of BG TFTs, the strategically positioned doped lines within the active channel facilitate the sharing of voltage drops across numerous reverse-biased junctions, thereby inhibiting HC degradation. Furthermore, BG TFTs exhibit potential in RSS-induced HC degradation [33].

Figs. (**8.32b** and **8.32c**) illustrate the electric field distributions at the conclusion of the t_r of RSS in normal TFTs and BG TFTs, respectively. A notable weakening of the electric field intensity within the channel, particularly at the source and drain sides, is evident in BG TFTs. As depicted in Fig. (**8.32a**), a marked reduction in the peak electric field values at both the source and drain sides is observed in BG TFTs compared to their normal counterparts. This suggests an improvement in the stability of BG TFTs under RSS conditions.

Subsequent application of the same RSS to BG TFTs reveals a more reliable transfer curve degradation profile, as shown in Fig. (**8.32d**). In comparison to normal TFTs, BG TFTs exhibit significantly more robust characteristics under identical RSS conditions, as anticipated. This substantial improvement can be primarily attributed to the reduction in the electric field achieved through the BG lines, as previously discussed [26]. Additionally, it is noteworthy that upon the removal of RSS, the I_{on} in BG TFTs recovers by approximately 4.51% within 30 seconds of recovery time, subsequently reaching saturation within 10,000 seconds of recovery [33], further underlining their enhanced stability and reliability.

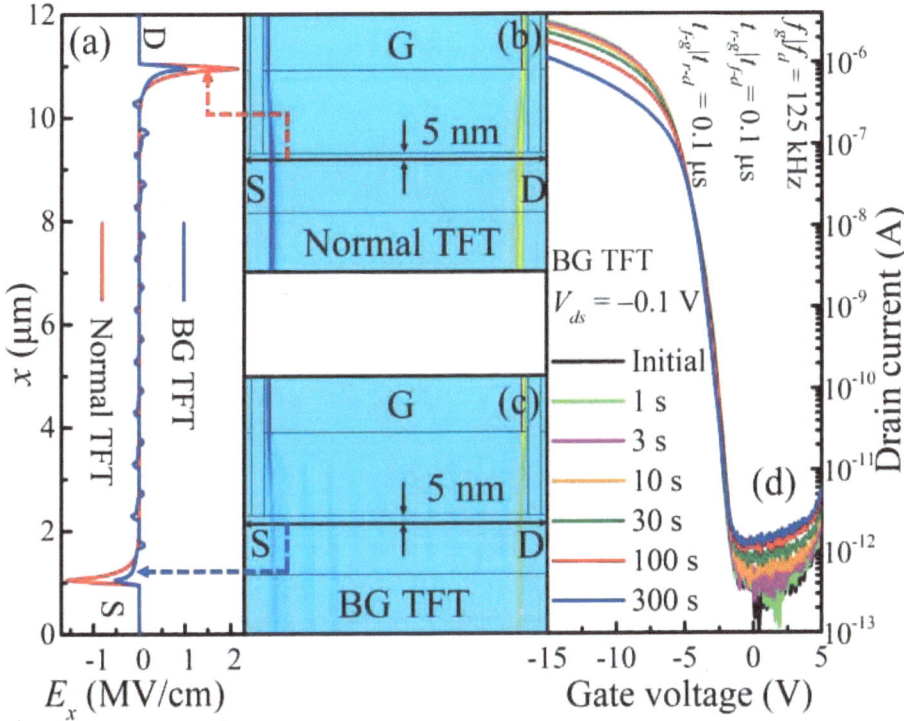

Fig. (8.32). (a) E_x in TFTs at the end of t_r of RSS. E_x distribution in the (b) normal and (c) BG TFT at the end of t_r. (d) Time evolution of transfer characteristics of BG TFTs [33].

8.4.4. Noise Analysis in Bridge-Grain Structure

Fig. (**8.33**) displays the normalized V_d PSD for both BG TFTs and normal TFTs across a range of V_{od}s. The S_{Id}s for both types of TFTs conform to the $1/f^{\gamma}$ noise relationship over different V_{od}s, with mean γ values of 1.02 for normal TFTs and 1.00 for BG TFTs from 2 to -10 V. This conformity suggests adherence to classical $1/f$ noise theory [40]. The normalized PSD diminishes as the absolute V_{od} increases for both BG TFTs and normal TFTs, signifying greater noise fluctuations in the subthreshold region compared to the on state. Notably, BG TFTs show a lower normalized PSD than normal TFTs at equivalent V_{od}, suggesting a reduced level of low-frequency noise (LFN). Additionally, the LFN was investigated for both TFT types, with L ranging from 8 to 20 μm, increasing by 2 μm. The drain current PSD decreases with increasing L for both TFT types, aligning with previous findings [41].

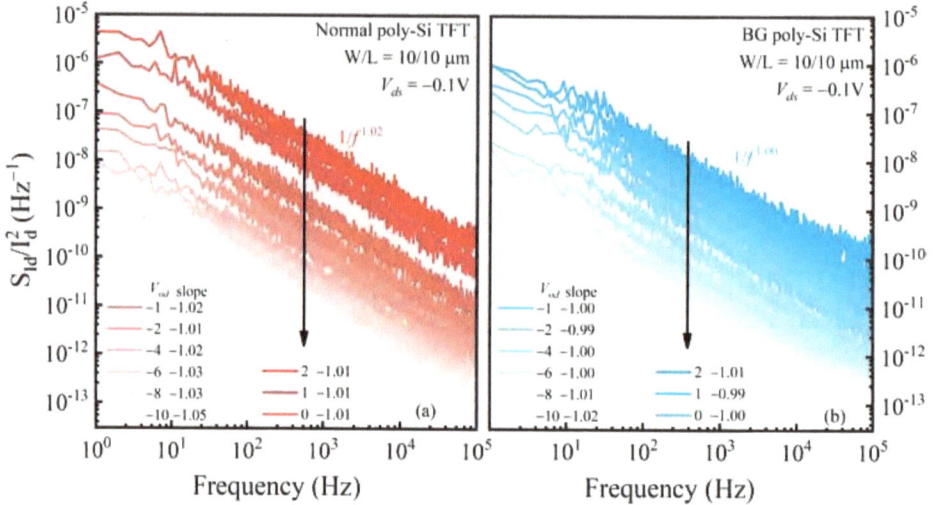

Fig. (8.33). (a) and (b) illustrate the normalized S_{Id} as a function of f for BG TFTs and normal TFTs, respectively, at a range of V_{od}s.

Fig. (**8.34**) shows the $1/f$ noise mechanism by presenting plots of the normalized PSD versus V_d and plots of $(g_m/I_d)^2$ versus drain current for both BG TFTs and normal TFTs. The data for both TFT types are well-described by the ΔN-$\Delta\mu$ model [42], suggesting it as the predominant factor influencing their low-frequency noise characteristics. The normalized PSD of BG TFTs is consistently lower than that of normal TFTs, aligning with the observations in Fig. (**8.33**) and suggesting superior bulk quality and/or interface in BG TFTs. The higher defect density in normal TFTs [34] leads to increased trap-assisted carrier capture and release, which in turn results in a larger drain current PSD compared to BG TFTs.

As detailed in Chapter 3.4, the αc for BG TFTs and normal TFTs are 3.43×10^4 and 4.66×10^4 Vs/C, respectively, with higher αc values indicating increased Coulombic scattering. The S_{vfb} values for BG TFTs and normal TFTs are 2.70×10^{-8} and 6.24×10^{-8} V^2/Hz, respectively. The BG TFTs exhibit reduced Coulombic scattering compared to normal TFTs, a trend that aligns with the observed transfer and output characteristics as well as the derived electrical parameters.

Additionally, the N_T for BG TFTs and normal TFTs is calculated to be 1.84×10^{18} and 8.47×10^{18} cm^{-3}eV^{-1}, respectively, with BG TFTs showing a significantly lower N_T than normal TFTs.

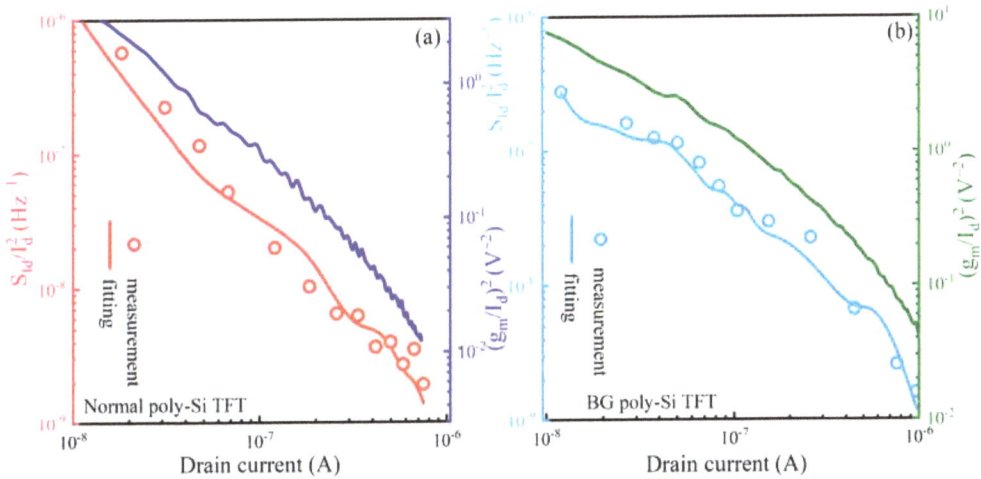

Fig. (8.34). The variation of normalized S_{Id} and $(g_m/I_d)^2$ with respect to drain current for both (a) normal TFTs and (b) BG TFTs.

In contrast to *c*-Si MOSFETs, the LFN in polycrystalline silicon TFTs stems from both detrapping/trapping at the boundary between the GI and the active channel, as well as from GBs within the active channel [43]. The significant variation in N_T between BG TFTs and normal TFTs can likely be ascribed to disparities in interface quality and/or GB trap densities.

The primary distinction between BG TFTs and normal TFTs lies in the additional doping of the polycrystalline silicon active channel by BG lines in the latter. Consequently, the LFN in BG TFTs is predominantly influenced by the undoped channel areas, the heavily boron-doped BG lines, and the interfaces between these regions. In BG TFTs, similar to normal TFTs, the undoped channel regions are the primary contributors to LFN due to fluctuations in carrier concentration and mobility. The heavily boron-doped BG lines in the channel, being highly conductive, minimally contribute to LFN. However, they facilitate carrier transport by providing more conductive pathways, thereby reducing the effective number of barriers and trap sites along the current path, which in turn lowers the LFN.

Furthermore, during fabrication, boron dopants can diffuse into the contact areas between the undoped regions and the boron-doped BG lines. The formation of B-H bonds or B-H-B bonds may lead to hydrogen localization, particularly at the junctions and the channel-insulator interface, as depicted in the inset of Fig. (**8.35**). The B-H bond has a higher dissociation energy than the Si-H bond, suggesting that a lower proportion of hydrogen is depassivated in BG TFTs under similar conditions. This results in a reduced trap density and, consequently, a lower LFN level. It can also be deduced that narrower spacing between BG lines can further decrease LFN in BG TFTs, as the relative area of boron diffusion rises with reduced spacing.

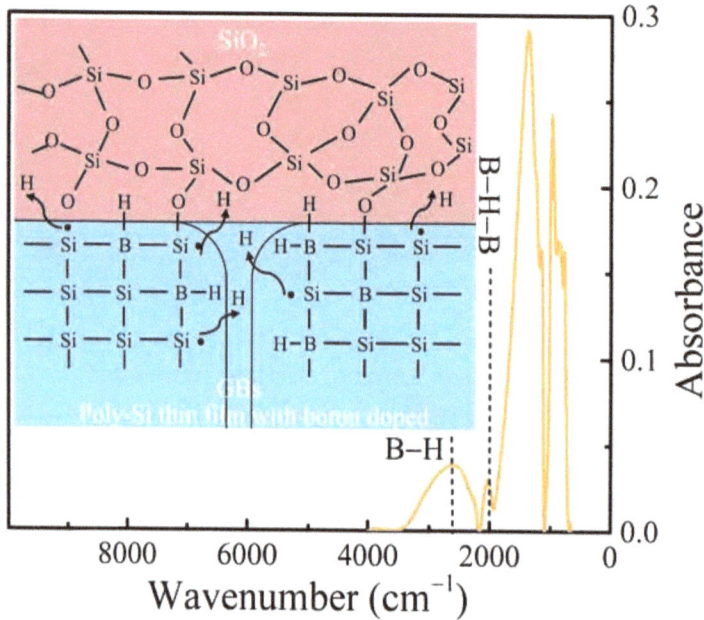

Fig. (8.35). FTIR spectrum of a boron-doped polycrystalline silicon thin film, with an inset showing a cross-sectional atomic-scale diagram of the interface between the boron-doped polycrystalline silicon film and the GI.

To confirm the presence of hydrogen near boron in the structure, an FTIR spectrum analysis was conducted on a boron-doped poly-Si layer, which was processed identically to the steps used in the BG TFT fabrication. Fig. (**8.35**) displays the FTIR spectrum, revealing two absorption bands at approximately 2000 and 2560 cm-1, associated with the B-H-B and B-H bonds, respectively [28]. This observation of hydrogen localization near boron suggests a probable reduction in the *NT* within the BG TFTs.

8.5. OTHER IMPROVEMENT METHOD

8.5.1. Reliability Improvement Method in Metal Oxide Thin-Film Transistor

As previously discussed, the hot-carrier effect significantly impacts the performance of metal oxide (MO) TFTs, regardless of the specific material used.

Consequently, reducing the HC effect in MO TFTs is one of the key issues in reliability research.

HC degradation is attributed to the presence of a large E_x within the active layer, a certain concentration of carriers, and an adequate acceleration distance. The suppression of HC effects in MO TFTs is primarily achieved through two approaches: the first is by altering the device structure to reduce the E_x in the active layer and the second is by introducing other degradation mechanisms during stress application to suppress or counteract the impact of HC degradation (Fig. **8.36**) [44].

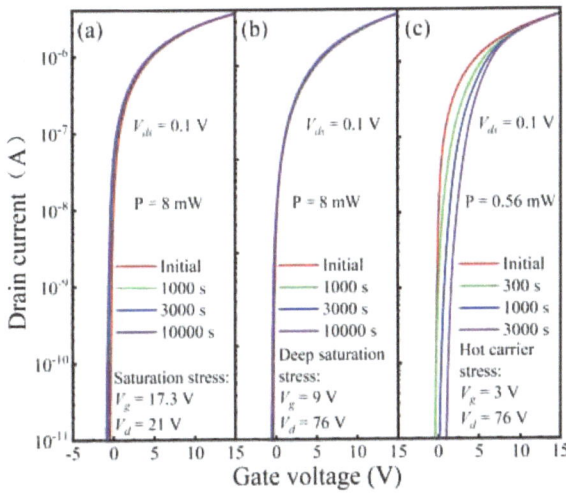

Fig. (8.36). Time evolution of transfer curves of IGZO TFTs under (a) saturation stress of $V_g = 17.3$ V and stress $V_d = 21$ V; (b) deep saturation stress of $V_g = 9$ V and $V_d = 76$ V; and (c) typical HC stress.

Currently, there are few reports on the suppression of HC degradation in MO TFTs through device structural modifications. However, observations of HC degradation in elevated-metal metal oxide (EMMO) IGZO TFTs suggest that this structure can significantly mitigate the impact of HC effects.

Additionally, introducing other degradation mechanisms can help suppress or offset the adverse effects of HC degradation. As mentioned earlier, HC degradation is highly sensitive to temperature; elevated temperatures can suppress HC degradation to a certain extent. This is because, as the temperature increases, lattice vibrations in the active layer of the TFT become more intense. The intense lattice vibrations cause carriers to lose energy through collisions during acceleration, making it difficult for them to become high-energy HCs [45, 46].

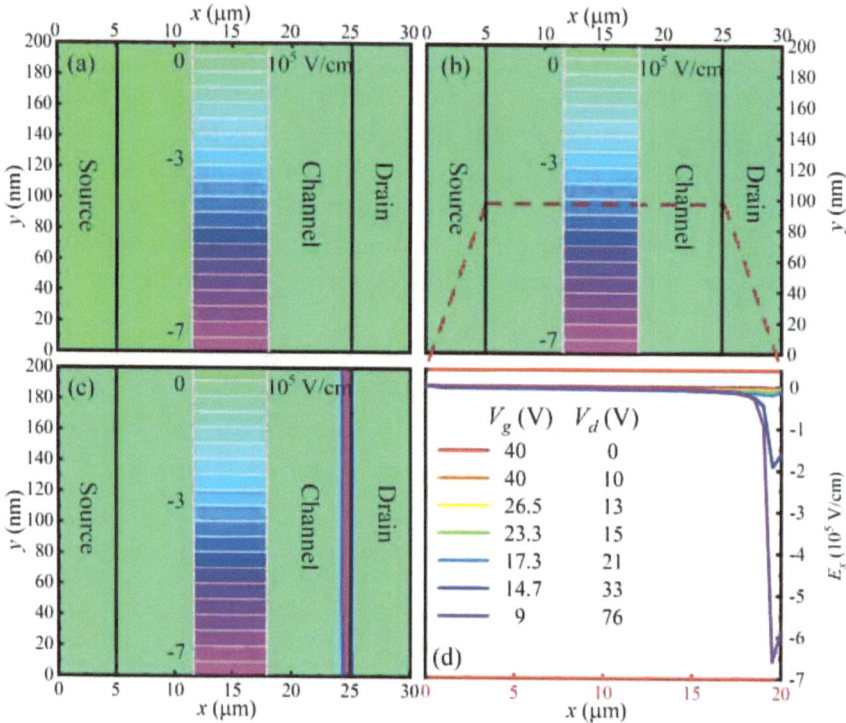

Fig. (8.37). Calculated E_x distribution in the active layer under (a) PBS of $V_g = 40$ V; (b) linear stress of $V_g = 40$ V and $V_d = 10$ V; and (c) saturation stress of $V_g = 9$ V and $V_d = 76$ V. (d) Extracted E_x in the middle of the channel from the source to the drain.

Based on this understanding, SH effects can be utilized to suppress HC degradation in TFTs. As shown in Fig. (**8.37**), when an HC stress of $V_d = 76$ V and $V_g = 3$ V is applied to an IGZO TFT, the transfer characteristics curve shifts positively, and the *SS* increases. At this time, the device's power consumption is 0.56 mW, and the heat generated by SH is insufficient to cause a significant temperature change in the active layer. Subsequently, when a stress of $V_d = 76$ V and $V_g = 9$ V is applied to the IGZO TFT, simulations reveal a significant E_x near the drain. However, the transfer characteristics curve shows little degradation, which is attributed to the device's power consumption of 8 mW at this stress. The SH effect induces a significant temperature change in the active layer, with thermal simulations indicating that the heat is concentrated near the drain, reaching a temperature of 218 °C [44]. The localized high temperature near the drain effectively suppresses HC degradation caused by the large E_x, resulting in minimal degradation of the transfer characteristics curve. The SH effect effectively mitigates the HC effect induced by the large electric field (Fig. **8.38**) [44].

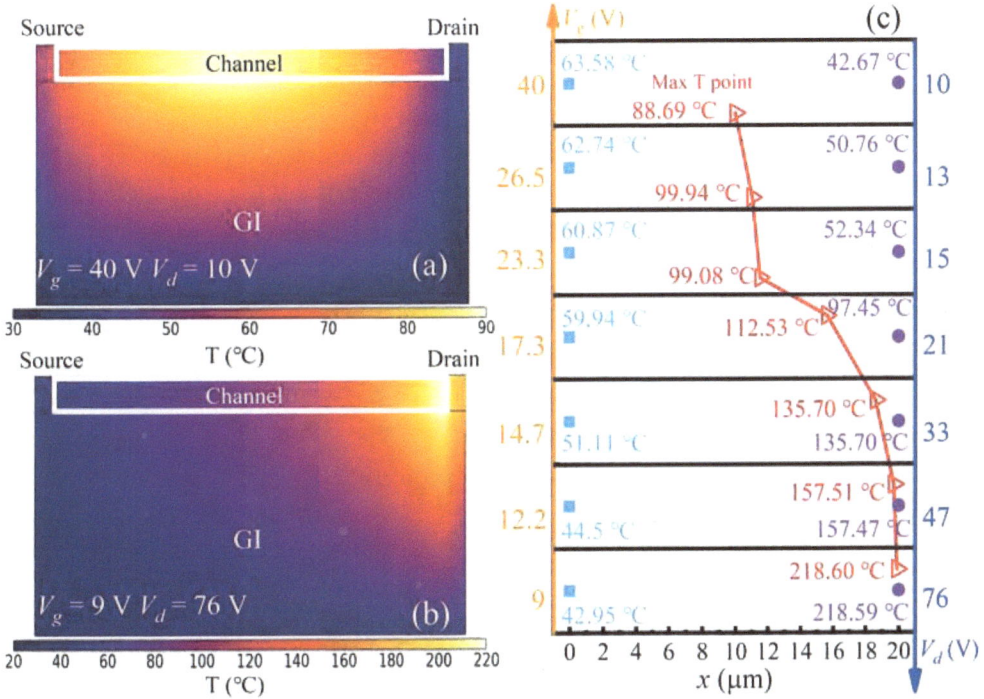

Fig. (8.38). T distribution in IGZO TFTs: (a) under a linear stress of $V_g = 40$ V and $V_d = 10$ V and (b) under a saturation stress of $V_g = 9$ V and $V_d = 76$ V. (c) Statistics of Tmax, TS, and TD under various combinations of stress Vg and stress V_d (red triangles: Tmax, blue squares: TS, and purple circles: TD).

8.5.2. Additional Reliability Improvement Methods

Aside from the two methods discussed earlier for improving device reliability through device structure modifications, reliability can also be enhanced by altering the internal materials of the TFT devices. Changing the internal materials typically involves modifications to the active layer or the gate-insulating layer, which can significantly improve the inherent reliability of the device. For example, replacing the MO active layer with a poly-Si thin film can enhance reliability. However, different active layers have their own strengths and weaknesses in terms of performance, and TFTs for specific application scenarios often require the use of particular materials.

Improving the interface state defects or bulk defects of the thin film can also improve device reliability to a certain extent. For instance, as discussed in Section 4.2, the common degradation mechanism under positive/negative bias stress

(PBS/NBS) is often due to carriers being trapped at interface states, leading to device V_{th} shift and degradation. This type of degradation can be mitigated by improving interface contact methods, such as plasma treatment [28] or replacing the GI layer with a denser material [47], thereby significantly enhancing reliability. Additionally, in the HC degradation of MO TFTs and some poly-Si TFTs, carrier injection into interface states is also a primary cause of device degradation, making the improvement of device interface contacts an important approach for enhancing reliability [48].

CONCLUSION

This chapter presents a comprehensive examination of various strategies designed to improve the reliability of both poly-Si and MO TFTs. It explores innovative structural designs, such as LDD configurations and bridged-grain structures, highlighting their fabrication processes and the significant impact they have on mitigating degradation mechanisms that affect device performance. The chapter delves into how the LDD structure enhances reliability through optimized doping levels and drain lengths while the BG architecture effectively suppresses both direct-current and alternating-current degradation, ultimately leading to enhanced noise performance. In addition to structural innovations, the chapter discusses signal modulation and material optimizations that further contribute to the long-term stability of TFTs. It emphasizes the importance of integrating these advancements into existing manufacturing processes to develop more robust electronic devices capable of withstanding varying operational conditions. Overall, the chapter not only outlines effective methods for enhancing TFT performance and reliability but also provides critical insights into the underlying mechanisms responsible for these improvements, paving the way for future advancements in TFT technology.

REFERENCES

[1] J.J. Chen, T-C. Chang, H-C. Chen, K-J. Zhou, C-W. Kuo, W-C. Wu, H-C. Li, M-C. Tai, Y-F. Tu, Y-L. Tsai, P-Y. Wu, and S.M. Sze, "Enhancing Hot-Carrier Reliability of Dual-Gate Low-Temperature Polysilicon TFTs by Increasing Lightly Doped Drain Length", *IEEE Electron Device Lett.,* vol. 41, no. 10, pp. 1524-1527, 2020.
http://dx.doi.org/10.1109/LED.2020.3018196

[2] S.W. Son, C.W. Byun, Y.W. Lee, S.J. Yun, A.V. Takaloo, J.H. Park, and S.K. Joo, "Effect of Dopant Concentration in Lightly Doped Drain Region on the Electrical Properties of N-Type Metal Induced Lateral Crystallization Polycrystalline Silicon Thin Film Transistors", *Jpn. J. Appl. Phys.,* vol. 52, no. 10S, p. 10MC13, 2013.
http://dx.doi.org/10.7567/JJAP.52.10MC13

[3]   L. Xu, R. Liu, B. Yuan, Z. Du, Y. Liu, and J. Li, "6.4: The Effect of LDD Structure on the Characteristics of TFT Devices", *SID Symposium Digest of Technical Papers,* vol. 50, 2019pp. 55-56

http://dx.doi.org/10.1002/sdtp.13385

[4]   M. Mativenga, M.H. Choi, W. Choi, J.W. Choi, J. Jang, R. Mruthyunjaya, T.J. Tredwell, E. Mozdy, and C. Kosik-Williams, "Reduction of Hot Carrier Effects in Silicon-on-Glass TFTs", *J. Electrochem. Soc.,* vol. 158, no. 6, pp. J169-J174, 2011.

http://dx.doi.org/10.1149/1.3573769

[5]   H.C. Casey, and G.L. Pearson, Diffusion in Semiconductors *in Point Defects in. Solids: Volume 2 Semiconductors and Molecular Crystals.* J. H. Crawford and L. M. Slifkin, Eds., Boston, MA: Springer US, 1975, pp. 163-255.

http://dx.doi.org/10.1007/978-1-4684-0904-8_2

[6]   K. Nomura, T. Kamiya, and H. Hosono, "Effects of Diffusion of Hydrogen and Oxygen on Electrical Properties of Amorphous Oxide Semiconductor, In-Ga-Zn-O", *ECS J. Solid State Sci. Technol.,* vol. 2, no. 1, pp. P5-P8, 2013.

http://dx.doi.org/10.1149/2.011301jss

[7]   G. Zhu, M. Zhang, L. Lu, M. Wong, and H.S. Kwok, "Hot Carrier Degradation Reduction in Metal Oxide Thin-Film Transistors by Implementing a Lightly Doped Drain-Like Structure", *IEEE Electron Device Lett.,* vol. 45, no. 9, pp. 1602-1605, 2024.

http://dx.doi.org/10.1109/LED.2024.3424473

[8]   S.-D. Wang, W.-H. Lo, and T.-F. Lei, "CF4 Plasma Treatment for Fabricating High-Performance and Reliable Solid-Phase-Crystallized Poly-Si TFTs", *Journal of The Electrochemical Society..*

[9]   C. Wu, Z. Meng, S. Zhao, S. Xiong, M. Wong, and H.S. Kwok, "Effects of high temperature post-annealing on the properties of solution-based metal-induced crystallized polycrystalline silicon films", *J. Mater. Sci. Mater. Electron.,* vol. 18, no. S1, pp. 355-358, 2007.

http://dx.doi.org/10.1007/s10854-007-9245-1

[10]   Mingxiang Wang, and Zhiguo Meng, "The effects of high temperature annealing on metal-induced laterally crystallized polycrystalline silicon", *IEEE Trans. Electron Devices,* vol. 47, no. 11, pp. 2061-2067, 2000.

http://dx.doi.org/10.1109/16.877167

[11]   W. Zhou, Z. Meng, S. Zhao, M. Zhang, R. Chen, M. Wong, and H-S. Kwok, "Bridged-Grain Solid-Phase-Crystallized Polycrystalline-Silicon Thin-Film Transistors", *IEEE Electron Device Lett.,* vol. 33, no. 10, pp. 1414-1416, 2012.

http://dx.doi.org/10.1109/LED.2012.2210019

[12]   Meng Zhang, Wei Zhou, Rongsheng Chen, Man Wong, and Hoi-Sing Kwok, "Characterization of DC-Stress-Induced Degradation in Bridged-Grain Polycrystalline Silicon Thin-Film Transistors", *IEEE Trans. Electron Dev.,* vol. 61, no. 9, pp. 3206-3212, 2014.

http://dx.doi.org/10.1109/TED.2014.2341676

[13]   M. Zhang, M. Wang, X. Lu, M. Wong, and H.S. Kwok, "Analysis of Degradation Mechanisms in Low-Temperature Polycrystalline Silicon Thin-Film Transistors under Dynamic Drain Stress", *IEEE Trans. Electron Dev.,* vol. 59, no. 6, pp. 1730-1737, 2012.

http://dx.doi.org/10.1109/TED.2012.2189218

[14]   M. Xue, M. Wang, Z. Zhu, D. Zhang, and M. Wong, "Degradation Behaviors of Metal-Induced Laterally Crystallized n-Type Polycrystalline Silicon Thin-Film Transistors Under DC Bias Stresses", *IEEE Trans. Electron Dev.*, vol. 54, no. 2, pp. 225-232, 2007.

http://dx.doi.org/10.1109/TED.2006.888723

[15]   Y.U. Yukiharu Uraoka, T.H. Tomoaki Hatayama, T.F. Takashi Fuyuki, T.K. Tetsuya Kawamura, and Y.T. Yuji Tsuchihashi, "Hot Carrier Effects in Low-Temperature Polysilicon Thin-Film Transistors", *Jpn. J. Appl. Phys.*, vol. 40, no. 4S, p. 2833, 2001.

http://dx.doi.org/10.1143/JJAP.40.2833

[16]   Ching-Fang Huang, Hung-Chang Sun, Ying-Jhe Yang, Yen-Ting Chen, Chun-Yuan Ku, Yuan-Jun Hsu, Ching-Chieh Shih, and Jim-Shone Chen, "Dynamic Bias Instability of p-Channel Polycrystalline-Silicon Thin-Film Transistors Induced by Impact Ionization", *IEEE Electron Device Lett.*, vol. 30, no. 4, pp. 368-370, 2009.

http://dx.doi.org/10.1109/LED.2009.2013644

[17]   H. Wang, M. Wang, Z. Yang, H. Hao, and M. Wong, "Stress Power Dependent Self-Heating Degradation of Metal-Induced Laterally Crystallized n-Type Polycrystalline Silicon Thin-Film Transistors", *IEEE Trans. Electron Dev.*, vol. 54, no. 12, pp. 3276-3284, 2007.

http://dx.doi.org/10.1109/TED.2007.908907

[18]   A. Valletta, P. Gaucci, L. Mariucci, A. Pecora, L. Maiolo, and G. Fortunato, "Downscaling effects on self-heating related instabilities in p-channel polycrystalline silicon thin film transistors", *Appl. Phys. Lett.*, vol. 99, no. 5, p. 053503, 2011.

http://dx.doi.org/10.1063/1.3621874

[19]   S. Inoue, H. Ohshima, and T. Shimoda, "Analysis of Degradation Phenomenon Caused by Self-Heating in Low-Temperature-Processed Polycrystalline Silicon Thin Film Transistors", *Jpn. J. Appl. Phys.*, vol. 41, no. Part 1, No. 11A, pp. 6313-6319, 2002.

http://dx.doi.org/10.1143/JJAP.41.6313

[20]   M. Kimura, S. Inoue, T. Shimoda, S.W-B. Tam, O.K.B. Lui, P. Migliorato, and R. Nozawa, "Extraction of trap states in laser-crystallized polycrystalline-silicon thin-film transistors and analysis of degradation by self-heating", *J. Appl. Phys.*, vol. 91, no. 6, pp. 3855-3858, 2002.

http://dx.doi.org/10.1063/1.1446238

[21]   P. Gaucci, A. Valletta, L. Mariucci, A. Pecora, L. Maiolo, and G. Fortunato, "Analysis of Self-Heating-Related Instability in Self-Aligned p-Channel Polycrystalline-Silicon Thin-Film Transistors", *IEEE Electron Device Lett.*, vol. 31, no. 8, pp. 830-832, 2010.

http://dx.doi.org/10.1109/LED.2010.2051137

[22]   L. Maiolo, M. Cuscunà, L. Mariucci, A. Minotti, A. Pecora, D. Simeone, A. Valletta, and G. Fortunato, "Analysis of self-heating related instability in n-channel polysilicon thin film transistors fabricated on polyimide", *Thin Solid Films*, vol. 517, no. 23, pp. 6371-6374, 2009.

http://dx.doi.org/10.1016/j.tsf.2009.02.105

[23]   K. Takechi, M. Nakata, H. Kanoh, S. Otsuki, and S. Kaneko, "Dependence of self-heating effects on operation conditions and device structures for polycrystalline silicon TFTs", *IEEE Trans. Electron Dev.*, vol. 53, no. 2, pp. 251-257, 2006.

http://dx.doi.org/10.1109/TED.2005.861729

[24] S. Hashimoto, Y. Uraoka, T. Fuyuki, and Y. Morita, "Suppression of Self-Heating in Low-Temperature Polycrystalline Silicon Thin-Film Transitors", *JJAP,* vol. 46, no. 4R, p. 1387, 2007.
http://dx.doi.org/10.1143/JJAP.46.1387

[25] J. Zhou, M. Wang, and M. Wong, "Two-Stage Degradation of p-Channel Poly-Si Thin-Film Transistors Under Dynamic Negative Bias Temperature Stress", *IEEE Trans. Electron Dev.,* vol. 58, no. 9, pp. 3034-3041, 2011.
http://dx.doi.org/10.1109/TED.2011.2158582

[26] C.Y. Chen, J-W. Lee, S-D. Wang, M-S. Shieh, P-H. Lee, W-C. Chen, H-Y. Lin, K-L. Yeh, and T-F. Lei, "Negative Bias Temperature Instability in Low-Temperature Polycrystalline Silicon Thin-Film Transistors", *IEEE Trans. Electron Dev.,* vol. 53, no. 12, pp. 2993-3000, 2006.
http://dx.doi.org/10.1109/TED.2006.885543

[27] M. Zhang, W. Zhou, R. Chen, M. Wong, and H.S. Kwok, "Water-enhanced negative bias temperature instability in p-type low temperature polycrystalline silicon thin film transistors", *Microelectron. Reliab.,* vol. 54, no. 1, pp. 30-32, 2014.
http://dx.doi.org/10.1016/j.microrel.2013.07.082

[28] M. Saß, A. Annen, and W. Jacob, "Hydrogen bonding in plasma-deposited amorphous hydrogenated boron films", *J. Appl. Phys.,* vol. 82, no. 4, pp. 1905-1908, 1997.
http://dx.doi.org/10.1063/1.365997

[29] M. Zhang, Z. Xia, W. Zhou, R. Chen, and M. Wong, "Significant Reduction of Dynamic Negative Bias Stress-Induced Degradation in Bridged-Grain Poly-Si TFTs", *IEEE Electron Device Lett.,* vol. 36, no. 2, pp. 141-143, 2015.
http://dx.doi.org/10.1109/LED.2014.2377040

[30] M. Zhang, W. Zhou, R. Chen, M. Wong, and H.S. Kwok, ""Driving"-Stress-Induced Degradation in Polycrystalline Silicon Thin-Film Transistors and Its Suppression by a Bridged-Grain Structure", *IEEE Electron Device Lett.,* vol. 38, no. 1, pp. 52-55, 2017.
http://dx.doi.org/10.1109/LED.2016.2626481

[31] M. Zhang, Y. Yan, G. Li, S. Deng, W. Zhou, R. Chen, M. Wong, and H-S. Kwok, "OFF-State-Stress-Induced Instability in Switching Polycrystalline Silicon Thin-Film Transistors and Its Improvement by a Bridged-Grain Structure", *IEEE Electron Device Lett.,* vol. 39, no. 11, pp. 1684-1687, 2018.
http://dx.doi.org/10.1109/LED.2018.2872350

[32] M. Zhang, X. Ma, S. Deng, W. Zhou, Y. Yan, M. Wong, and H-S. Kwok, "Degradation Induced by Forward Synchronized Stress in Poly-Si TFTs and Its Reduction by a Bridged-Grain Structure", *IEEE Electron Device Lett.,* vol. 40, no. 9, pp. 1467-1470, 2019.
http://dx.doi.org/10.1109/LED.2019.2931007

[33] M. Zhang, S. Deng, W. Zhou, Y. Yan, M. Wong, and H.S. Kwok, "Reversely-Synchronized-Stress-Induced Degradation in Polycrystalline Silicon Thin-Film Transistors and Its Suppression by a Bridged-Grain Structure", *IEEE Electron Device Lett.,* vol. 41, no. 8, pp. 1213-1216, 2020.
http://dx.doi.org/10.1109/LED.2020.3005046

[34]   Wei Zhou, Shuyun Zhao, Rongsheng Chen, Meng Zhang, J.Y.L. Ho, Man Wong, and Hoi-Sing Kwok, "Study of the Characteristics of Solid Phase Crystallized Bridged-Grain Poly-Si TFTs", *IEEE Trans. Electron Dev.,* vol. 61, no. 5, pp. 1410-1416, 2014.
http://dx.doi.org/10.1109/TED.2014.2308579

[35]   M. Zhang, W. Zhou, R. Chen, M. Wong, and H.S. Kwok, "High-performance polycrystalline silicon thin-film transistors integrating sputtered aluminum-oxide gate dielectric with bridged-grain active channel", *Semicond. Sci. Technol.,* vol. 28, no. 11, p. 115003, 2013.
http://dx.doi.org/10.1088/0268-1242/28/11/115003

[36]   D. Zhang, M. Wang, and X. Lu, "Two-Stage Degradation of p-Type Polycrystalline Silicon Thin-Film Transistors Under Dynamic Positive Bias Temperature Stress", *IEEE Trans. Electron Dev.,* vol. 61, no. 11, pp. 3751-3756, 2014.
http://dx.doi.org/10.1109/TED.2014.2359299

[37]   Y.H. Tai, S.C. Huang, and C.K. Chen, "Analysis of Poly-Si TFT Degradation Under Gate Pulse Stress Using the Slicing Model", *IEEE Electron Device Lett.,* vol. 27, no. 12, pp. 981-983, 2006.
http://dx.doi.org/10.1109/LED.2006.886416

[38]   D.R. Khanal, and J. Wu, "Gate coupling and charge distribution in nanowire field effect transistors", *Nano Lett.,* vol. 7, no. 9, pp. 2778-2783, 2007.
http://dx.doi.org/10.1021/nl071330l PMID: 17718588

[39]   M. Kimura, "Degradation Analysis of p-Type Poly-Si Thin-Film Transistors Using Device Simulation," IEEE Transactions on Electron Devices, vol. 58, no. 11, pp. 4106-4110, 2011.
http://dx.doi.org/10.1109/ted.2011.2163801

[40]   Y. Liu, H. He, R. Chen, Y.F. En, B. Li, and Y.Q. Chen, "Analysis and Simulation of Low-Frequency Noise in Indium-Zinc-Oxide Thin-Film Transistors", *IEEE J. Electron Devices Soc.,* vol. 6, pp. 271-279, 2018.
http://dx.doi.org/10.1109/JEDS.2018.2800049

[41]   Y. Liu, S-T. Cai, C-Y. Han, Y-Y. Chen, L. Wang, X-M. Xiong, and R. Chen, "Scaling Down Effect on Low Frequency Noise in Polycrystalline Silicon Thin-Film Transistors", *IEEE J. Electron Devices Soc.,* vol. 7, pp. 203-209, 2019.
http://dx.doi.org/10.1109/JEDS.2018.2890737

[42]   G. Ghibaudo, O. Roux, C. Nguyen-Duc, F. Balestra, and J. Brini, "Improved Analysis of Low Frequency Noise in Field-Effect MOS Transistors", *Phys. Status Solidi, A Appl. Res.,* vol. 124, no. 2, pp. 571-581, 1991. [a].
http://dx.doi.org/10.1002/pssa.2211240225

[43]   C. A. Dimitriadis, F. V. Farmakis, G. Kamarinos, and J. Brini, "Origin of low-frequency noise in polycrystalline silicon thin-film transistors", 2014.

[44]   G. Zhu, M. Zhang, Z. Jiang, J. Huang, Y. Huang, S. Deng, L. Lu, M. Wong, and H-S. Kwok, "Significant Degradation Reduction in Metal Oxide Thin-Film Transistors via the Interaction of Ionized Oxygen Vacancy Redistribution, Self-Heating Effect, and Hot Carrier Effect", *IEEE Trans. Electron Dev.,* vol. 70, no. 8, pp. 4198-4205, 2023.
http://dx.doi.org/10.1109/TED.2023.3283940

[45] S. Tyaginov, M. Jech, J. Franco, P. Sharma, B. Kaczer, and T. Grasser, "Understanding and Modeling the Temperature Behavior of Hot-Carrier Degradation in SiON nMOSFETs", *IEEE Electron Device Lett.,* vol. 37, no. 1, pp. 84-87, 2016.

http://dx.doi.org/10.1109/LED.2015.2503920

[46] P. Heremans, G. Van den Bosch, R. Bellens, G. Groeseneken, and H.E. Maes, "Temperature dependence of the channel hot-carrier degradation of n-channel MOSFET's", *IEEE Trans. Electron Dev.,* vol. 37, no. 4, pp. 980-993, 1990.

http://dx.doi.org/10.1109/16.52433

[47] J. Huang, "Enhanced Visible Light Response of Amorphous InZnO Thin-Film Transistors by Hydrogen Doping via Al2O3/SiO2 Gate Dielectric", *Digest of Technical Papers-SID International Symposium,* 2023pp. 1802-1805

http://dx.doi.org/10.1002/sdtp.16955

[48] T. Kawamura, M. Matsumura, T. Kaitoh, T. Noda, M. Hatano, T. Miyazawa, and M. Ohkura, "A Model for Predicting On-Current Degradation Caused by Drain-Avalanche Hot Carriers in Low-Temperature Polysilicon Thin-Film Transistors", *IEEE Trans. Electron Dev.,* vol. 56, no. 1, pp. 109-115, 2009.

http://dx.doi.org/10.1109/TED.2008.2008376

CHAPTER 9

Summary

9.1. SUMMARY OF RELIABILITY OF THIN-FILM TRANSISTORS

This book primarily uses silicon-based (Si-based) thin-film transistors (TFTs) and metal oxide (MO) TFTs as examples to thoroughly explain the materials, structures, and working principles of TFT devices. It also provides a detailed discussion of the degradation phenomena, mechanisms, and improvement methods for TFT reliability across different types.

The book begins with an introductory chapter on the history of TFT development, followed by a brief overview of TFT classification and applications. It then delves into several common TFT structures, detailing the typical structures of Si-TFTs and MO-TFTs, as well as their fabrication methods. The book proceeds to discuss various common applications of TFTs, with an emphasis on the impact of reliability in these applications.

Subsequently, Chapter 2 focuses on reliability issues in TFTs, starting with the characterization of basic TFT performance, emphasizing measurement and analysis methods for transfer and output characteristic curves. It then highlights typical defects found within Si-TFTs and MO-TFTs, concluding with a summary of several typical degradation mechanisms in TFTs. Chapter 2 serves as the theoretical foundation for the entire book.

After an overall introduction to TFT reliability, Chapter 3 explores various methods for analyzing TFT reliability. It covers the characterization of basic transfer characteristics, including analysis methods for the on-state and subthreshold swing (SS) of the transfer curve before and after degradation. The chapter also introduces the analysis method for TFT's capacitance-voltage (CV) characteristics and highlights several methods for extracting defect density from CV curves. Additionally, it discusses low-frequency noise analysis methods and key application formulas, followed by an introduction to various TFT thin-film characterization techniques, including x-ray photoelectron spectroscopy, transmission electron microscopy, x-ray diffraction, scanning electron microscopy, atomic force microscopy, and Raman analysis. The chapter concludes with two simulation methods: semiconductor device simulation and thermal simulation.

Chapter 4 focuses on the degradation conditions and mechanisms of TFTs under direct current (DC) stress, detailing the typical degradation scenarios and special cases for Si-TFTs and MO-TFTs under gate bias stress, hot carrier (HC) stress, and self-heating (SH) stress.

Chapters 5 and 6 provide an in-depth look at the degradation of TFTs under different alternating current (AC) stress conditions. They begin with a brief overview of TFT degradation under typical single-ended pulse stress and analyze how changes in pulse parameters affect device degradation. A non-equilibrium PN junction AC reliability model is introduced to explain AC degradation under various conditions. Since TFTs often operate in more complex pulse environments, Chapter 7 discusses TFT degradation under DC bias AC stress, dual-end AC stress, and ultrafast AC stress, presenting an enhanced non-equilibrium PN junction AC reliability model.

Chapter 7 primarily examines the impact of different environmental factors on TFT reliability, discussing the effects of temperature, light exposure, and humidity on device reliability. Chapter 8 introduces several methods for improving TFT reliability, focusing on structural modifications, including the LDD and BG structures, as well as other improvement methods.

Overall, this book provides a comprehensive introduction to the degradation phenomena and principles of TFT reliability, supplemented by an overview of necessary foundational knowledge. The content presented serves as a theoretical basis for research and production work related to TFT reliability studies.

9.2. FUTURE WORK ON THE RELIABILITY OF THIN-FILM TRANSISTORS

This book primarily focuses on the reliability analysis of mainstream Si-based TFT (Si-TFT) devices and MO TFT (MO-TFT) devices. As a new generation of TFTs, MO-TFTs offer advantages such as low-temperature fabrication and extremely low leakage current that are difficult for polycrystalline silicon (poly-Si) TFTs to achieve. However, reliability research for MO-TFTs is still not comprehensive. In the case of DC gate bias stress, the same stress voltage leads to both positive and negative shifts in different types of MO-TFTs. Although numerous mechanisms have been proposed to explain these phenomena, there is still a lack of a unified model to predict when MO-TFTs will exhibit negative or positive shifts under gate bias stress.

Similarly, the reliability of MO-TFTs is a major obstacle to their practical application. For basic gate bias stress, poly-Si TFTs typically require high gate voltages or high-temperature treatment to cause significant device degradation, while MO-TFTs often show noticeable degradation at gate voltages above 20 V. Additionally, due to the conduction mechanisms of MO-TFTs, those with higher mobility often suffer from poorer reliability. This is why, despite the availability of high-mobility MO-TFTs, IGZO TFTs still dominate commercial applications. Improving the reliability of MO-TFTs is one of the key research areas for future MO-TFT development.

It is also worth noting that most current reliability analyses are qualitative, which can guide the improvement of TFT device reliability in commercial applications to some extent. However, to further expand on these reliability analyses, such as predicting the working life of TFTs or the overall display, or to compensate for TFT degradation specifically, quantitative research on device degradation is required. This includes the study of analytical models for device performance and the development of degradation analysis models. While there is a substantial amount of work on the initial analytical models for TFTs, which can well fit the performance changes of devices in on-state, off-state, and subthreshold state, research on analytical models for degraded devices is relatively scarce, with most focusing on DC studies. There is almost no research on analytical models for AC reliability

SUBJECT INDEX

www.ingramcontent.com/pod-product-compliance
Lightning Source LLC
Chambersburg PA
CBHW050805220326
41598CB00006B/124